Clay Mineralogy: Spectroscopic and Chemical Determinative Methods

Clay Mineralogy: Spectroscopic and Chemical Determinative Methods

Edited by

M. J. Wilson FRSE

Head, Division of Soils
Macaulay Land Use Research Institute
Craigiebuckler
Aberdeen, UK

CHAPMAN & HALL

London · Glasgow · Weinheim · New York · Tokyo · Melbourne · Madras

Published by Chapman & Hall, 2-6 Boundary Row, London SE1 8HN, UK

Chapman & Hall, 2-6 Boundary Row, London SE1 8HN, UK

Blackie Academic & Professional, Wester Cleddens Road, Bishopbriggs, Glasgow G64 2NZ, UK

Chapman & Hall GmbH, Pappelallee 3, 69469 Weinheim, Germany

Chapman & Hall USA., 115 Fifth Avenue, New York, NY 10003, USA

Chapman & Hall Japan, ITP-Japan, Kyowa Building, 3F, 2-2-1 Hirakawacho, Chiyoda-ku, Tokyo 102, Japan

Chapman & Hall Australia, 102 Dodds Street, South Melbourne, Victoria 3205, Australia

Chapman & Hall India, R. Seshadri, 32 Second Main Road, CIT East, Madras 600 035, India

First edition 1994
Reprinted 1995

Typeset in 10/12pt Palatino by Pure Tech Corporation, Pondicherry, India
Printed in Great Britain at the Alden Press, Oxford

ISBN 0 412 53380 4

A Catalogue record for this book is available from the British Library

Library of Congress Catalog Card Number: 93—74207

∞ Printed on permanent acid-free text paper, manufactured in accordance with ANSI/NISO Z39.48-1992 (Permanence of Paper).

Contents

Preface

A knowledge of clay is important in many spheres of scientific endeavour, particularly in natural sciences such as geology, mineralogy and soil science, but also in more applied areas like environmental and materials science. Over the last two decades research into clay mineralogy has been strongly influenced by the development and application of a number of spectroscopic techniques which are now able to yield information about clay materials at a level of detail that previously would have seemed inconceivable. This information relates not only to the precise characterization of the individual clay components themselves, but also to the ways in which these components interact with a whole range of absorbate molecules. At present, however, the fruits of this research are to be found principally in a somewhat widely dispersed form in the scientific journals, and it was thus considered to be an appropriate time to bring together a compilation of these spectroscopic techniques in a way which would make them more accessible to the non-specialist. This is the primary aim of this book. The authors of the various chapters first describe the principles and instrumentation of the individual spectroscopic techniques, assuming a minimum of prior knowledge, and then go on to show how these methods have been usefully applied to clay mineralogy in its broadest context. The book concludes with two chapters describing a purely chemical approach to the characterization of clays which, although somewhat traditional, is nevertheless considered to be a useful adjunct to the sophisticated physical techniques described beforehand. The philosophy throughout the book has been to try to strike a balance between an introductory text and a research compendium so that it may be useful to scientists with no previous experience of clays as well as to those who are familiar with the problems of investigating these difficult materials. This approach is similar to that adopted in a previous book of this kind (*A Handbook of Determinative Methods in Clay Mineralogy*, 1987, edited by M. J. Wilson, Blackie, Glasgow) which is now out of print, and in fact Chapters 2, 8 and 9 are updated versions from this work. All other chapters are, however, completely new.

Most of the authors are, or have been, associated with the Macaulay Land Use Research Institute in Aberdeen, which has had a long and distinguished history of research into soil mineralogy. Despite recent difficulties occasioned by financial constraints, soil mineralogy is identified in the Institute's mission statement as an area which it is regarded as important to develop in an international context. It is hoped that this book may be of some assistance in the fulfilment of this aspiration.

Finally, we are most grateful to the Director of MLURI, Professor T. J. Maxwell, for allowing us to undertake and complete this book with the minimum of formalities. We also thank Mrs. Aileen Stewart for the time and effort she has devoted to processing the manuscript.

M. J. Wilson
Macaulay Land Use Research Institute
Aberdeen

Contributors

D. C. Bain, Macaulay Land Use Institute, Craigiebuckler, Aberdeen AB9 2QJ.

J. A. Chudek, Department of Chemistry, University of Dundee, Dundee DD1 4HN

A. R. Fraser, Macaulay Land Use Institute, Craigiebuckler, Aberdeen AB9 2QJ.

B. A. Goodman, Scottish Crops Research Institute, Invergowrie, Dundee DD2 5DA.

P. L. Hall, Schlumberger, PO Box 153, Cambridge CB2 3BE.

E. E. Lachowski, Department of Chemistry, University of Aberdeen AB9 2UE

W. J. McHardy, Macaulay Land Use Institute, Craigiebuckler, Aberdeen AB9 2QJ.

E. Paterson, Macaulay Land Use Institute, Craigiebuckler, Aberdeen AB9 2QJ.

J. D. Russell, Macaulay Land Use Institute, Craigiebuckler, Aberdeen AB9 2QJ.

B. F. L. Smith, Macaulay Land Use Institute, Craigiebuckler, Aberdeen AB9 2QJ.

R. Swaffield, Macaulay Land Use Institute, Craigiebuckler, Aberdeen AB9 2QJ.

CHAPTER 1

Molecular spectroscopy: introduction and general principles

B. A. Goodman

1.1 BACKGROUND

Molecular spectroscopy involves the interaction of electromagnetic radiation with materials in order to produce an absorption pattern (i.e. a spectrum) from which structural or compositional information can be deduced. Electromagnetic radiation is conventionally, and rather arbitrarily, divided into a number of distinct regions, each of which covers a range of energies that corresponds to a different type of molecular process (Figure 1.1). In order of increasing frequency these are:

1. Radiofrequency (*c.* $10^5 - 10^9$ Hz). This corresponds to the energy involved in changing the direction of spin of a nucleus (nuclear magnetic resonance, NMR).
2. Microwave (*c.* 10^9 to 3×10^{10} Hz). This range covers the energy involved in changing the direction of electron spin (electron paramagnetic resonance, EPR, also known as electron spin resonance, ESR) and the separations between rotational energy levels of gaseous molecules.
3. Infrared (*c.* 3×10^{10} to 3×10^{14} Hz). Separations between vibrational energy states occur in this region, in which the techniques of infrared absorption and Raman spectroscopy occur.
4. Visible and ultraviolet (*c.* 3×10^{14} to 3×10^{16} Hz). This region corresponds to the separations between the energy levels of valence electrons.

Clay Mineralogy: Spectroscopic and Chemical Determinative Methods.
Edited by M. J. Wilson. Published in 1994 by Chapman & Hall, London.
ISBN 0 412 53380 4.

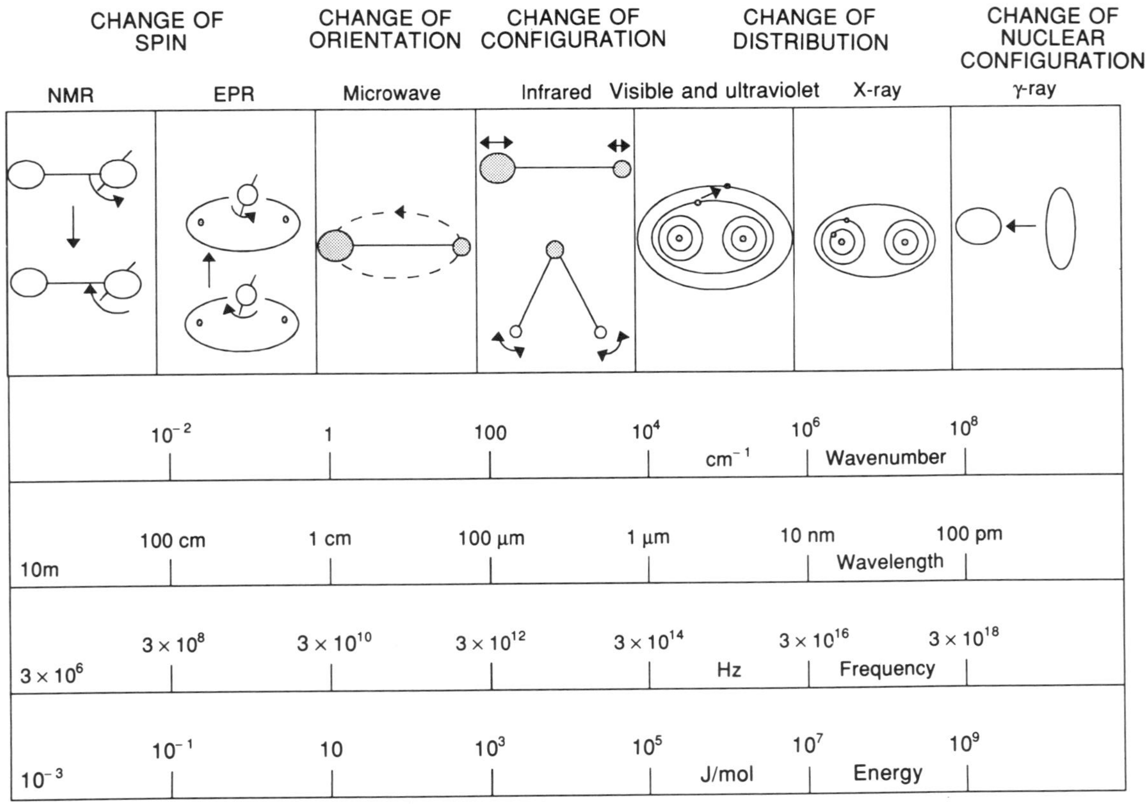

Figure 1.1 The electromagnetic spectrum and corresponding spectroscopic techniques. NMR, nuclear magnetic resonance; EPR, electron paramagnetic resonance.

5. X-rays (*c.* 3×10^{16} to 3×10^{18} Hz). The energies of transitions involving inner-shell electrons occur in this region (photoelectron spectroscopy).
6. Gamma (γ)-rays (*c.* 3×10^{18} to 10^{21} Hz). These correspond to the energies required to change the energy states of nuclear particles (Mössbauer spectroscopy).

All of the above energy states are quantized, i.e. they occur as discrete values. Any discussion of quantum mechanics is beyond the scope of this text, except to say that the energy states of all atoms and molecules are determined by sets of quantum numbers which determine the number and relative energies of the states in the various regions of the electromagnetic spectrum. These will be discussed further in the relevant sections, which are concerned primarily with the establishment of criteria for the existence of transitions between the various energy states.

1.2 ABSORPTION SPECTROSCOPY

In absorption spectroscopy the absorption of radiation is monitored as it is swept through a range of frequencies; absorption at a frequency, ν, indicates that the specimen contains energy levels separated by $h\nu$, where h is Planck's constant (6.626×10^{-34} J s).

The basic components of an absorption spectrometer (Figure 1.2) are a radiation source, which is directed on to the sample being investigated, and an analyser system which detects the amount of radiation transmitted through the specimen. A spectrum of absorbance versus radiation frequency is produced by synchronizing the recording device with the analyser system. In infrared, visible and ultraviolet spectroscopy a typical spectrometer contains a 'white' radiation source (i.e. it emits radiation that covers equally a wide range of frequencies) and incorporates a modulator between the sample and the analyser. This interrupts the radiation beam at a fixed number of times per second and causes the detector to send an alternating current signal to the recorder. The amplifier is tuned to accept only radiation at the frequency imposed by the modulator, thus removing stray radiation and other extraneous signals from the spectrum. In Mössbauer spectroscopy modulation of the

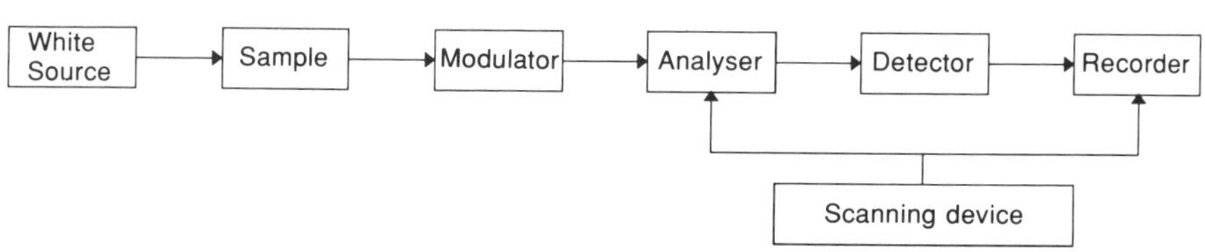

Figure 1.2 Basic block diagram for an absorption spectrometer.

radiation is usually performed at the source, which consists of a single-energy radiation beam. The resulting signal-to-noise ratio is generally low and a Mössbauer spectrum typically consists of the time-average of many spectral scans. With continuous wave magnetic resonance techniques a single radiation frequency is used and spectra are obtained by synchronizing the recorder to varying magnetic fields from electromagnets.

1.3 EMISSION SPECTROSCOPY

In emission spectroscopy transitions occur from higher to lower energy states with the emission of radiation. The sample is excited by a radiation source and the radiation emitted as it returns to thermal equilibrium is detected and analysed (Figure 1.3). This free induction decay (FID) can be converted by a Fourier transform to produce a spectrum in the frequency domain that contains information that is essentially analogous to that obtained by absorption techniques. Fourier transform spectroscopy is generally more sensitive and versatile than the corresponding absorption methods and now dominates the NMR field and is rapidly taking over the infrared and Raman areas.

1.4 SPECTRAL INTENSITY

The intensity of a spectral peak, which corresponds to a transition between one state and another, is determined by three factors: the transition probability, the population of the states and the concentration of the sample.

The intensity of a peak that arises from a transition from a state with energy E is dependent upon the population of that state, $N(E)$, and is given by the Boltzmann distribution function

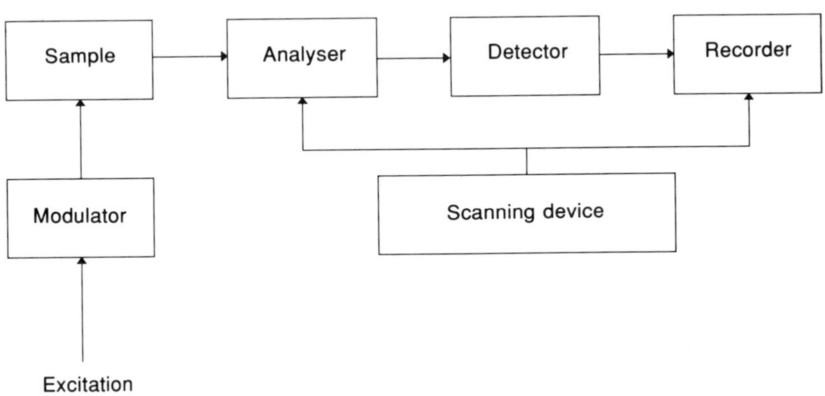

Figure 1.3 Basic block diagram for an emission spectrometer.

$$N(E) \propto N_0 \exp\left(-\frac{E}{KT}\right),$$

where N_0 is the population of the lowest energy state, T is the temperature, and k is the Boltzmann constant (1.38×10^{-23} J/K^{-1}).

Calculation of the probability of a transition occurring involves detailed quantum mechanical calculations, which are not usually possible for mineral systems. However, it is usually possible to decide whether a particular transition is forbidden or allowed by the application of empirically derived selection rules for the various techniques. These will be discussed in more detail in the relevant sections.

The amount of radiation absorbed by a sample is proportional to the amount of the absorbing species present in the beam. The intensity, I, of the transmitted radiation is given by the Beer–Lambert law,

$$I = I_0 \exp(-\varepsilon cl),$$

where I_0 is the intensity of the incident radiation, c is the concentration of the absorbing species, l is the length of the sample and ε (known as the extinction or absorption coefficient) is a constant characteristic of the transition and is closely related to the transition probability.

1.5 SPECTRAL PEAK WIDTHS

Various processes contribute to the shape and width of spectral lines. No excited state has an infinite lifetime and it can be shown by quantum mechanics that if a system is changing with time, it is impossible to specify its energy levels exactly (the Heisenberg uncertainty principle). Thus, the shorter the lifetimes of the states involved in a transition the broader the peaks in the resulting spectrum. The natural linewidth of a transition is determined by the rate at which the excited state emits a photon. It is strongly frequency dependent, so that low-frequency techniques, such as NMR spectroscopy, have much smaller natural linewidths than higher frequency methods, such as photoelectron spectroscopy. In practice other processes contribute to the spectral linewidth and in only a few cases are linewidths close to the natural linewidths observed.

1.6 SPECTRAL RESOLUTION

The setting of the various types of spectrometer for the optimization of spectral quality is covered in the individual sections, but some general principles will be covered briefly here.

Because absorption takes place over a spread of frequencies the ability to resolve two separate overlapping peaks depends on the relative magnitudes of the separation between the peaks and the magnitude of the window of observation. This is illustrated diagramatically in Figure 1.4.

The resolution of spectra can sometimes be improved by recording the first or higher derivatives of the absorption instead of the absorption itself (Figure 1.5). It should be noted that an absorption peak corresponds to the point of cross-over of the baseline in a first-derivative recording. In a second-derivative recording, however, it corresponds to a peak, making this a useful level for spectral display – because many spectroscopists find it more convenient to look for peaks.

This approach is very useful for the separation of shoulders on the sides of peaks and for the deconvolution of multicomponent peaks. It should be appreciated, however, that with successively higher derivatives, the heights of the narrowest peaks are enhanced relative to those with broader linewidths. Thus it is possible in just a second-derivative recording to fail to identify the presence of a broad peak that may account for much of the total spectral area.

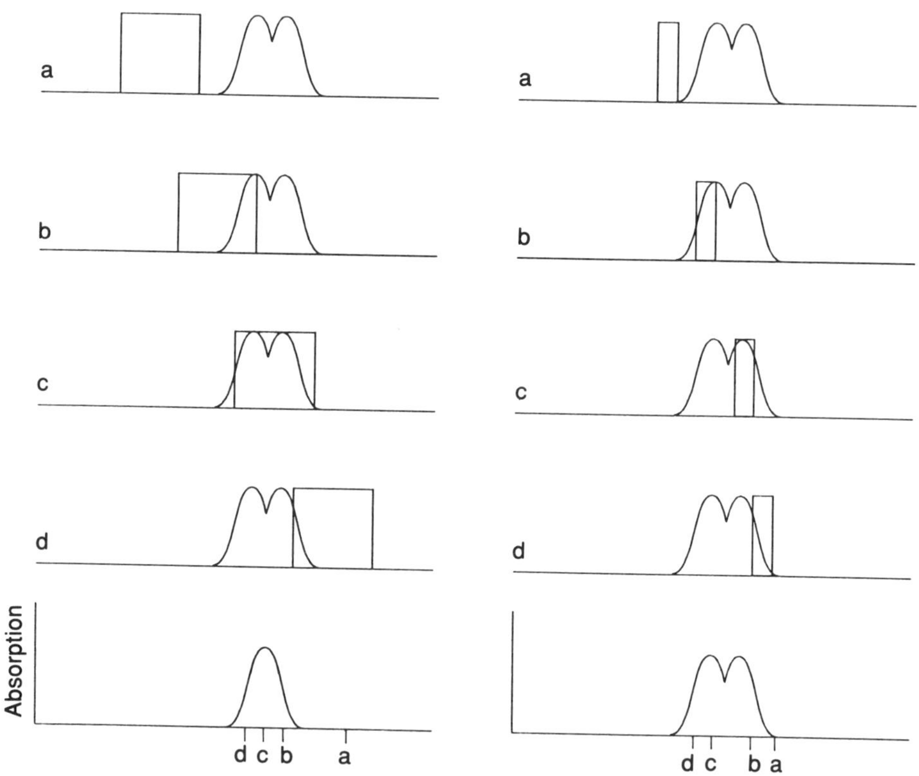

Figure 1.4 The relationship between the size of the window of observation and the ability to resolve overlapping peaks. The broad window of observation on the left side fails to resolve the overlapping peaks. The narrow window on the right side resolves the peaks successfully.

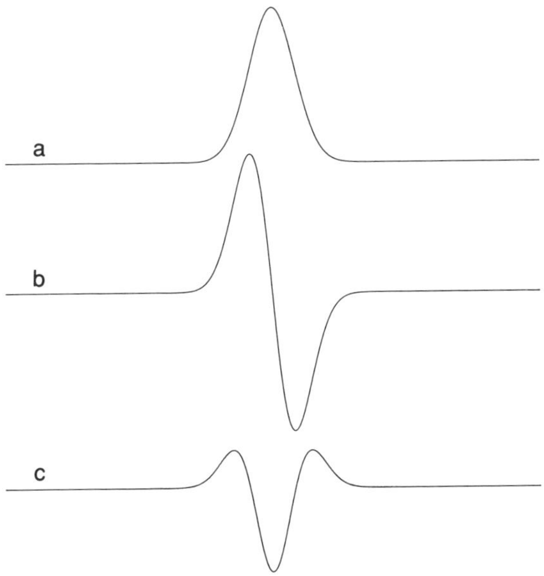

Figure 1.5 (a) Absorption, (b) first and (c) second derivative line shapes.

In recent years numerical filters have been developed which are able to effectively narrow the widths of experimental absorption peaks. A combination of absorption and even-order derivatives can be used to produce a pseudo-absorption spectrum with considerably narrower widths than the original.

1.7 SIGNAL-TO-NOISE OPTIMIZATION

In all forms of spectroscopy the signal produced by the detector is amplified electronically before display. As a result all spectra contain a background of random fluctuations generated from spurious signals ('noise') in the detector or the amplifying equipment. For positive identification a spectral peak should have an intensity at least twice that of the background noise, and preferably greater than this.

The procedures for selection of instrumental parameters for optimization of spectral quality will be dealt with in the individual spectroscopy chapters. However, there are various computer techniques for the improvement of signal-to-noise ratios that have general application. The most common of these will be described here.

1.7.1 Computer averaging of transients (CAT)

This involves recording a spectrum stepwise into a computer, where the intensity of each point is stored in a separate memory location. In the

CAT procedure this process is repeated many times, with the new data being added to the existing data each time. After n summed scans the signal will be n times larger than on a single scan, but the noise intensity, because it is random, will have increased by only $n^{1/2}$. This is a very useful procedure for the resolution of weak peaks in all forms of spectroscopy, but it is very expensive in time because 100 scans are required to give a 10-fold increase in the signal-to-noise ratio. The CAT procedures are used routinely on Fourier transform spectrometers, such as NMR spectrometers, where time-domain scans are of the order of a second, but are much less common in frequency-swept spectroscopy. However, in Mössbauer spectroscopy, which is a very insensitive technique, a CAT procedure is normally used on all spectrometers and as a result spectra typically take many hours, and sometimes days, to acquire.

1.7.2 Numerical noise reduction procedures

A large amount of computer software exists for the improvement of spectral signal-to-noise ratios. The most common procedures concentrate on the removal of high-frequency noise, i.e. noise peaks with widths much smaller than those of the spectral lines. The simplest of these is the moving average, in which each point in a spectrum is replaced by the average of that point and n points on either side of it. In other related routines a weighting is given such that there is a progressive decrease in the contribution of points as their distance from the central point increases. This decrease can be linear, as in a triangular smoothing function, or curved, as in a Lorentzian or Gaussian function. The main problem with this approach is that the residual noise peaks are of comparable width to the spectral peaks and could be mistaken for genuine peaks if strict intensity criteria are not adhered to.

An alternative approach to noise reduction is through the use of a Fourier transform. After Fourier transformation, peaks of small width produce a slowly decaying envelope, whereas those of large width produce an envelope that decays rapidly (Figure 1.6). Therefore, by editing the Fourier transform of a spectrum and then performing a reverse Fourier transform, it is possible to remove numerically either very narrow or very broad spectral components.

1.8 SPECTRAL EDITING AND ANALYSIS

Conventional procedures for spectral analysis vary greatly from one technique to another and will be dealt with in some detail in their individual sections. With some techniques extensive libraries have been built up as aids for spectral interpretation. Software for computer-assisted

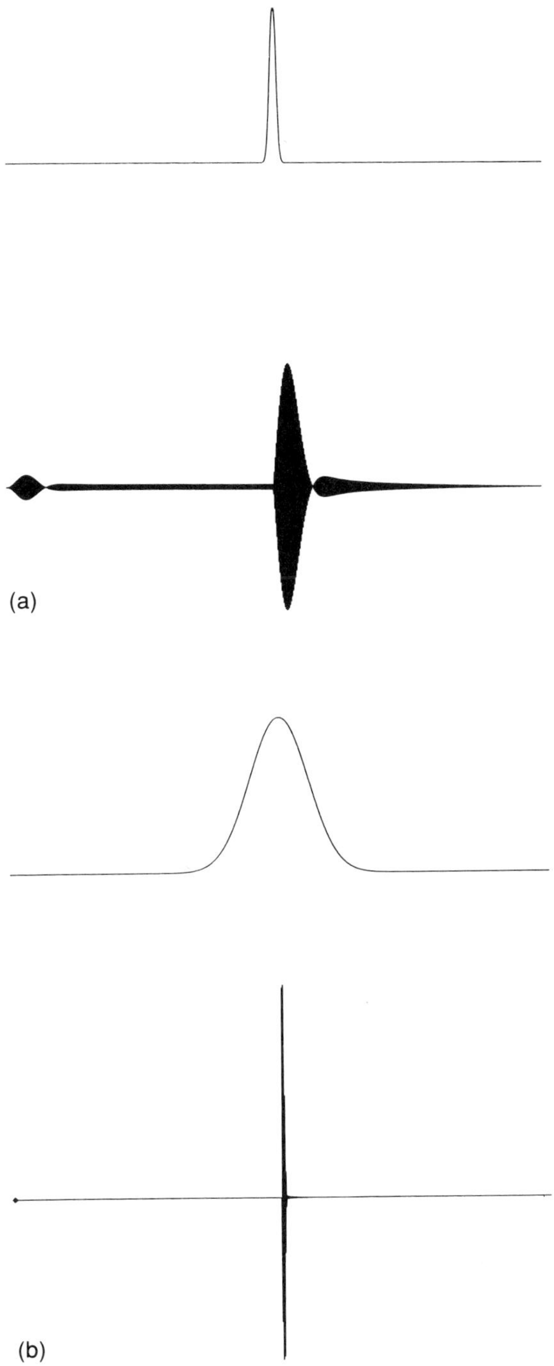

Figure 1.6 Absorption peak and Fourier transform of (a) narrow and (b) broad peaks.

spectral assignment has been developed for techniques such as IR (infrared) and NMR spectroscopy. These are used extensively in chemical applications of the techniques, but are less valuable in mineralogy, where samples may be non-uniform and peaks tend to be broader.

There are some numerical procedures for spectral deconvolution that have general application. The most obvious is the subtraction of a background, such as might arise from a matrix that is used in sample preparation. An extension of this approach involves the successive subtraction of the spectral contributions from identifiable components in the spectrum of a complex mixture. In this way it is possible to obtain spectra of minor components, whose individual peaks may be partially obscured in the original spectrum by peaks from the major components.

One of the problems with the subtraction method outlined above is the progressive increase in noise in the derived spectrum with successive subtractions. It may be possible to overcome this problem with some techniques if the individual spectra can be simulated. Apart from the justification of the spectral parameters that are presented for such species, the subtraction of a simulated component from a complex spectrum does not increase the background noise level. Extreme care must be used when operating such a procedure because, if there are any deviations between the actual and simulated spectral envelopes, artefacts will be produced in the spectrum obtained by the subtraction routine. Nevertheless, spectral simulation is becoming increasingly important with a number of spectroscopic techniques as an aid to interpretation, especially where there is extensive peak overlap.

An extension of the spectral simulation approach is to use a least-squares fitting routine to obtain the optimum values of the parameters for different simulation models and to use statistical goodness-of-fit criteria to differentiate between unsatisfactory and potentially satisfactory models. This approach has not yet gained wide acceptance in most fields of spectroscopy because of the large amount of computer time that would be required to fit complex spectra. It is, however, used routinely in Mössbauer spectroscopy, where the combination of a comparatively small number of peaks and poor spectral sensitivity make the use of statistical methods for spectral interpretation a cost-effective approach.

ACKNOWLEDGEMENT

The Scottish Office Agriculture and Fisheries Department is gratefully acknowledged for funding for this work.

CHAPTER 2

Infrared methods

J. D. Russell and A. R. Fraser

2.1 INTRODUCTION

Infrared absorption spectroscopy is a rapid, economical and non-destructive physical method universally applicable to structural analysis. The technique is so versatile that it can be used both as a source of the physical parameters of crystal lattice determinations, and as a means of eliciting purely empirical qualitative relationships between specimens. It is an intrinsically simple technique that deserves to be more widely used in clay mineralogy and soil science, and this chapter aims to provide some basic information to help bring this about. It is proposed to indicate briefly the basic principles of IR, describe the instrumentation and methods, present some typical spectra of standard minerals, and finally to show spectra of a few representative soil clays, indicating how they may be used to assess the constituent minerals.

2.2 PRINCIPLES

The absorption of infrared radiation by clay minerals depends critically on atomic mass, and the length, strength and force constants of inter-atomic bonds in the structures of these minerals. It is also controlled by the constraints of the overall symmetry of the unit cell, and the local site symmetry of each atom within the unit cell. The total number of potentially active internal vibrations is given by $3n - 6$, where n is the number of atoms in the unit cell. Not all of these vibrations are active in the infrared, only those that undergo a change in dipole moment during the absorption process. More detailed theoretical treatments of vibration spectroscopy as applied to inorganic solids may be found in the books by Hadni (1967) and Turrell (1972). The monograph by

Clay Mineralogy: Spectroscopic and Chemical Determinative Methods.
Edited by M. J. Wilson. Published in 1994 by Chapman & Hall, London.
ISBN 0 412 53380 4.

Farmer (1974) provides one of the most appropriate texts relating specifically to clay minerals.

The absorption of infrared radiation is also strongly influenced by the degree of crystalline order (Lazarev, 1974) and the size and shape of the mineral particles (Farmer and Russell, 1966; Rendon and Serna, 1981; Serna *et al.*, 1982). It is crucial also to be aware of the effect of particle orientation relative to the intercepting beam of IR radiation. Absorption of radiation by a vibrating interatomic bond will be at a maximum when the axis of the bond is parallel to the electric vector of the beam, and will be at a minimum when the axis and vector are mutually perpendicular. Thus, the extent to which clay-mineral particles adopt a preferred orientation – for example in the sedimentation plane for platy particles of a layer silicate – will be important in the identification and structural orientation of specific interatomic groupings in the mineral. The infrared spectrum of a clay mineral is, therefore, sensitive to chemical composition, isomorphous substitution and crystallinity, and provides fundamental information not only on mineral identification, but also on surface properties and reactions of minerals with chemicals in their environment. This chapter aims to provide a representative collection of IR spectra of the most important mineral groups likely to be encountered in clay-size fractions isolated from soils, mineral deposits, etc., and to indicate methods of spectral interpretation that can identify these minerals and the types of isomorphous substitution present.

2.3 INSTRUMENTAL REQUIREMENTS

To provide adequate characterization of a clay mineral by infrared spectroscopy, the spectrum should be recorded over the range 4000 to $250\,cm^{-1}$, and if possible to $200\,cm^{-1}$. Such coverage ensures that most of the useful vibrations active in the infrared will be included, although some minerals, such as micas, carbonates and sulphides, exhibit a few characteristic bands in the region below $200\,cm^{-1}$. Facilities should also be available to expand the wavenumber scale of particular spectral regions to provide good resolution of closely spaced absorption bands, for example in the hydroxyl stretching region of kaolinite.

The spectrometers commonly in use are basically of two types, dispersive and Fourier transform (FT).

2.3.1 Dispersive instruments

In these instruments, the detector monitors the intensity of narrow bandwidths of radiation, dispersed by diffraction gratings and/or prisms in a monochromator. The operational mode is double beam, and may be either null balance, in which a servo system equalizes the intensity of radiation in the two beams by attenuating the reference beam, or ratio recording, in which the absolute value of the radiation intensity ratio in both the reference beam and the sample beam is continuously monitored. Both types have advantages, the major one for the latter being that in regions of strong absorption the detection system can still measure with high response, whereas the former loses response with increasing absorption.

The power and versatility of a dispersive instrument can be greatly enhanced by interfacing with a microcomputer and, with appropriate programming (Adams and Black, 1983), can acquire a potential approaching that of an FT instrument.

2.3.2 Fourier transform instruments

In non-dispersive FTIR spectrometers, the detector continuously monitors the full wavenumber range of radiation emitted by the IR source, providing an inherently more sensitive system than that in a dispersive instrument. Fourier transform instruments are interferometers, and they require a dedicated computer to transform their output – an interferogram – into an absorption spectrum. Fourier transform infrared spectrometers may be either single beam, in which the sample spectrum must be ratioed against a background, or double beam, in which ratioing against background is carried out continuously. Martin (1980) and Durig (1980) have provided excellent texts on the details of FTIR spectrometers.

Such is the power, convenience, versatility and speed of FTIR instruments that they are greatly enhancing the application of the IR technique in a wide range of endeavours, including field studies such as archaeology (Weiner and Goldberg, 1990). The FTIR method derives its power from the ready mathematical manipulation of spectral data in operations such as enhancement of weak absorption bands on intense backgrounds, subtraction of component spectra and component mixture simulation. Examples of the use of a computer operating on the output of a conventional dispersive spectrometer are given in section 2.8.

One relatively recent innovation is the use of FTIR in conjunction with an analytical microscope, which enables infrared spectra to be obtained from sample areas as small as 10 μm diameter. Despite the need for care in minimizing problems associated with contamination of small samples, and the limitation of the spectral range to about 600 cm^{-1} imposed by low energy through such small apertures, applications of the FTIR microscope technique are proliferating in mineral studies, gemology, forensic science, industry and medicine.

2.3.3 Spectral format adopted

Samples were dispersed, usually in KBr, or occasionally CsI, pressed discs and the spectra illustrated in this chapter were recorded in percent

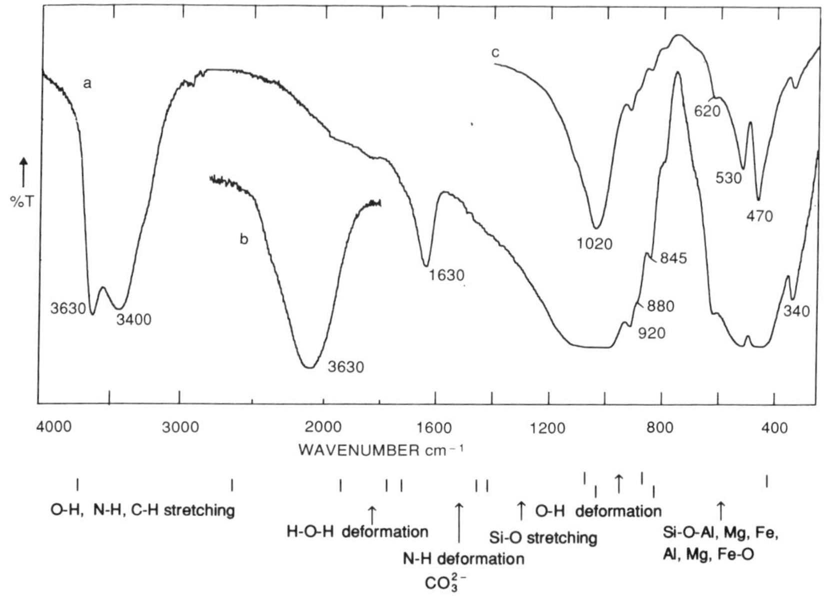

Figure 2.1 Infrared spectra of untreated Camp Berteaux montmorillonite in KBr discs: (a) unheated, 3 mg, (b) 3 mg, heated 16 h, 150°C, (c) 0.25 mg, T = transmission.

transmission (%T) over the range 4000 to 250 (KBr) or 200 cm^{-1} (CsI) on a *Perkin-Elmer 580 B* spectrometer. Three types of spectra are shown, but not for every sample:

- high sample concentration, 1–3 mg (Figure 2.1(a)), usually showing total absorption of radiation in the 1000–1100 and 430 – 530 cm^{-1} silicate regions;
- high concentration, 1–3 mg, after heating the disc to 150°C for 16 h to minimize the absorption of adsorbed water, with the wavenumber scale expanded by a factor of five to provide higher resolution (Figure 2.1(b));
- low concentration, 0.25–0.5 mg (Figure 2.1(c)), showing the spectral detail in the regions of intense silicate absorption near 1000 cm^{-1}.

Absorption bands most characteristic of each particular mineral will be presented, as shown in Figure 2.1 for montmorillonite (Wyoming type, see later for fuller description). Absorption bands of adsorbed molecular water appearing at 3400 and 1630 cm^{-1} will not normally be identified, and neither will C–H absorption bands at 2850–2960 cm^{-1} which arise from organic contaminants.

Also shown in Figure 2.1 are wavenumber ranges for some of the characteristic absorption bands likely to be encountered in spectra of soil clays.

2.4 PREPARATIVE METHODS AND PRETREATMENTS

2.4.1 Particle size

By definition, soil clay minerals are less than 2 μm equivalent spherical diameter, a size that is ideal for examination by IR spectroscopy. Samples with particle sizes much greater than 2 μm must be preground thoroughly in order to minimize scattering of incident IR radiation and the distortion and broadening of absorption bands. Pregrinding is most effectively achieved, with minimal damage to the mineral structure, by keeping the sample moist using a volatile, inert and organic liquid, such as isopropyl alcohol. Grinding may be done mechanically in a stainless steel capsule with ball-bearings, or by hand with a conventional agate mortar and pestle, usually for 1–5 min. The alcohol must be allowed to evaporate completely before proceeding further.

The effect of pregrinding a coarse-grained montmorillonite (Camp Berteaux) can be clearly seen in Figure 2.2, where the broad distorted bands of the original sample (a) become narrower, on a flatter background, and (b) show greatly improved resolution. A small amount of

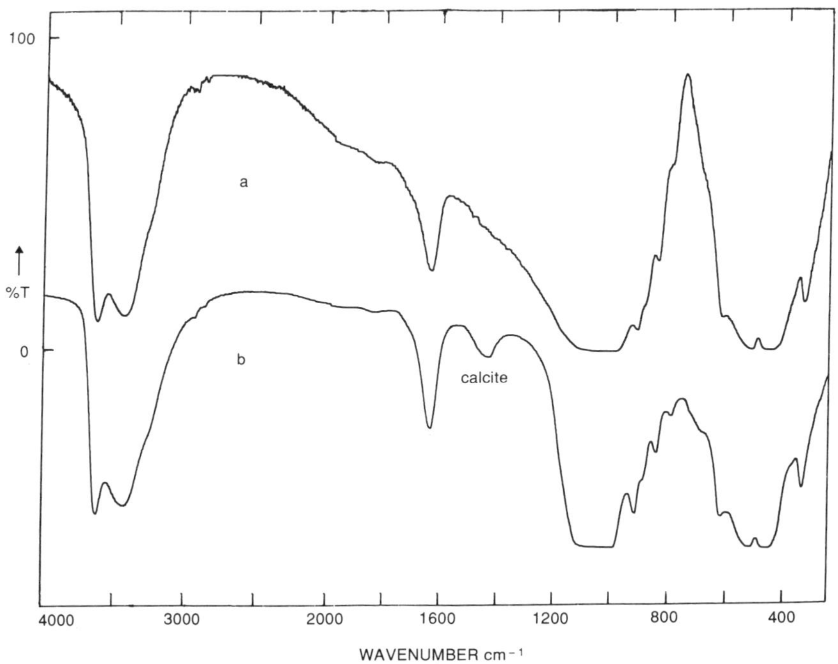

Figure 2.2 Infrared spectra of Camp Berteaux montmorillonite in KBr discs: (a) untreated, 3 mg; (b) preground, 2 mg.

coarse-grained calcite absorbing near 1430 cm^{-1} is scarcely detectable before pregrinding, but is very obvious after.

2.4.2 Removal of organic matter

If a soil clay contains significant organic matter it may be advantageous to remove it, so as to minimize complications in the IR spectrum. This may be done by boiling with H_2O_2, or by treatment with alkaline hypobromite. Although these methods are generally effective for oxidizing organic matter, the former has the disadvantage of producing complex oxalates which may not be completely removed by washing with water (Farmer and Mitchell, 1963). The effect of such contamination will be shown later.

2.4.3 Alkali halide pressed discs

The required weight of clay mineral (1–2 mg) and 170 mg alkali halide are thoroughly mixed for 1–2 min in a vibrating stainless steel capsule containing three ball-bearings (2–3.5 mm diameter). The mixture is then pressed in an evacuated die to give a 13 mm diameter disc. The discs may be heated overnight at 150°C to remove adsorbed water, whose absorption bands overlap some of the spectral regions that are important for clay-mineral identification. Heated discs must be allowed to cool to room temperature in a desiccator before recording the spectra. For routine survey work on soil clays, the KBr pressed disc is the most appropriate method, and is used throughout this chapter to illustrate the utility of IR. For more demanding investigations, such as sorption of organic molecules by smectites or determination of intrinsic structural properties of minerals, other methods of sample preparation may be used.

2.4.4 Sedimented films and deposits

In this method, which is analogous to the oriented aggregate method used in X-ray diffraction studies, the clay mineral is sedimented, usually from 0.5 to 1% w/v aqueous suspension, on to either an IR-transparent plate (AgCl, CaF_2, Irtran 2, Irtran 4, Ge, Si) to give a uniform deposit, or on to a thin plastic sheet from which a self-supporting film of the clay mineral may be peeled after evaporation to dryness. Further details on other aspects of sample preparation for IR analysis have been described by Russell (1974, 1987), White (1964) and van der Marel and Beutelspacher (1976). For the platy-layer silicates, such preparations behave like quasi-two-dimensional crystals and have important properties that help to identify certain clay minerals. To illustrate this concept, a comparison is made between the spectra of talc, from randomly oriented

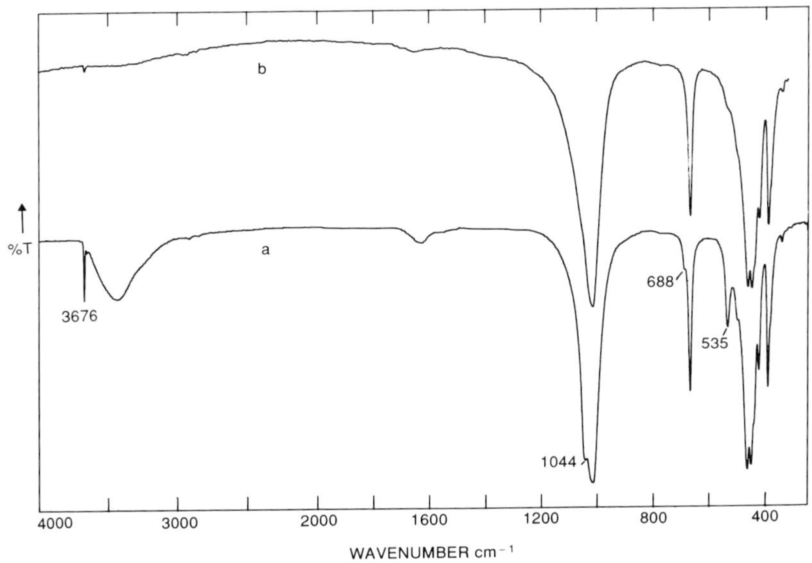

Figure 2.3 Infrared spectra of talc: (a) random particles (KBr disc), (b) oriented particles with c^* parallel to the IR beam, i = 0°.

particles in a KBr pressed disc and from a well-oriented sedimented deposit with the sedimentation plane at 0° incidence to the IR beam (Figure 2.3). Only the bands at 3676, 1044, 688 and 535 cm^{-1} show variation in intensity between the two orientations. They are clearly very much weaker with the film at normal incidence (Figure 2.3(b)) and, therefore, originate in vibrational modes that are perpendicular to the electric vector of the beam (i.e. parallel to the beam direction). The assignments of these bands have been discussed in detail by Russell *et al.* (1970), but the OH-stretching band occurring at 3676 cm^{-1} has the most diagnostic value, because this intensity behaviour, called pleochroism, is exhibited only by triocta- hedral layer silicates. Dioctahedral minerals such as muscovite, pyrophyllite and montmorillonite, in which the O–H dipole lies at 75–80° to the c^* axis, show no change in OH intensity between random orientation and normal incidence. Bands such as those listed above for talc are designated as perpendicular bands, because their principal dipole moment change is perpendicular to the electric vector of the IR beam. Their intensity, in keeping with the relationship described above for random and oriented particles, increases when the angle of incidence between the film and the beam is increased by rotating the film. This is an important and rapid means of distinguishing di- and trioctahedral minerals.

2.5 IDENTIFICATION AND CHARACTERIZATION OF CLAY MINERALS AND ASSOCIATED MINERALS

Spectra are presented of a selection of the layer silicates and associated minerals most likely to be encountered in IR analysis of clay materials. Where possible, the minerals have been selected for their high purity and authenticity. Characteristic absorption features in the spectra suitable for the identification and, if possible, quantification of particular mineral phases are highlighted, along with any other relevant information. The minerals selected are ordered and disordered, crystalline and non-crystalline, the last-named being of considerable importance in some soils. A few examples of how to analyse and interpret soil-clay spectra for their mineral components are then presented, and some guidance is given on keeping misinterpretation to a minimum. Important compilations of spectra have been published by van der Marel and Beutelspacher (1976), Moenke (1962, 1966), Ferraro (1982) and Kodama (1985), and the collection of spectra of inorganic salts by Nyquist and Kagel (1971) is extremely useful. It should be remembered, however, that none of these collections is perfect, incorrect identification or impurities being not uncommon.

2.5.1 Clay minerals

2.5.1.1 Kaolins

Minerals of this group are frequently encountered in clay materials. Kaolinite of variable order and crystallinity and halloysite are the most common, whereas dickite is rare. Nacrite, the other polymorph of kaolinite is even rarer. Spectra of all members show a strong family resemblance (Figure 2.4), particularly over the 1200–250 cm^{-1} range, with variations in absorption pattern in the 1150–1100 cm^{-1} range arising from effects of particle size and shape (Farmer and Russell 1966; Rendon and Serna, 1981). The three pure kaolinite polymorphs, kaolinite, dickite and nacrite, can be readily distinguished by differences in position and relative intensity of their OH-stretching bands (Figure 2.4(a), (b), (c)). The complete assignment of these bands is still in dispute, although it is generally accepted that the band near 3620 cm^{-1} arises from internal OH groups, and that near 3700 cm^{-1} arises from internal surface OH groups. Disorder in kaolinite is detectable mainly in the OH-stretching region, although some general broadening of all bands in the spectrum may also occur. Whereas the 3620 and 3700 cm^{-1} bands are essentially unchanged, the 3669, 3652 cm^{-1} doublet is replaced by a single broad band at 3653 cm^{-1} (Figure 2.4(d)). This pattern is typical of disordered kaolinites and a comparison with the dickite and nacrite spectra (Figure 2.4(b), (c)) suggests that such disorder could arise from a small amount of dickite-

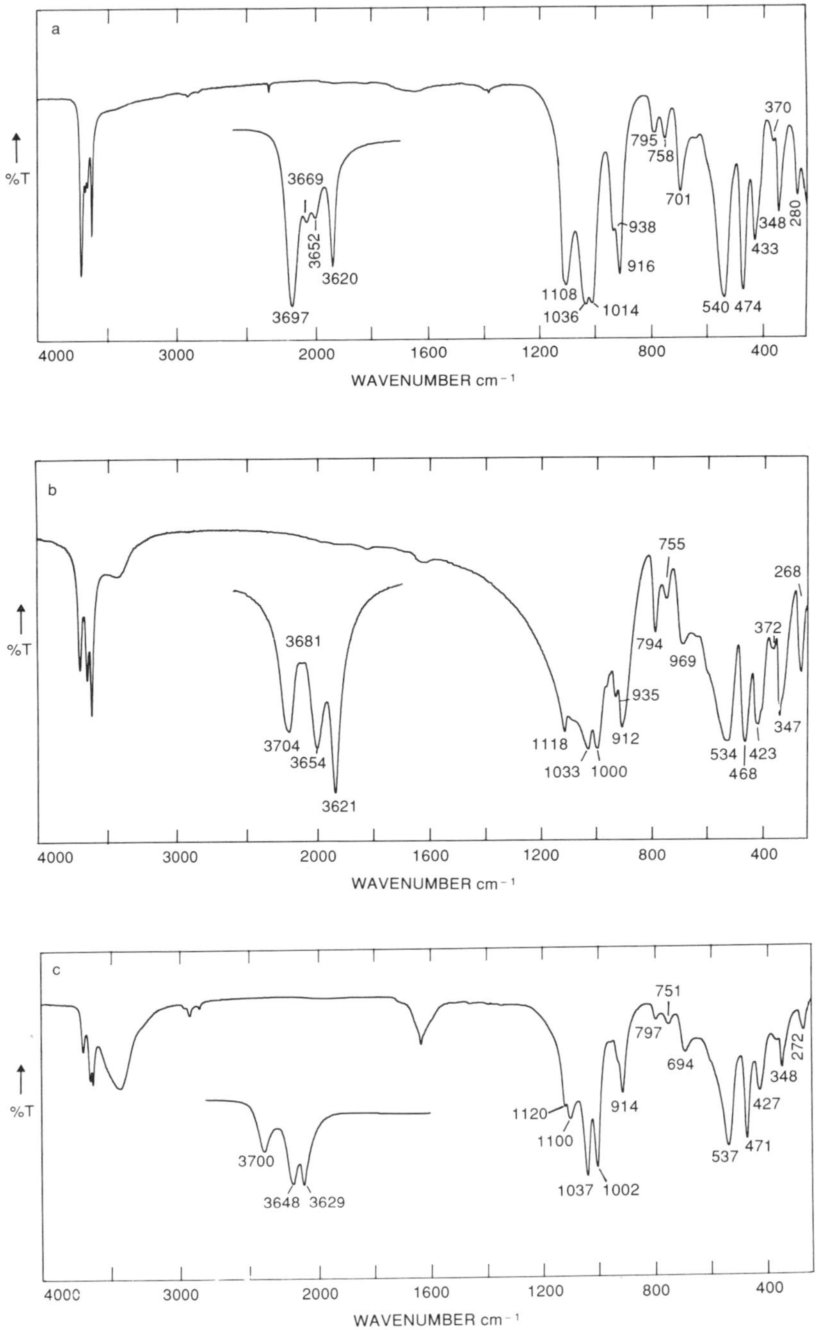

Figure 2.4 Infrared spectra of kaolin minerals: (a) kaolinite, well crystallized (International), (b) dickite (Anglesey), (c) nacrite (Germany), (d) disordered kaolinite (Frantex B), (e) halloysite (Eureka).

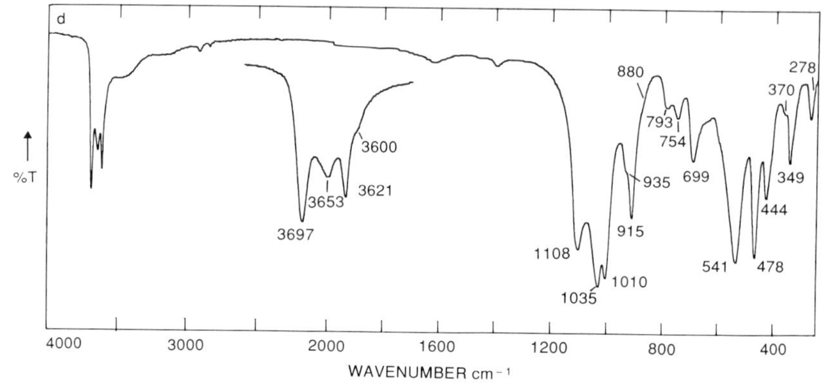

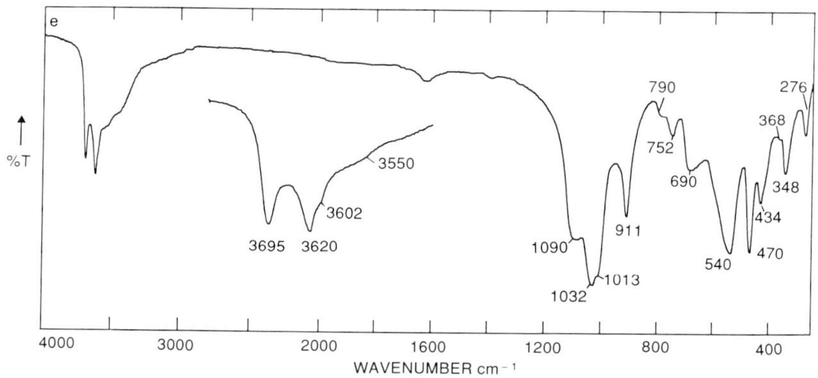

Figure 2.4 (contd).

and/or nacrite-like stacking in the kaolinite structure. This interpretation, originally proposed by Farmer and Russell (1964) and amplified by Farmer (1964), is supported by recent results from the low-temperature studies of Prost *et al.* (1989). Halloysite shows significant broadening of bands throughout its spectrum (Figure 2.4(e)), particularly for the 3700, $3620\,cm^{-1}$ OH-stretching bands because of structural distortion caused by variable hydration (Theng *et al.*, 1982). The weak band near $3550\,cm^{-1}$ is not found in all halloysites and may arise from H-bonding between surface OH groups and interlayer water (Kodama and Oinuma, 1963).

The 3700, $3620\,cm^{-1}$ doublet is characteristic for the kaolin group in general, and a semiquantitative estimate of kaolin content can be made from the absorbance of the 3700 cm^{-1} band, which usually lies well separated from those of most other mineral bands. The OH deformation bands of kaolinite at 938, 916 cm^{-1} (Figure 2.4(a)) are also typical for the kaolin group, and arise from vibrations of the inner and inner surface OH groups, respectively. These assignments have been made following deuterium-exchange studies (Ledoux and White, 1964; Rouxhet *et al.*, 1977). A useful spectral region to help to distinguish between well-crystallized kaolinite and halloysite is between 750 and 800 cm^{-1}. The two weak bands found here at 795 and 758 cm^{-1} are of about equal intensity for kaolinite (Figure 2.4(a)), whereas for halloysite, the 795 cm^{-1} band is reduced to a very weak inflection (Figure 2.4(e)). The relative intensity of this band in disordered kaolinite (Figure 2.4(d)) has an intermediate value.

Another potentially useful means of assessing the crystallinity of kaolinites is based on the ratio of the 3700 and 915 cm^{-1} hydroxyl band absorbancies following the method of Neal and Worrall (1977), who incorrectly but fortuitously used a linear measure of the bands instead of absorbancy. The magnitude of the absorbancy ratio of the bands and the crystallinity of the kaolinites are inversely related. Problems are likely to arise with this method, however, if significant amounts of fine-grained dioctahedral illite and/or dioctahedral smectite are present. Both minerals contribute OH bending absorption at 915 cm^{-1} but relatively little OH stretching absorption at 3700 cm^{-1}, thereby decreasing the ratio of 3700 to 915 absorbancies and indicating an erroneously high value for the crystallinity of the kaolinite.

A weak inflexion at 3600 cm^{-1} and a corresponding weak feature near 880 cm^{-1} (Fig. 2.4(d)) have been correctly interpreted by Mendelovici, Yariv and Villalba (1979) as arising from OH stretching and bending vibrations of $AlFe^{3+}OH$ groupings resulting from Fe^{3+}-for-Al substitution in the octahedral sheet of the kaolinite structure. This type of substitution has been previously associated only with poorly crystallized kaolins, as illustrated here, and Mendelovici, Yariv and Villalba (1979) further concluded that the use of the more exotic alkali halides CsBr or CsI was necessary before the $AlFe^{3+}OH$ bands were fully exhibited. However, these bands have been observed recently in KBr disc spectra of well-crystallized kaolinites of lateritic origin (Russell and Fraser, unpublished).

2.5.1.2 *Serpentine minerals*

Minerals in this group include chrysotile and antigorite, trioctahedral analogues of kaolinite with very different morphologies. The spectra (Figure 2.5(a), (b)) show fundamental differences from those of the dioctahedral minerals, namely, a lower frequency for the principal Si–O stretching band at or below 1000 cm^{-1}, a single lower-frequency band for

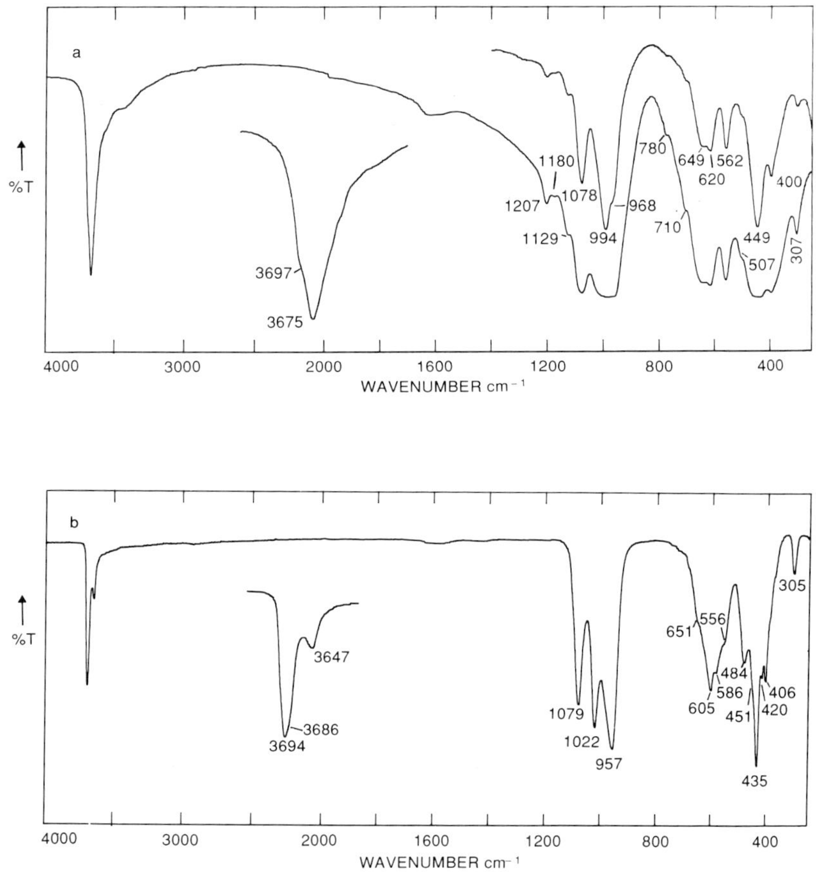

Figure 2.5 Infrared spectra of serpentine minerals: (a) antigorite (Aberdeenshire), (b) chrysotile (Canada).

Si–O–Mg near 460–450 cm^{-1} compared with two or three strong bands in this area for the dioctahedral minerals, and an OH deformation band at a much lower frequency near 700–600 cm^{-1}, compared with 950–900 cm^{-1}. These di–tri differences are also well illustrated by pyrophyllite and talc (Figure 2.6). Chrysotile and antigorite spectra are significantly different from those of the kaolin minerals and are easily distinguished from each other, particularly in the OH-stretching region. Neither pattern, however, shows clear resolution of bands that might be assigned to inner and inner-surface OH groups, as in kaolinite. These serpentine minerals are readily identified in clays by their characteristic absorption bands, but as with all minerals, the whole spectrum must be considered for reliable identification.

2.5.1.3 *Pyrophyllite and talc*

Although these non-expanding layer silicates occur only rarely in clay materials, they are worthy of inclusion here because they can be regarded as the parent unsubstituted structures from which the smectites and mica minerals are derived by isomorphous substitution. Spectral shifts resulting from this substitution have been considered in detail by Farmer and Russell (1964) and Farmer (1974). Spectra of pyrophyllite and talc (Figure 2.6) show the differences in peak positions already reported for di- and trioctahedral minerals. The absorption bands are

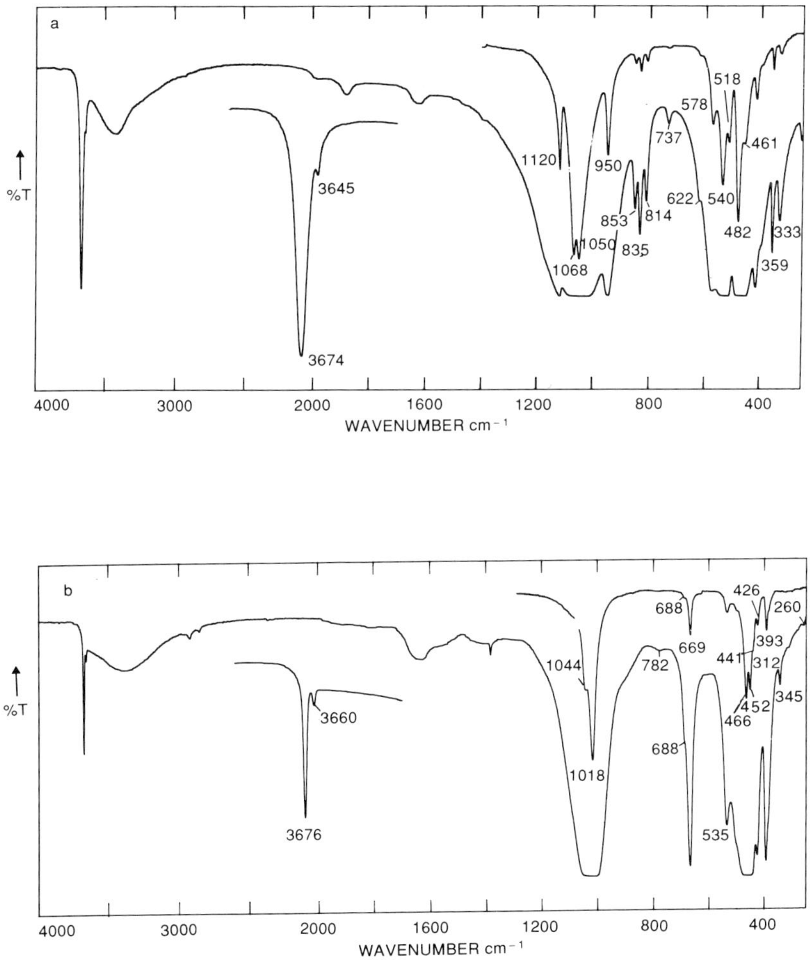

Figure 2.6 Infrared spectra of (a) pyrophyllite, (b) talc.

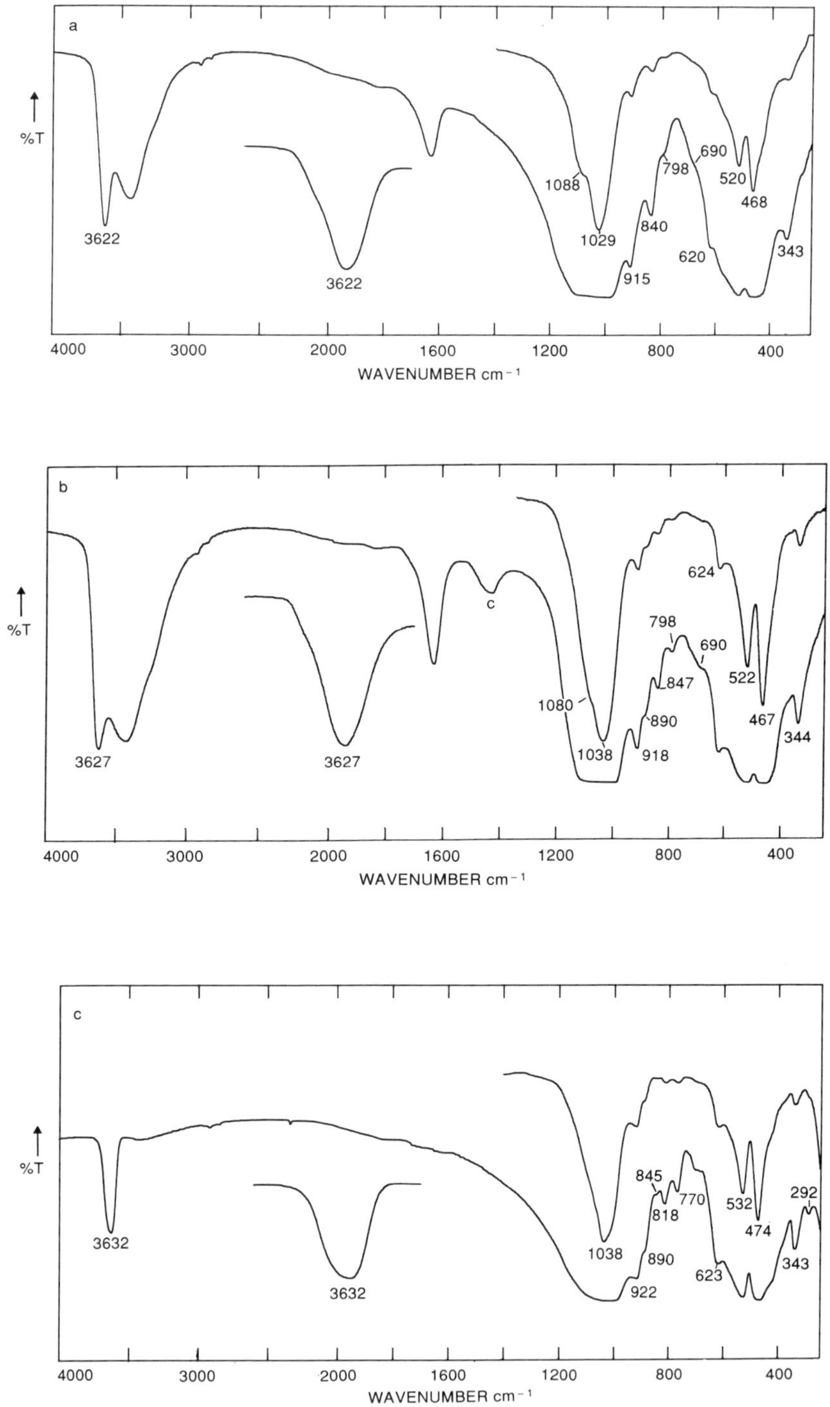
a
%T
3622
3622
1088
1029
840
915
798
690
620
520
468
343
4000
3000
2000
1600
1200
800
400
WAVENUMBER cm⁻¹
b
%T
3627
3627
c
1080
1038
890
918
847
798
690
624
522
467
344
4000
3000
2000
1600
1200
800
400
WAVENUMBER cm⁻¹
c
%T
3632
3632
1038
845
818
770
890
922
623
532
474
292
343
4000
3000
2000
1600
1200
800
400
WAVENUMBER cm⁻¹

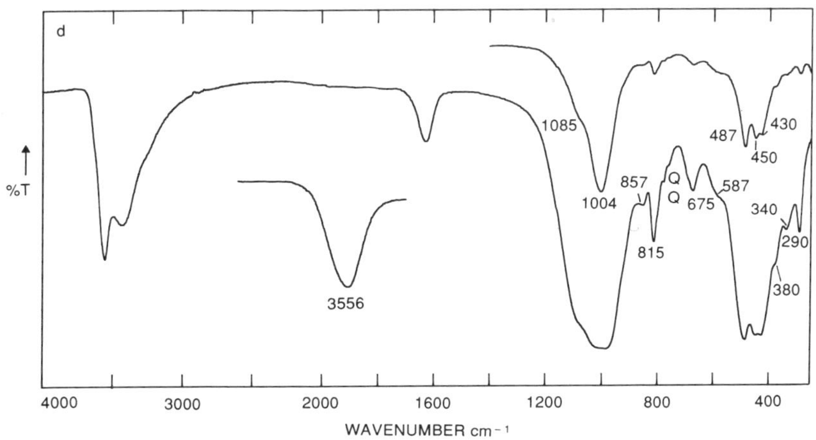

Figure 2.7 Infrared spectra of dioctahedral smectites: (a) montmorillonite (Otay), (b) montmorillonite (Camp Berteaux[1]), (c) beidellite-rich clay (Unter-rupsroth[2]), (d) nontronite (California[3]).
[1]1430 cm^{-1} band due to calcite. [2]Nadeau *et al.* (1985). [3]Goodman *et al.* (1976).

sharp, as anticipated for well-crystallized, ordered minerals, but it can be seen that the OH-stretching band for the dioctahedral pyrophyllite (Figure 2.6(a)) is considerably broader than that of talc. In talc, a much more symmetrical environment exists for OH groups because of full octahedral occupancy by Mg ions, higher symmetry and crystalline order. The OH groups in talc are pleochroic, as discussed in section 2.4 (Figure 2.3), whereas in the dioctahedral pyrophyllite the OH groups show no such effect. The weak, low-frequency satellite bands at 3645 cm^{-1} for pyrophyllite and at 3660 cm^{-1} for talc arise from small amounts of iron in these minerals, producing the groupings $AlFe^{3+}OH$ and $Mg_2Fe^{2+}OH$ respectively. A very weak inflection, just discernible near 900 cm^{-1} (Figure 2.6(a)) may be the deformation vibration of the former grouping, although an alternative explanation is also possible (Russell, Farmer and Velde, 1970).

2.5.1.4 *Smectites*

(i) Dioctahedral smectites

Montmorillonites contain both tetrahedral and octahedral isomorphous substitution, Al (and occasionally Fe^{3+}) for Si in the former case, and Fe^{3+} and Mg for Al in the latter. As a result of these substitutions, crystalline order is reduced and structural imperfections arise. These leading to considerable broadening of the IR absorption bands compared with those of the essentially unsubstituted pyrophyllite. Low-iron montmorillonite, so-called Cheto-type (Figure 2.7(a)), shows a typically broad OH-stretching band at 3622 cm^{-1}, which is the overall envelope for a wide range of AlAlOH and AlMgOH environments in the highly substituted and distorted structure. This band occurs close to that of the inner OH groups in kaolinite (Figure 2.4(a)) but can be distinguished by its much greater breadth. More characteristic, however, are the well-resolved OH deformation bands at 915 cm^{-1} (AlAlOH) and 840 cm^{-1} (AlMgOH), assignments that were confirmed by deuteration (Russell, Farmer and Velde, 1970). The spectrum of a so-called Wyoming-type montmorillonite (Figure 2.7(b)) in which there is some replacement of octahedral Al by Fe^{3+}, is very similar to that of the iron-free type (Figure 2.7(a)), apart from some relatively minor shifts in peak positions. The most significant difference between the two is the presence of the OH deformation band of $AlFe^{3+}OH$ groupings at 890 cm^{-1} in the Wyoming-type, this being diagnostic of iron substitutions in the smectite structure. The small amount of calcite (which absorbs near 880 cm^{-1}, Figure 2.21(b)) in the montmorillonite would make little or no contribution to the intensity of the 890 cm^{-1} band. It is uncertain whether the weak band at 798 cm^{-1} arises from montmorillonite, a band at this frequency having been ascribed to a platy form of tridymite in a montmorillonite (Wilson, Russell and Tait, 1974).

In contrast to montmorillonite, substitution in beidellite is exclusively Al for Si in the tetrahedral sheet. This type of substitution gives rise to a quite different pattern of absorption bands. Although the beidellite spectrum generally has a strong montmorillonite character (Figure 2.7(c)), the two bands at 818 and 770 cm^{-1} are uniquely characteristic. Their origins are Al–O out-of-plane, and Al–O–Si in-plane, vibrations, and are analogous to similar tetrahedral sheet vibrations in muscovite (Farmer, 1974). The sample of beidellite illustrated contains some montmorillonite (Nadeau *et al.*, 1985), as shown by the OH deformation bands at 890 and 845 cm^{-1}, and the lower OH-stretching frequency at 3632 cm^{-1} compared with 3661 cm^{-1} in pure beidellite.

With increasing substitution of Al by Fe^{3+} in montmorillonite, the $AlFe^{3+}OH$ deformation band shifts from 890 to 870 cm^{-1} and, when sufficient iron is present to give $Fe^{3+}Fe^{3+}OH$ groupings as in nontronite, a band appears at 815 cm^{-1} (Figure 2.7(d)). This grouping is dominant in nontronite, and gives an OH-stretching frequency of 3556 cm^{-1}, some 100 cm^{-1} lower than the equivalent AlAlOH band in beidellite. This low-frequency shift of OH-stretching and deformation bands with increasing Fe^{3+} content is a very important criterion for the determination of ferruginous smectites in clay materials. Other characteristic bands for nontronite are found at 675 cm^{-1}, which is an Fe^{3+}–O out-of-plane vibration, and 290 cm^{-1}. A series of iron-rich smectites and nontronites has been studied by Goodman *et al.* (1976).

(ii) Trioctahedral smectites

Hectorite is formed when about one in nine of the octahedral Mg cations in talc is replaced by Li; it is the trioctahedral equivalent of montmorillonite.

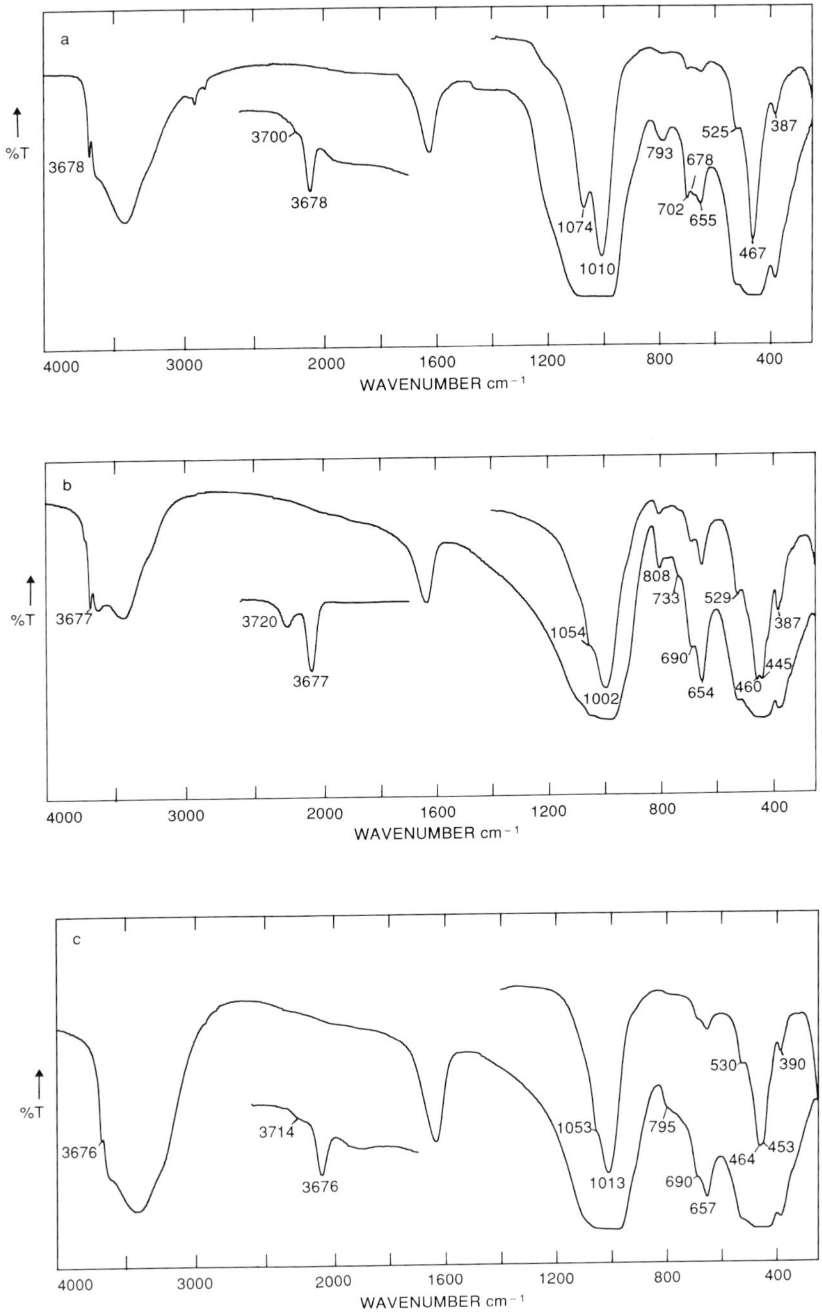

Figure 2.8 Infrared spectra of trioctahedral smectites: (a) hectorite (Hector, USA), (b) saponite (Allt Ribhean, Scotland); (c) stevensite (Arizona).

Its spectrum (Figure 2.8(a)) shows marked perturbation from that of talc (Figure 2.6(b)), i.e. the bands are broadened and shifted. Similarly, for saponite, where about 1 in 12 of the tetrahedral Si is replaced by Al, the spectrum (Figure 2.8(b)) shows significant displacement from that of talc but less than observed for hectorite. Both saponite and hectorite show the typical trioctahedral spectrum, but their OH-stretching bands are characteristic. These bands occur within a few wavenumbers of that of talc, but are markedly broadened, the result of lattice distortion and crystal imperfection caused by the isomorphous substitutions. Again, substitution of Mg by Fe^{2+} produces one or more of a group of three bands at lower frequency, depending on iron content. These bands are analogous to those in iron-substituted talc, and occur at 3663 (MgMgFe), 3646 (MgFeFe) and 3642 cm^{-1} (FeFeFe) (Wilkins and Ito, 1967). Saponite and hectorite are potentially difficult to distinguish from their IR spectra, but a reaction occurs in KBr pressed discs after heating to 150°C to remove absorbed water, and this is diagnostic. Under these conditions, saponite produces a high-frequency satellite at 3720 cm^{-1} on the OH-stretching band (Figure 2.8(b)). It is an effect produced by interlayer K^+ ions (from the KBr), electrostatically repelling the proton of the OH group, thereby raising its frequency. The effect occurs in saponite because of the highly localized negative charge on the Si–O surface, resulting from Al-for-Si substitution. It is a much weaker effect in hectorite (Figure 2.8(a)), where the Li substitution for octahedral Mg results in a more delocalized surface charge. The effect is also very weak in stevensite (Figure 2.8(c)), where the layer charge is derived from a deficiency of octahedral Mg ions, again resulting in the same delocalization of charge at the Si–O interlayer surface. The spectrum of stevensite is otherwise very similar to those of saponite and hectorite, but the weakness of the K-effect band distinguishes it from saponite. Chemical analysis would also make this distinction because stevensite contains no Al, only Si and Mg, as for talc.

2.5.1.5 *Vermiculite, biotite and phlogopite*

Identification of trioctahedral vermiculite is difficult by IR spectroscopy, because there are no sharp, characteristic absorption bands that might be recognized in the complex pattern of a mineral mixture. This can be seen from its spectrum (Figure 2.9(a)), which, although obviously of the trioctahedral type, shows broad, rounded absorption bands that generally resemble saponite. The only recognizably different feature is the OH-stretching band, which is broad, strong and at low frequency. These properties indicate that, although vermiculite is nominally trioctahedral, some of the OH groups are associated with vacancies, and are in a locally dioctahedral environment. Such vacancies are produced by ejection of octahedral Fe ions during oxidation and vermiculitization of biotite

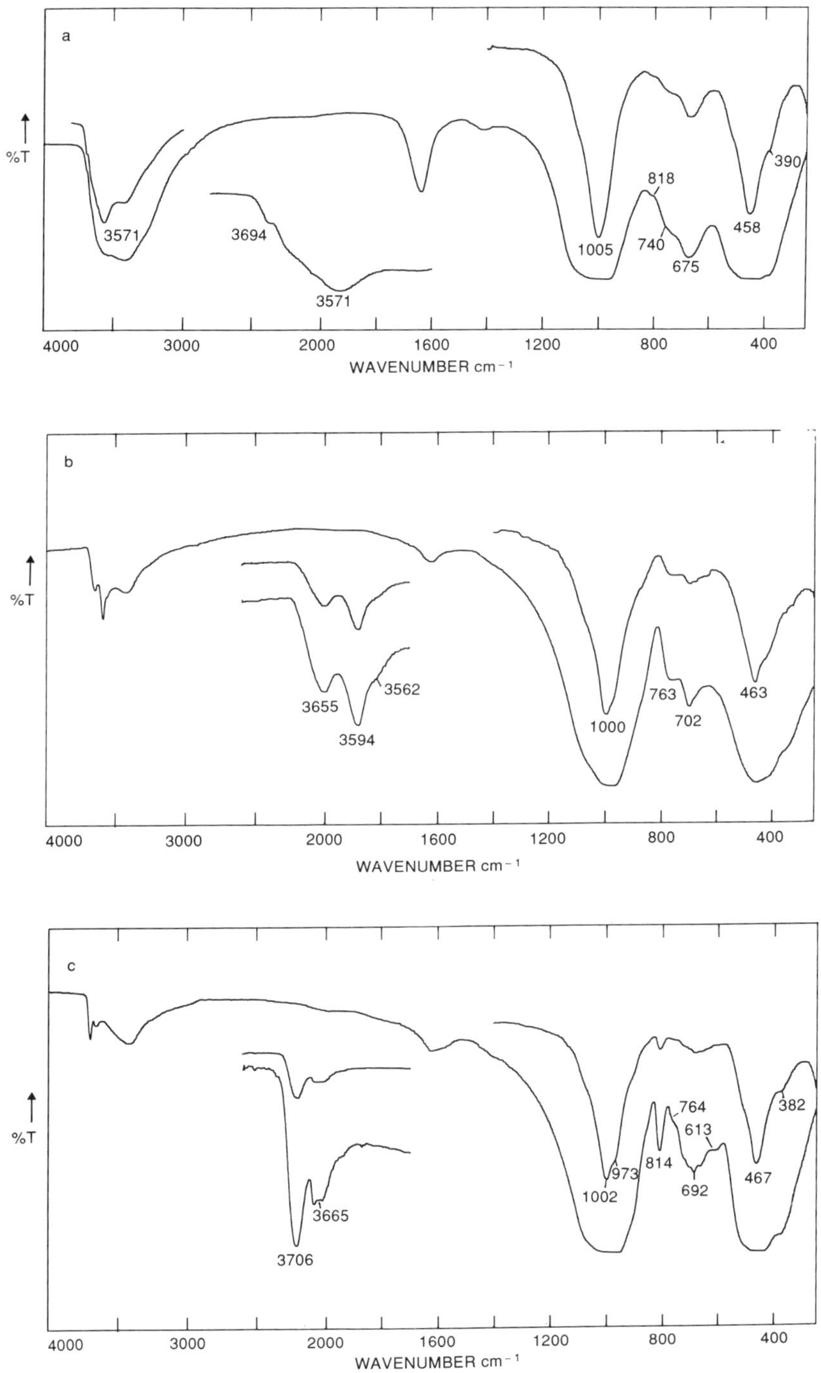

Figure 2.9 Infrared spectra of (a) vermiculite (G. Walker), (b) biotite, (c) phlogopite.

(Farmer *et al.*, 1971). The resulting band at 3571 cm^{-1} (Figure 2.9(a)) is close to that for nontronite (Figure 2.7(d)).

Spectra of biotite, which vermiculite closely resembles, and phlogopite (Figure 2.9(b), (c)) show a pattern of OH-stretching bands that vary more or less continuously from the iron-free phlogopite to the iron-rich biotite shown here. Detailed treatments of these spectra can be found in Vedder (1964), Juo and White (1969) and Farmer and Russell (1964).

2.5.1.6 *Illite*

The composition of illite is quite variable, as is its IR spectrum. Nevertheless, two main types of spectrum have been identified, being associated with muscovite-like (Figure 2.10(a)) and phengite-like (Figure 2.10(b))

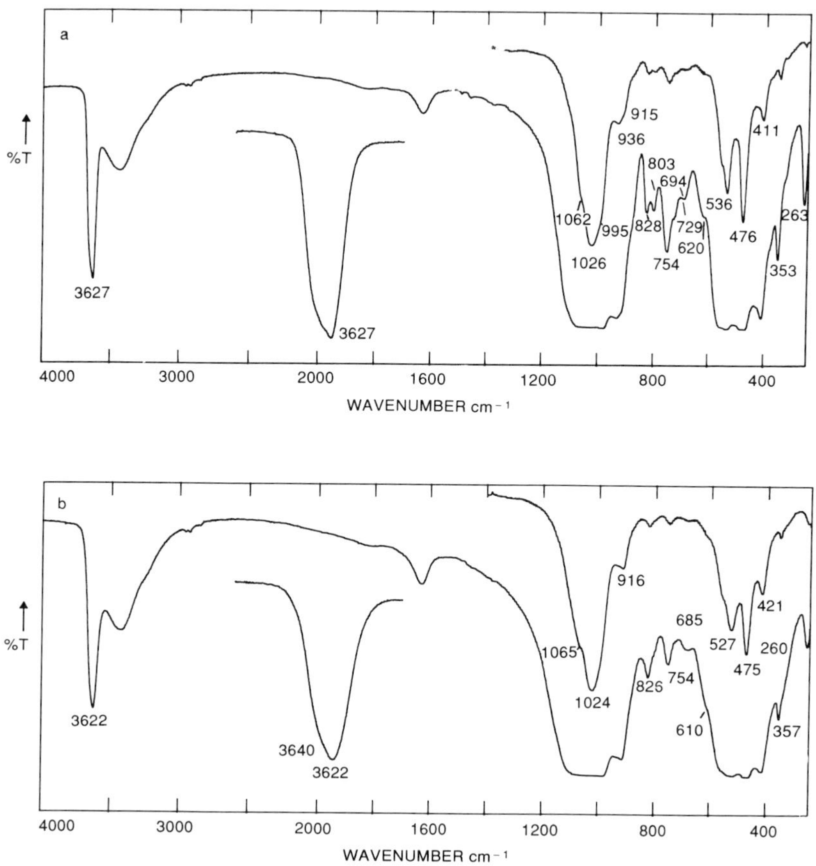

Figure 2.10 Infrared spectra of illite: (a) 2 M_1 illite (muscovite-like); (b) 1M illite (phengite-like).

illites. An important chemical difference is that phengite has a much higher Mg content than muscovite. The illite spectra can range between these two types and may show even weaker absorption bands near 825 and 750 cm^{-1} than in the spectra illustrated. Consequently, it can be difficult to identify the type of illite, but determination of the presence of the mineral usually presents little difficulty. The broad OH-stretching band near 3625 cm^{-1}, ranging down below 3620 cm^{-1} in some instances, coupled with the 825, 750 cm^{-1} doublet is almost certainly diagnostic. The variety most frequently encountered in soil clays is the phengite type, with relatively weak bands at 825 and 750 cm^{-1}. These bands originate from Al–Mg–OH deformation and, by analogy with muscovite

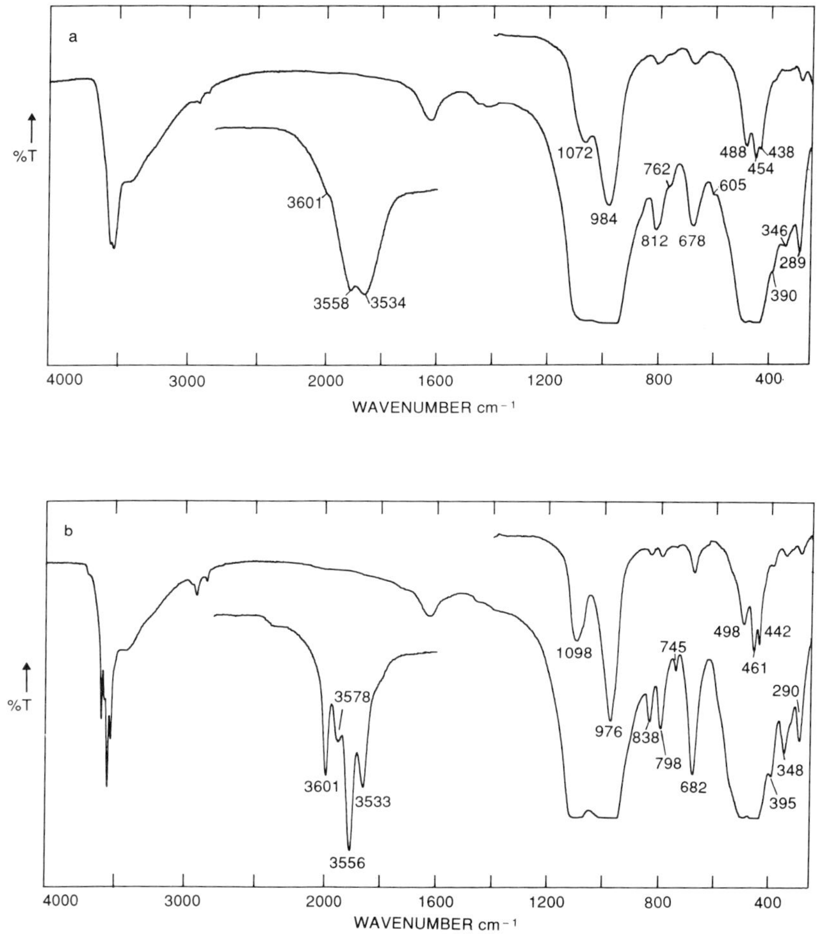

Figure 2.11 Infrared spectra of (a) glauconite; (b) celadonite. Buckley *et al.* (1978).

(Farmer, 1974), Al–O–Si in-plane vibration. Diagenetic illites sometimes contain interlayer NH_4^+ ions replacing K^+ (Nadeau and Bain, 1986), the NH_4^+ ions absorbing at 3260 and 1430 cm^{-1}.

2.5.1.7 Glauconite and celadonite

These iron-rich micas derive most of their layer charge from octahedral replacement of trivalent (Al, Fe^{3+}) by divalent cations (Mg, Fe^{2+}). Their structures are characterized by pairing of a trivalent and a divalent cation in the octahedral sheet. In glauconite there is also significant tetrahedral substitution and this can lead to disorder and broadening of bands, as can be seen in Figure 2.11(a). Nevertheless, the poorly resolved OH-stretching bands can be distinguished at 3601 (AlMgOH), 3558 (Fe^{3+}MgOH) and 3534 cm^{-1} ($Fe^{3+}Fe^{2+}$OH) (Farmer *et al.*, 1967). The spectrum of the celadonite shown here is much sharper (Figure 2.11(b)) and testifies to a more highly ordered structure. The four clearly resolved OH-stretching bands correspond with all possible tri–divalent pairings from an octahedral population containing Al, Mg, Fe^{2+}, Fe^{3+}. The assignments are those detailed for glauconite, the band at 3578 cm^{-1} being attributed to (AlFe^{2+}OH). This is a very characteristic pattern, which varies with composition but which is easily recognized, as is the Si–O pattern. The OH deformation bands for the major octahedral pairings occur at 838 (AlMgOH) and 798 cm^{-1} (Fe^{3+}MgOH) in celadonite (Figure 2.11(b)), whereas in glauconite the unresolved doublet at 812 cm^{-1} (Fe^{3+}MgOH and $Fe^{3+}Fe^{2+}$OH) is characteristic (Figure 2.11(a)). Buckley *et al.* (1978) proposed that celadonite was ordered (sharp spectrum), glauconite was disordered (broad spectrum), and that they were two distinct mineral types. Slonimskaya *et al.* (1986), however, have shown from IR spectroscopy that both minerals can be ordered or disordered, that the sharpness of the spectrum is not diagnostic, and that there is probably a continuous series between the two minerals.

2.5.1.8 Chlorites

The compositions and, therefore, the IR spectra of chlorites vary widely. Figure 2.12(a) shows the spectrum of a fairly iron-rich type, also containing aluminium (brunswigite). The strong broad doublet at 3565, 3434 cm^{-1} arises from OH-stretching vibrations of the interlayer hydroxide sheet. The exact position of this doublet varies with composition, shifting to lower frequency with increasing aluminium and iron content, but is distinctive for a fully trioctahedral chlorite. The weak feature at 3662 cm^{-1}, sometimes sharp and well-resolved at higher frequency, arises from OH groups in the trioctahedral 2 : 1 layer, similar to that observed in the spectra of biotite and phlogopite (Figure 2.9(b), (c)). The strong band at 667 cm^{-1} (Figure 2.12(a)) comprises OH deformation

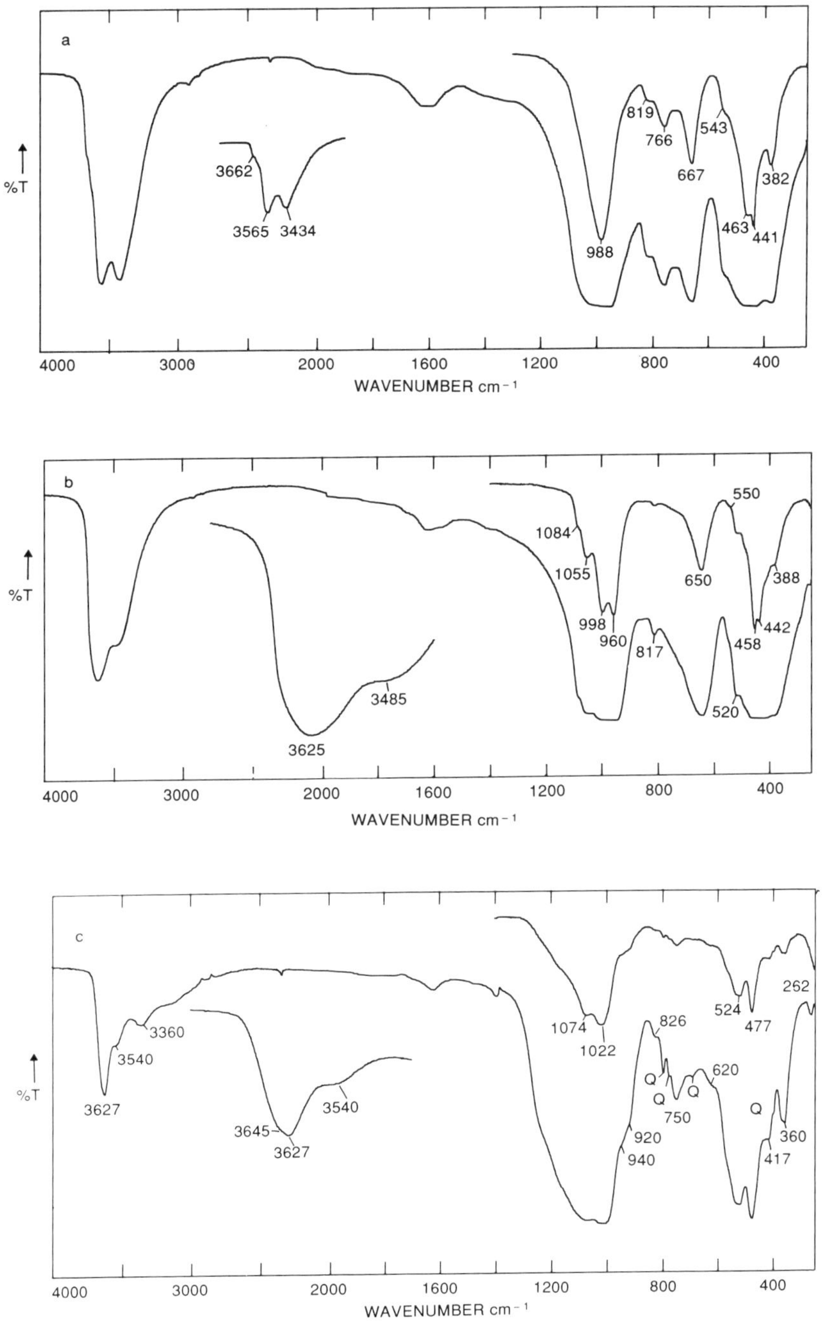

Figure 2.12 Infrared spectra of chlorites: (a) brunswigite, (b) penninite (Zermatt), (c) sudoite.

components from the 2 : 1 layer and the hydroxide sheet. The broad OH-stretching doublet of chlorite in clay materials is frequently masked by the stretching band of adsorbed water, removal of which is essential for unambiguous identification.

The IR spectrum of the magnesium-rich penninite (Figure 2.12(b)) is generally much sharper in the silicate region than that of brunswigite. This arises because of the lower levels of isomorphous substitution in penninite in both the 2 : 1 and the hydroxide sheet. The OH-stretching bands of the latter occur at 3625 and 3485 cm^{-1}, the higher frequency band moving towards the unsubstituted brucite position 3698 cm^{-1}. The marked splitting of the principal Si–O band at 998, 960 cm^{-1} reflects the relatively low level of Al-for-Si substitution in the 2 : 1 layer (Tuddenham and Lyon, 1959), and this can be an important diagnostic feature.

With increasing substitution of Mg by Al, chlorites become more dioctahedral. The aluminous chlorite, sudoite, shows three strong bands in the OH-stretching region (Farmer, 1974; Kodama, 1985), as shown in Figure 2.12(c). The band at 3627 cm^{-1} is much stronger than the other two bands, suggesting a stronger contribution from a more dioctahedral 2 : 1 layer. Another feature indicating that the 2 : 1 layer resembles that of muscovite is the band at 750 cm^{-1}. However, the weak 3540, 3360 cm^{-1} OH doublet (Figure 2.12(c)) suggests that the hydroxide sheet still retains some trioctahedral character.

2.5.1.9 *Interstratified minerals*

Infrared spectroscopy responds to interstratified clay minerals as if they were a simple mixture, their IR components being additive. The only exception known to the authors is rectorite, which has a spectrum showing unique bands not present in the spectra of its separate components.

The spectrum of an illite–smectite interstratification (Figure 2.13(a)) closely resembles that of a Cheto-type montmorillonite (Figure 2.7(a)), with a contribution from a phengite-like illite (Figure 2.10(b)). The weakness of the 756 cm^{-1} illite band should not necessarily be taken to indicate a minor amount of illite, because, as indicated previously, this band may be inherently weak in some illites.

Similarly, the spectrum of aliettite (Figure 2.13(b)) shows features that can be ascribed to its separate components, talc and saponite (Figures 2.6(b), 2.8(b)). Its absorption bands have intermediate breadth and resolution. Specifically, saponite is indicated by the 3715 cm^{-1} K^+-effect band and the relatively broad $MgFe^{2+}Fe^{2+}OH$ band at 3633 cm^{-1}. Talc is signalled by the overall sharpness of all bands, but particularly by the $MgMgFe^{2+}OH$ band at 3661 cm^{-1}, which has the same width at half height as the Mg_3OH 3676 cm^{-1} band (Figure 2.6(b)).

The spectrum of rectorite (Figure 2.13(c)), which is a regular alternation of the Na-mica paragonite and Na-saturated beidellite, obviously

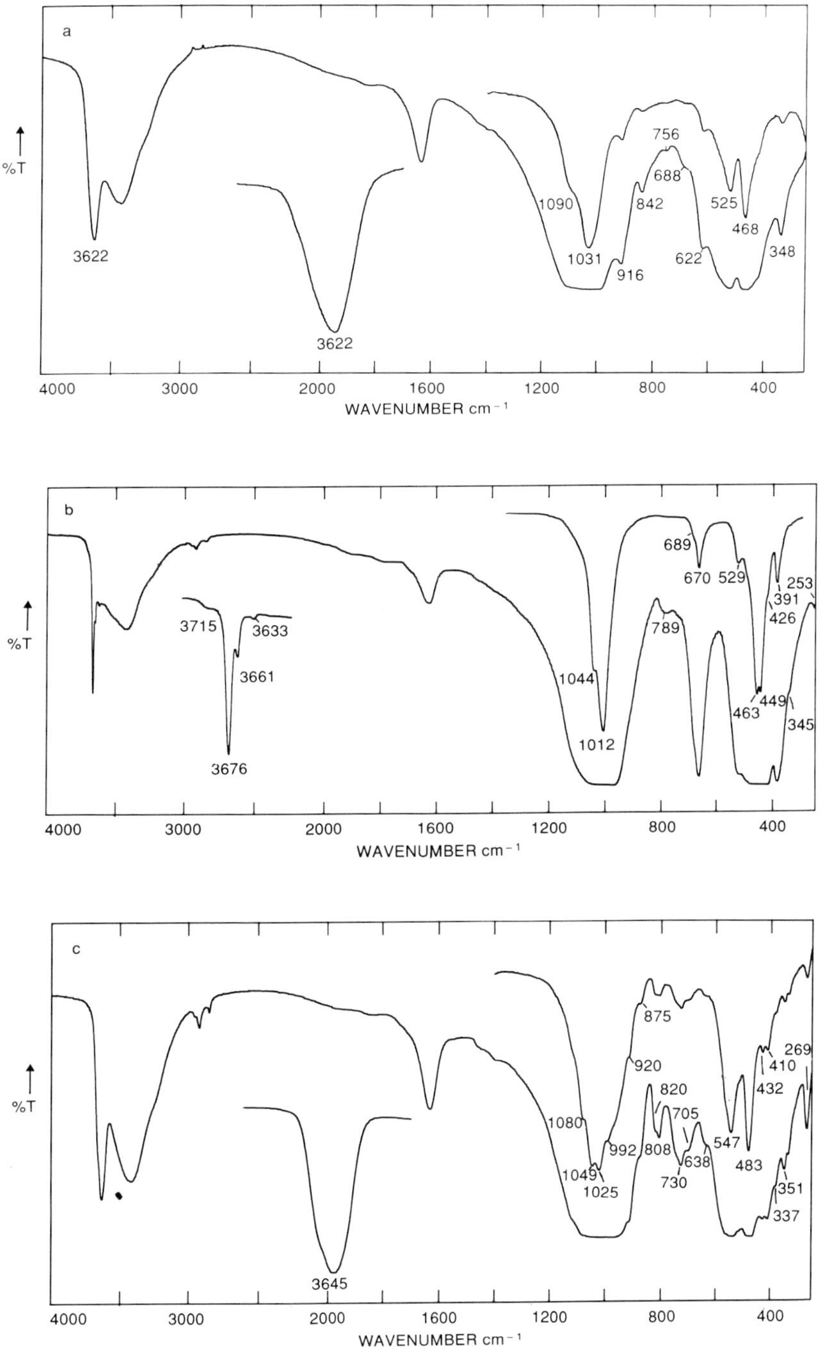

Figure 2.13 Infrared spectra of interstratified minerals: (a) illite–smectite (Nadeau, 1980), (b) aliettite, (c) rectorite (Baluchistan).

resembles those of muscovite (closely similar to paragonite) (Kodama, 1985) and beidellite (Figure 2.7(c)) (Farmer and Russell, 1964). However, there are many additional bands, and a spectral sharpness not shown by either component. For example there are bands at 1120, 992, 920, 808, 730 and 337 cm^{-1}, which suggest a structural integrity and order beyond that implied by simple interstratification.

In other interstratified minerals, for example kaolinite–smectite, both components are relatively easy to identify from their IR spectra. In chlorite–vermiculite, however, only one component can be recognized because the chlorite pattern dominates and completely overlies that of the vermiculite, to which it has many similarities. It should perhaps be stressed that IR cannot distinguish between regular and random interstratifications.

Other examples of spectra of interstratified minerals can be found in Kodama (1985), van der Marel and Beutelspacher (1976) and Oinuma and Hayashi (1968), which the interested reader should consult.

2.5.1.10 Palygorskite and sepiolite

Despite their unique channel structures, these minerals are similar to layer silicates intermediate between the di- and trioctahedral types respectively. In palygorskite, the structural OH-stretching bands at 3615 cm^{-1}, with a weak inflection at 3690 cm^{-1} (Figure 2.14(a)) originate in the aluminous octahedral layer, which must contain both vacancies and complete site filling. The dioctahedral character is dominant, as shown by the well-defined AlAlOH deformation band at 913 cm^{-1}. The unique nature of the 2 : 1 ribbon layers in the structure gives the complex Si–O stretching pattern, the most characteristic features of which are the high- and low-frequency components at 1195 and 990 cm^{-1}.

The more completely trioctahedral sepiolite shows a dominant OH-stretching band at 3686 cm^{-1} (Figure 2.14(b)), slightly higher than that of talc (Figure 2.6(b)) and antigorite (Figure 2.5(a)), with which it might be compared, and a typically trioctahedral OH deformation band within the 690, 646 cm^{-1} doublet. Again, the unique ribbon layer structure gives rise to the very characteristic Si–O vibration at 1208 cm^{-1}, one of the highest recorded for layer silicates. The bands near 1200 cm^{-1} for palygorskite and sepiolite may well have their origin in the virtually straight Si–O–Si bonds linking the 2 : 1 ribbons along their edges. The only other silicate mineral absorbing near 1200 cm^{-1} is sillimanite, the origin of the band being the straight Si–O–Al bond linking tetrahedral silicate and aluminate chains (Farmer, 1974).

A striking feature of palygorskite and sepiolite spectra is the complex nature of absorption bands in stretching and bending regions of water

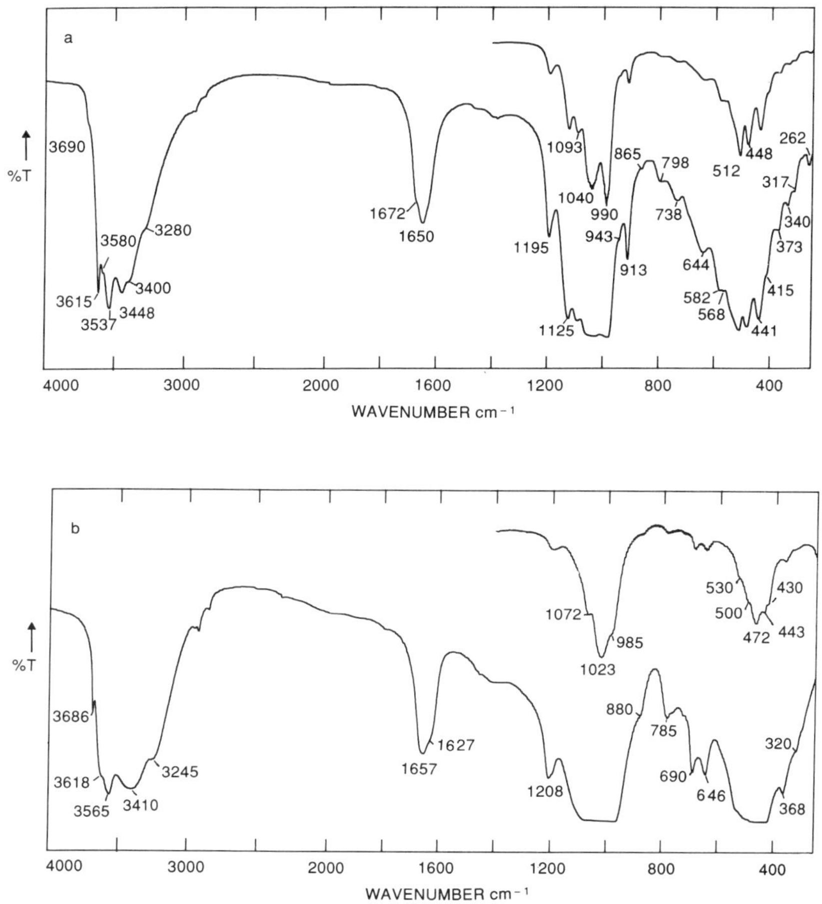

Figure 2.14 Infrared spectra of (a) palygorskite (Cabrach, Scotland), (b) sepiolite (Spain).

molecules. Both minerals contain water in channels in their structures, and even in KBr discs, as here, they display patterns of bands originating in bound molecular water directly coordinated to Al or Mg ions at the edges of the 2 : 1 ribbon layers (3580, 3537, 1650 cm^{-1} in palygorskite; 3618, 3565, 1627 cm^{-1} in sepiolite), and unbound zeolitic water molecules within the channels (3448, 3400, 3280, 1672 cm^{-1} for palygorskite; 3410, 3245, 1657 cm^{-1} for sepiolite). The characteristic IR pattern of the water molecules in these minerals has been studied by many workers, among them Hayashi, Otsuka and Imai (1969), Tarasevich (1970) and, more recently, Serna, Vanscoyoc and Ahlrichs (1977).

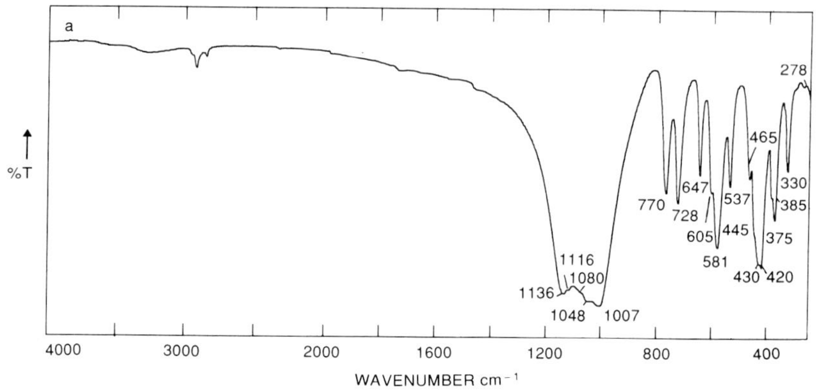
a
%T
278
465
330
385
375
647
537
770
728
605
445
581
430
420
1116
1080
1136
1048
1007
4000
3000
2000
1600
1200
800
400
WAVENUMBER cm⁻¹

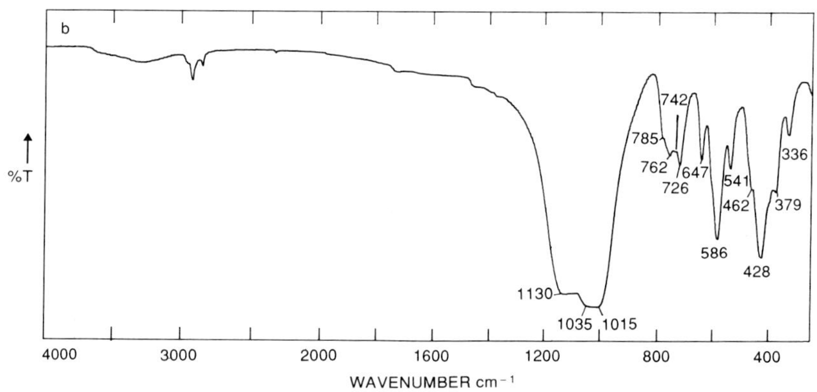
b
%T
742
785
762
647
726
541
462
336
379
586
428
1130
1035
1015
4000
3000
2000
1600
1200
800
400
WAVENUMBER cm⁻¹

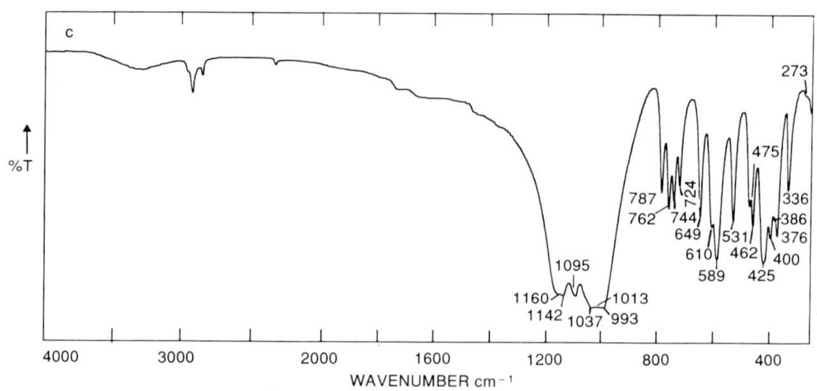
c
%T
273
475
787
724
762
744
649
610
531
462
336
386
376
400
589
425
1095
1160
1142
1013
1037
993
4000
3000
2000
1600
1200
800
400
WAVENUMBER cm⁻¹

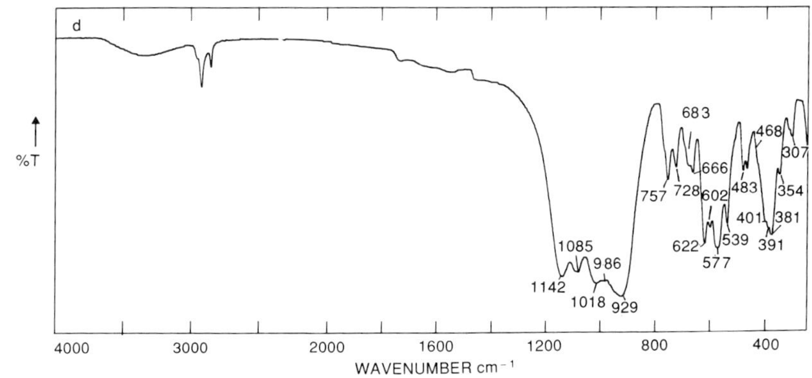

Figure 2.15 Infrared spectra of feldspars: (a) microcline, (b) orthoclase, (c) albite, (d) anorthite.

2.5.2 Other minerals

2.5.2.1 Feldspars

These framework silicate minerals are frequent constituents of clay materials. Despite their extensive mutual solid solution and complexity of composition, they can be distinguished by their IR spectra, at least at principal species level. The spectra of microcline and albite (Figure 2.15(a), (c)), which are close to the K and Na end-member compositions respectively, are closely related, but the quartet of bands between 700 and 800 cm^{-1} and the sharp feature at 400 cm^{-1} are particularly characteristic of albite (Figure 2.15(c)). These bands can also be clearly seen in the spectrum of orthoclase (Figure 2.15(b)), identifying the presence of an albite component in this predominantly K-feldspar. Anorthite, the Ca-plagioclase end-member, has a related but uniquely different spectrum (Figure 2.15(d)). It too has a very complex pattern of absorption bands in the 800–200 cm^{-1} region, but at 929 cm^{-1} there is an intense Si–O vibration not seen in the other feldspars. Kodama (1985) has presented a short series of plagioclase-feldspar spectra in which the steady enhancement of absorption at about 920 cm^{-1} correlates well with the increasing proportion of anorthite. Feldspars show wide variation in crystalline order, and this is reflected in the IR spectra, which can range from broad and diffuse to sharp and well defined.

2.5.2.2 Imogolite and allophane

Allophane is a general term referring to non-crystalline, hydrated aluminosilicates of variable composition. Infrared spectra of allophanes reflect this variation and indicate, from the position of the main Si–O absorption band, the degree of polymerization of the silicate anion and the likely coordination number of aluminium. Imogolite is a specific name for a cryptocrystalline aluminosilicate, with an Al : Si molar ratio of 2 : 1, and with a fibrous tubular morphology (Cradwick *et al.*, 1972). It contains orthosilicate anions coordinated to the inside surface of a tubular gibbsite sheet. Proto-imogolite allophane, a closely related non-crystalline aluminosilicate, has the same short-range structure as imogolite, but does not show its fibrous morphology. These minerals typically occur in acid soils and have very characteristic spectra.

The IR spectrum of imogolite (Figure 2.16(a)) shows well-resolved components at 993 and 946 cm^{-1} in the orthosilicate Si–O band, and clearly defined bands in the 800–200 cm^{-1} region, the one at 343 cm^{-1} (shifting to 348 cm^{-1} on heating to 150°C) being particularly useful for identification of imogolite structures (Farmer *et al.*, 1977). The spectrum of proto-imogolite allophane (Figure 2.16(b)) has substantially the same pattern as imogolite, but is more diffuse. The band at 348 cm^{-1} is nevertheless still well defined.

Allophanes formed at high pH have a framework type of structure, with Al in fourfold and sometimes in both fourfold and sixfold coordination. The spectrum of such an allophane, with the same 2 : 1 Al/Si ratio as proto-imogolite allophane (Figure 2.16(c)), shows the Si–O band at 1016 cm^{-1}, indicative of a polymerized silicate anion containing tetrahedral Al. The general pattern resembles that of a feldspathoid, and has been described as a hydrous feldspathoid (Farmer *et al.*, 1979). Other allophanes contain phyllosilicate anions which may be related to kaolins (Farmer *et al.*, 1979). Allophanes generally have broad, unresolved OH-stretching bands, and become impossible to distinguish in a mixture of minerals. However, the considerable adsorbed and hydration water associated with them can suggest their presence. A more positive indication of imogolite structures can be obtained from difference spectroscopy using computer subtraction, either for the clay before and after oxalate extraction of the allophane/imogolite, or before and after dehydroxylation of the allophane (Farmer *et al.*, 1984).

2.5.2.3 *Ortho- and pyrosilicates*

Olivines, of general formula $M_2^{2+}SiO_4$ (ortho), garnets, M_3^{2+}, $M_2^{3+}(SiO_4)_3$ (ortho), and epidotes, M_2^{2+}, $M_3^{3+}Si_2O_7SiO_4$ (pyro), are occasionally encountered in clay materials and in hand-picked specimens from weathered rocks. Their spectra are dominated by the influence of the SiO_4 and Si_2O_7 anions, but the nature of the cations present in the structure can have a profound effect on the appearance of the spectrum and on the positions of absorption bands.

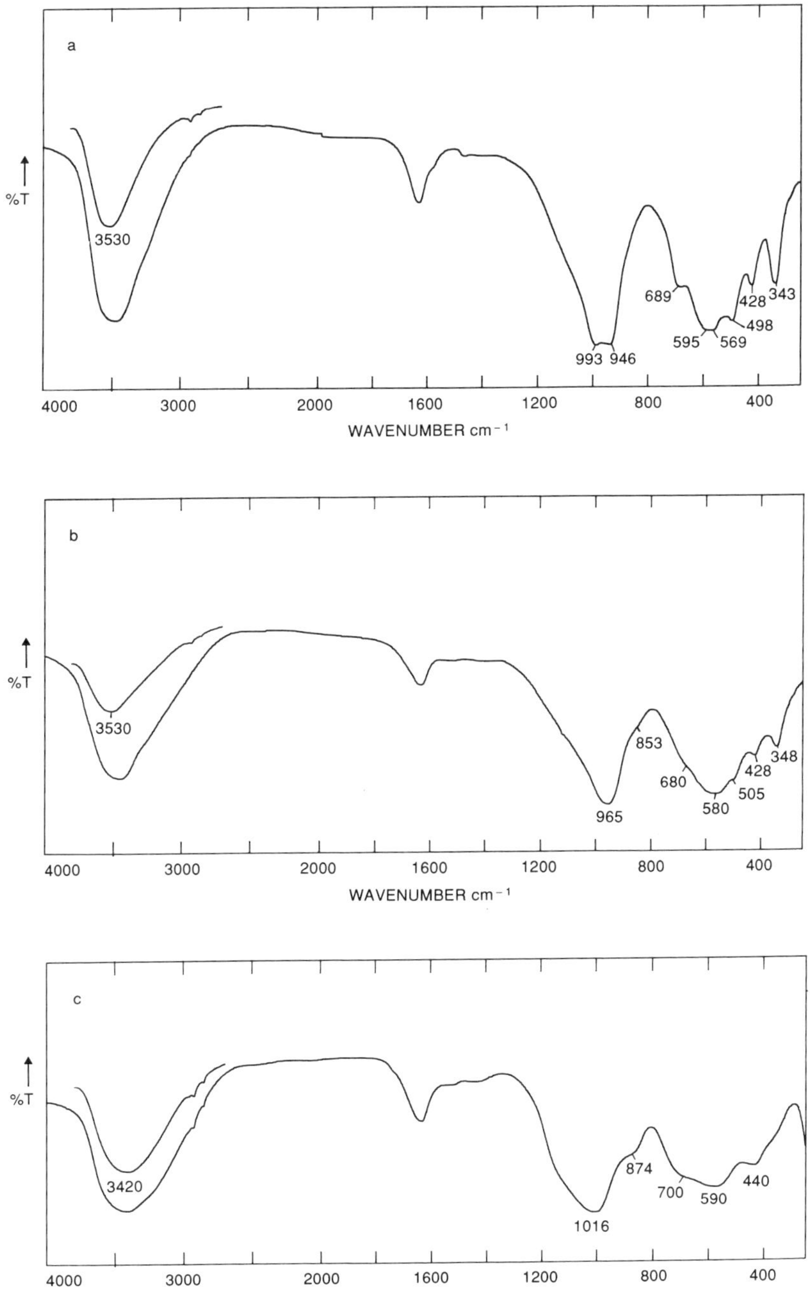

Figure 2.16 Infrared spectra of (a) imogolite (Kurayoshi), (b) proto-imogolite allophane (synthetic), (c) hydrous feldspathoid allophane (Silica Springs, New Zealand).

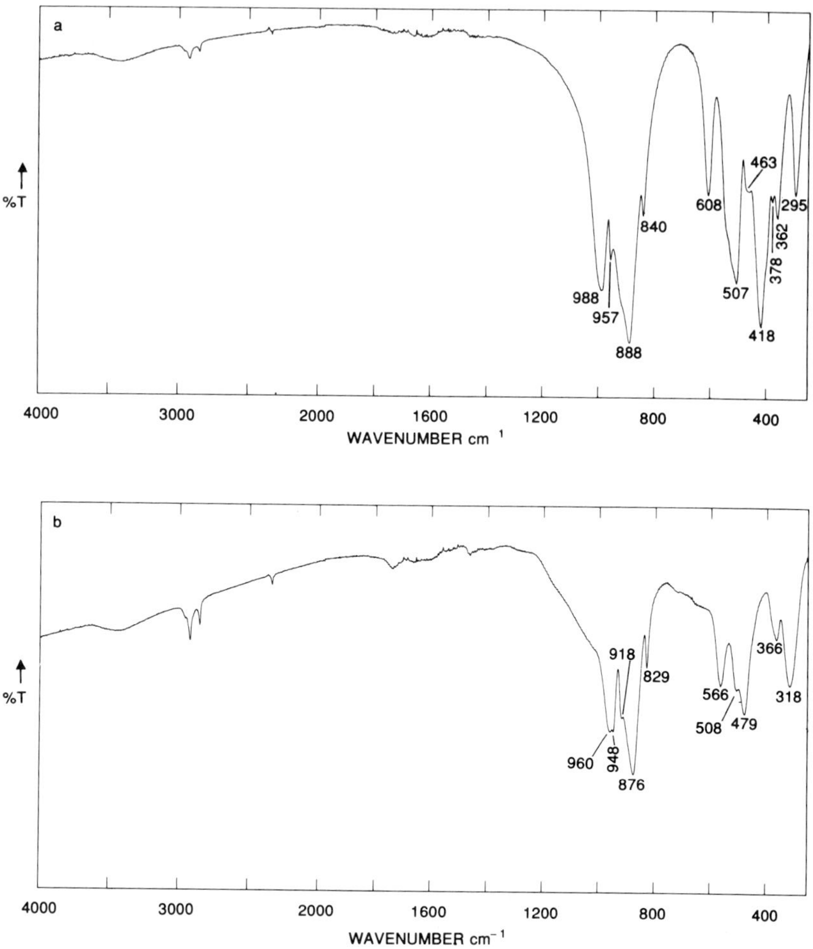

Figure 2.17 Infrared spectra of olivines: (a) forsterite (Norway), (b) fayalite (Aberdeen, Scotland). KBr discs heated 16 h at 150°C.

Spectra of the olivine-group minerals are generally similar except for the end-members. Spectra of forsterite (Mg) and fayalite (Fe^{2+}) (Figure 2.17(a), (b)) illustrate this, with that of the former displaying an intense unique absorption band near 420 cm^{-1}. Also evident is the general shift to lower frequencies of the fayalite bands under the influence of the larger, heavier ferrous cation. The forsterite examined here contains a trace (less than 1%) of Fe^{2+}, insufficient to affect the IR pattern significantly.

The typical orthosilicate pattern of absorption bands of the olivine minerals in the region 1000–800 cm^{-1} is also exhibited by garnets, but there are several notable differences, particularly in the 700–300 cm^{-1} region. Within the garnet group, spectra show a strong progressive pattern of shifts in band position dependent on the divalent–trivalent cation pairings in the structure. This is well shown in the work of Tarte (1965), illustrated by Farmer (1974). In the examples shown here (Figure 2.18(a), (b)), the higher frequency almandine (Fe^{2+}, Al) pattern of bands can be readily distinguished from the lower frequency grossular (Ca, Al) pattern. The pattern of absorption bands observed from the olivine minerals shifts steadily to lower frequencies for garnets of different chemical composition, from pyrope (Mg, Al) to almandine (Fe, Al), spessartine (Mn, Al), grossular (Ca, Al), uvarovite (Ca, Cr) and andradite (Ca, Fe),

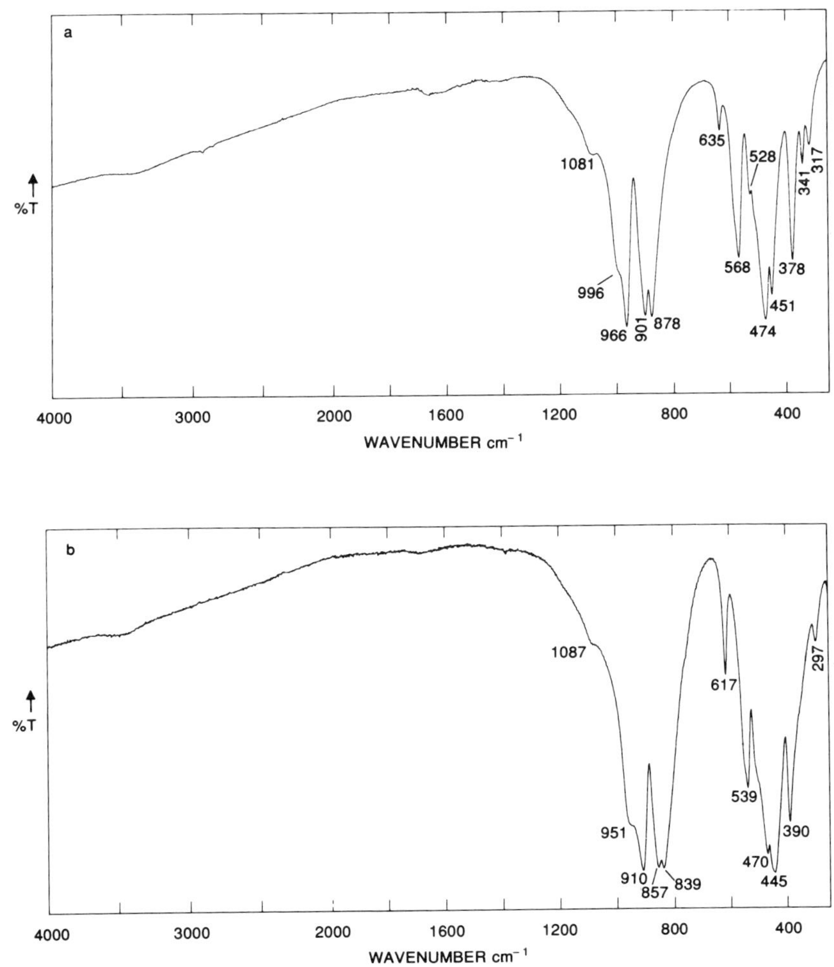

Figure 2.18 Infrared spectra of garnets: (a) almandine (Fauske, Norway), (b) grossular (Laguno de Jaco, Mexico). KBr discs heated 16 h at 150°C.

generally in the direction of increasing atomic weight and ionic radius of the divalent–trivalent pairs. Positive identification of the orthosilicates from their spectra can be achieved using the differentiation table of band frequencies published by Tarte (1965) and reproduced by Farmer (1974).

The spectra of pyrosilicates are readily distinguished from those of orthosilicates by their much higher frequency Si–O bands, which usually occur in two clearly differentiated groups, one centred on about 1170–1100 cm^{-1}, the other near 900–800 cm^{-1}. This can be seen for epidote (Figure 2.19(a)), a complex pyrosilicate, $Ca_2(Al, Fe)_3(Si_2O_7)(SiO_4)O(OH)$, containing in addition to the pyrosilicate anion, orthosilicate and hydroxyl. The orthosilicate anion may be responsible for some of the

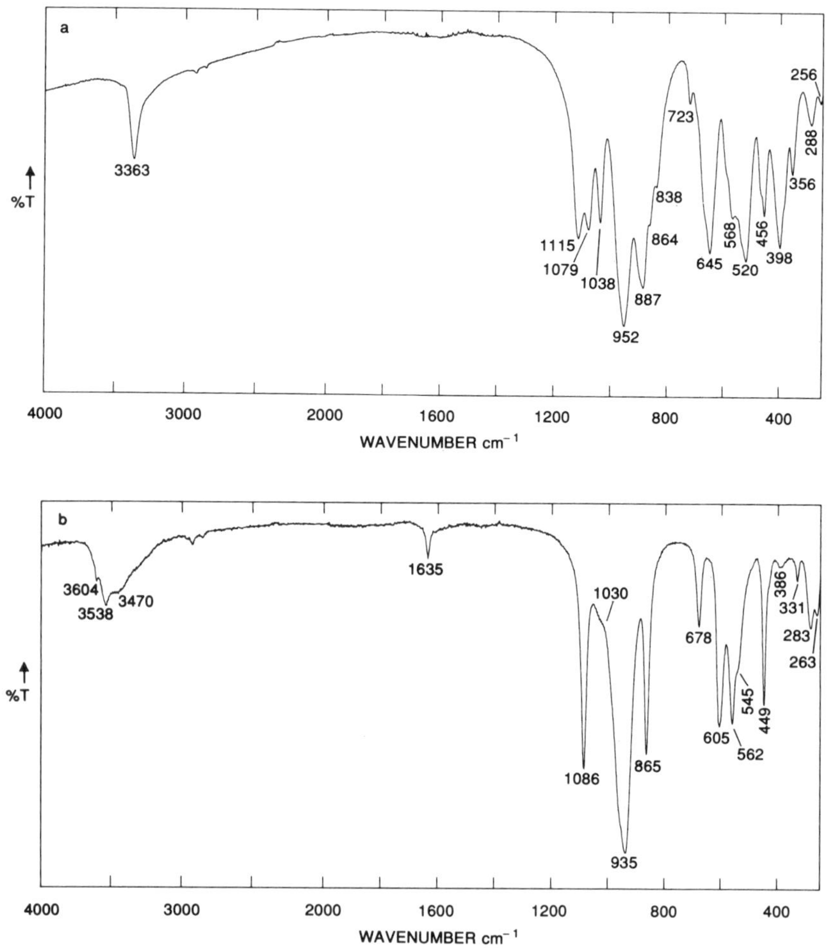

Figure 2.19 Infrared spectra of pyrosilicates: (a) epidote (Tyrol, Austria), (b) hemimorphite (Leadville, Colorado). KBr discs heated 16 h at 150°C.

enhanced intensity of the strong doublet near 890 and 950 cm^{-1}, and the hydroxyl group for the stretching band at 3363 cm^{-1}. The latter band, although distinctive and usefully characteristic for some pyrosilicates, e.g. epidotes and zoisites, is not diagnostic for the pyrosilicate group as a whole, which may contain no hydroxyl groups. Nevertheless, Langer and Raith (1974) have deduced several important correlations between band frequency and occupancy of octahedral cation sites in epidotes and zoisites; using their formulae for the epidote discussed here, the 3363 cm^{-1} OH band and the substitution-sensitive bands at 1038, 723, 398 and 356 cm^{-1} should correspond to a mean occupancy of one of the types of three octahedral sites (M_3) by Fe^{3+} of 70%, and be equivalent to an iron content of about 11.7% Fe_2O_3 (Jovic, Fraser and Russell, unpublished data).

Compared with that of epidote, the spectrum of hemimorphite, $Zn_4Si_2O_7(OH)_2.H_2O$ (Figure 2.19(b)) is probably more typical of a pyrosilicate. As in the orthosilicates, the influence of the larger and heavier cation, has resulted in a shift to lower frequencies of many of the absorption bands. However, the most characteristic feature of the hemimorphite spectrum is the absorption due to the strongly held water of crystallization at 1635 cm^{-1} and its contribution to the complex pattern of bands at 3604, 3538 and 3470 cm^{-1}. These bands must also include absorption due to the structural OH groups. Compared with those in epidote, these OH groups are much more weakly bonded, judging from the much higher frequencies of their absorption bands.

2.5.2.4 *Zeolites*

Among the zeolites most frequently found in association with clays are heulandite–clinoptilolite, analcime, stilbite and laumontite. All are significantly hydrated and show spectra with strong, discrete water absorption bands in the 3500 and 1640 cm^{-1} regions (Figures 2.20, 2.21). These water bands can display considerable complexity, both for the fully hydrated state as recorded for unheated KBr discs (full curves) and for the partly dehydrated state after heating the discs for 16 h at 150°C (inset curves). The water bands and their changes on heating are potentially characteristic, or even diagnostic, for some zeolites, e.g. laumontite. The marked differences between analcime and laumontite in this regard indicate the variety of bonding states of water molecules associated with Na and Ca, respectively, in the channels and cages in these minerals. There are also marked differences elsewhere in the spectra, as befits their different aluminosilicate structures and compositions.

According to the structural scheme described by Breck (1974), based on the type of secondary building unit (SBU) in the structure, analcime and laumontite have similar aluminosilicate structures and are classified as belonging to group 1, in which the SBU is a single four-membered Al–Si ring (S4R). This similarity is generally reflected in their infrared spectra (Figure 2.20), which display common regions of major absorption at 1040–1020, 760–740 and 450–430 cm^{-1}. There are clearly significant differences in detail, with the greater sharpness and multiplicity of bands in the spectrum of laumontite (Figure 2.20(b)), probably originating from its greater ordering and greater long-range order, compared with analcime (Figure 2.20(a)).

In contrast, heulandite–clinoptilolite and stilbite belong to group 7 in Breck's classification, where the SBU is a complex unit comprising ten interlinked Al–Si tetrahedra. Again their spectra (Figure 2.21) show

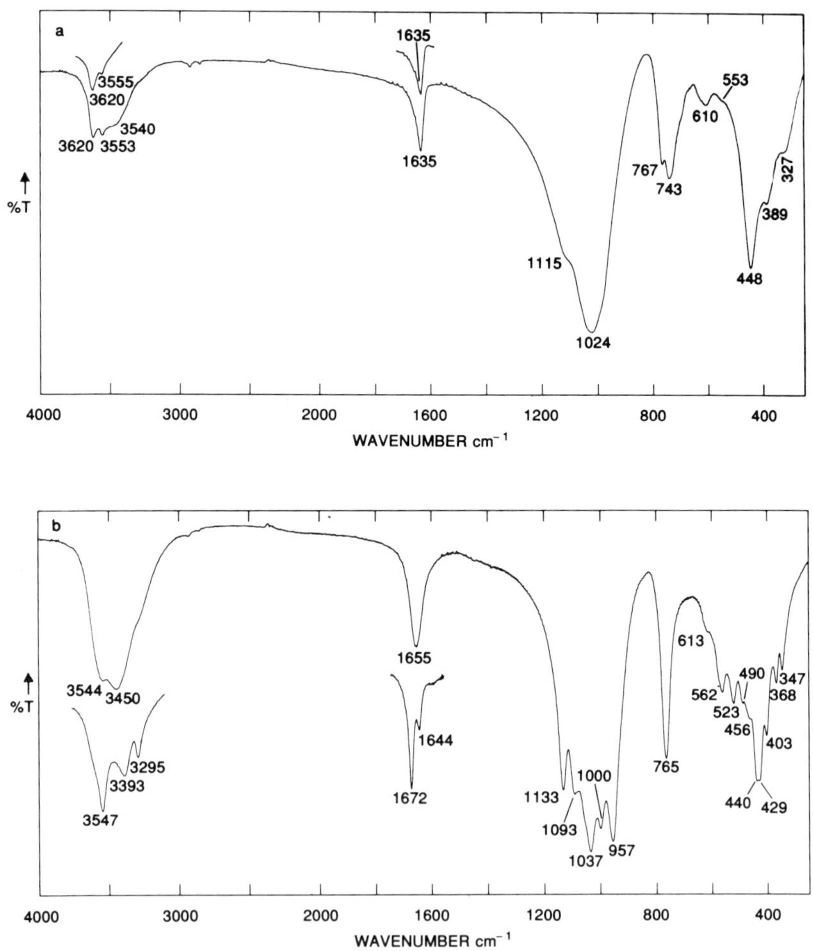

Figure 2.20 Infrared spectra of zeolites (group 1): (a) analcime (Barrhead, Scotland), (b) laumontite (Skye, Scotland). Inset curves, KBr discs heated 16 h at 150°C.

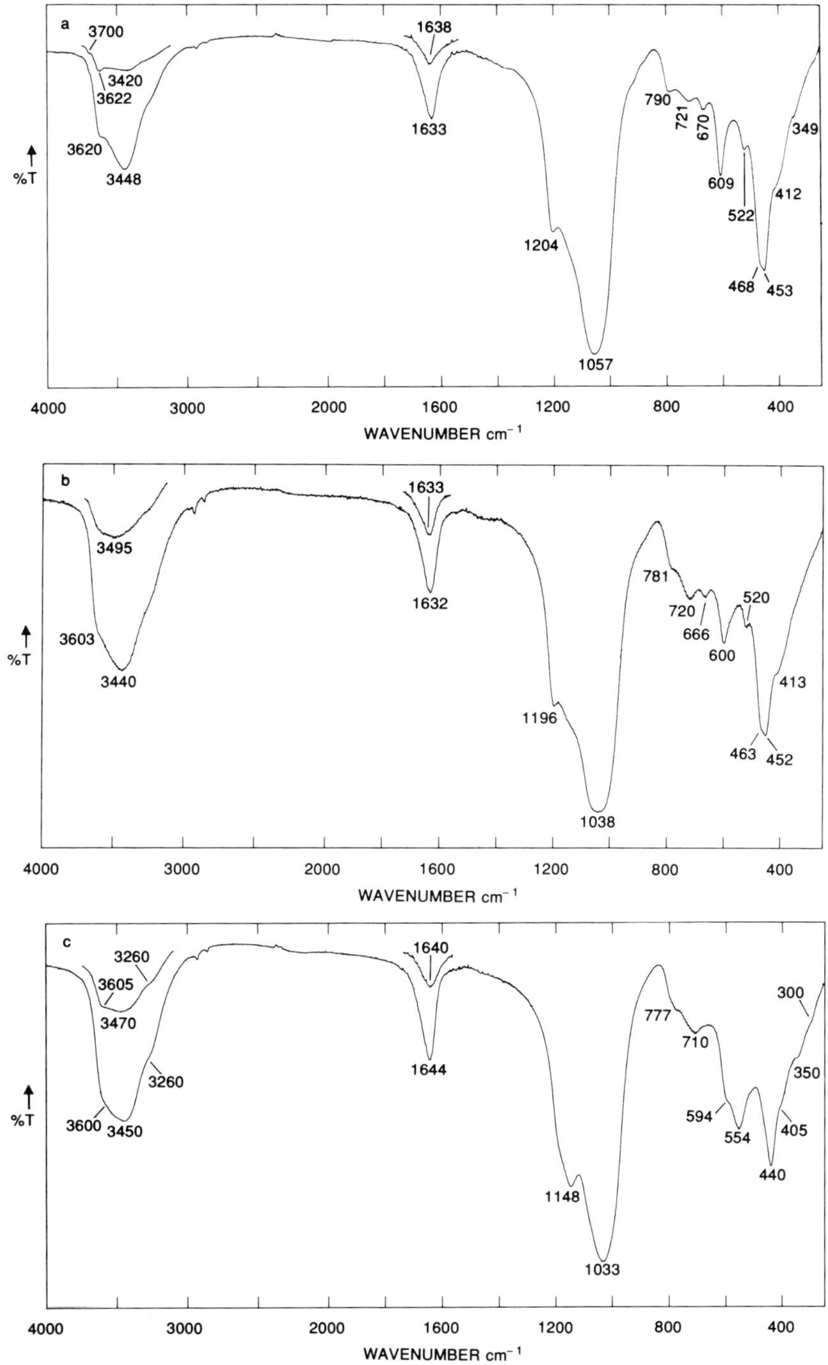

Figure 2.21 Infrared spectra of zeolites (group 7): (a) clinoptilolite (Sheuville, Oregon), (b) heulandite (Peyton Collection, Aberdeen University), (c) stilbite (Skye, Scotland). Inset curves, KBr discs heated 16 h at 150°C.

common regions of major absorption, which at 1060–1030, 610–550 and 470–440 cm^{-1} are appreciably different from those of the group 1 minerals.

Clinoptilolite and heulandite are isostructural, differing only in the occupancy of the four octahedral cation sites (Koyama and Takeuchi,

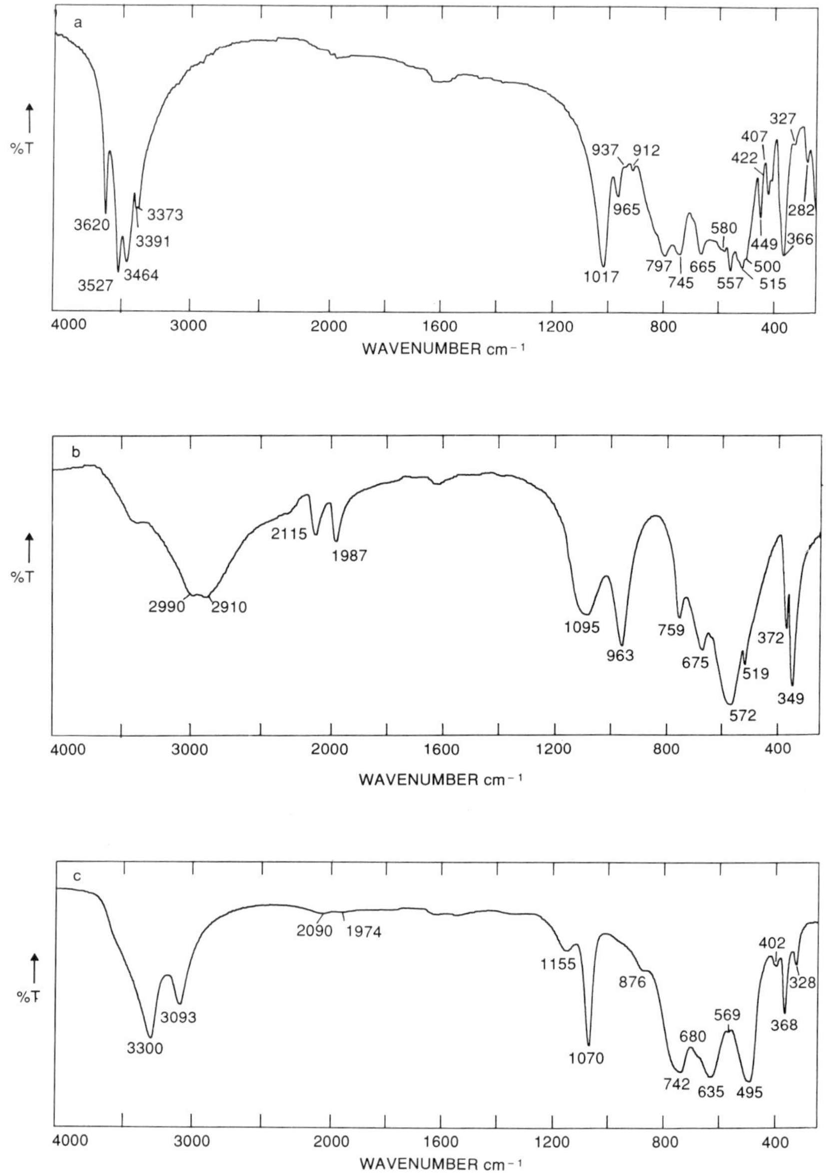

Figure 2.22 Infrared spectra of aluminium hydroxides: (a) gibbsite; (b) diaspore; (c) boehmite (synthetic).

1977), and would be expected to have similar IR spectra. However, as shown in Figure 2.21, while their spectra are obviously closely related, there are two important differences. The first is that virtually all of the metal-oxygen bands between 1400 and 200 cm^{-1} in heulandite are of lower frequency (by up to 20 cm^{-1}) than those in clinoptilolite. The second is that the water absorption pattern in the region 3800–3200 cm^{-1} in clinoptilolite is sharper and shows more band resolution, in both the fully and partly hydrated states (Figure 2.21(a)), than that in heulandite (Figure 2.21(b)), where the OH absorption envelope is always much more broad and diffuse. The minerals examined here are well authenticated by both X-ray diffraction (XRD) and electron-probe microanalysis (EPM) and conform to end-member compositions. These spectra do not confirm that clinoptilolite and heulandite are different minerals, but they do establish that these Na-rich and Ca-rich end-member compositions are readily differentiated.

2.5.2.5 *Aluminium minerals*

The spectra of the hydroxide mineral gibbsite and the oxyhydroxides diaspore and boehmite are shown in Figure 2.22. The oxyhydroxides do not occur extensively, but are included for comparison with their iron analogues goethite and lepidocrocite, both of which have a widespread occurrence. The most characteristic feature of gibbsite is its pattern of OH-stretching bands. The specimen illustrated is well crystallized and shows splitting of the 3391, 3373 cm^{-1} doublet. This is not always observed, but the quartet of bands at 3620, 3527, 3464 and 3380 cm^{-1}, and the 1017 OH deformation and 366 cm^{-1} bands are quite diagnostic. Disappearance of the bands after heating to 350°C will provide confirmation of gibbsite.

2.5.2.6 *Iron minerals*

Hematite, an anhydrous oxide, occurs in two morphological forms, namely, a platy form (kidney ore) and a more equant form (specularite). Their spectra (Figure 2.23(a), (b)) are generally similar, but show considerable differences in detail. Because of differences in crystal size, and particularly in shape, the bands at 532, 450 and 317 cm^{-1} in kidney ore shift to higher frequencies at 562, 480 and 352 cm^{-1} in specularite. Again, the weaker bands at 643 and 400 cm^{-1} in kidney ore shift to lower frequencies and merge with the stronger bands in specularite, the 378 cm^{-1} band being just resolved (Rendon and Serna, 1981; Serna *et al.*, 1982). The specularite form is the more common in clay materials. Solid-state dehydration of platy goethite also yields platy hematite, not a new form of hematite as was proposed by Yariv and Mendelovici (1979). This type of hematite morphology was also reported in the new mineral

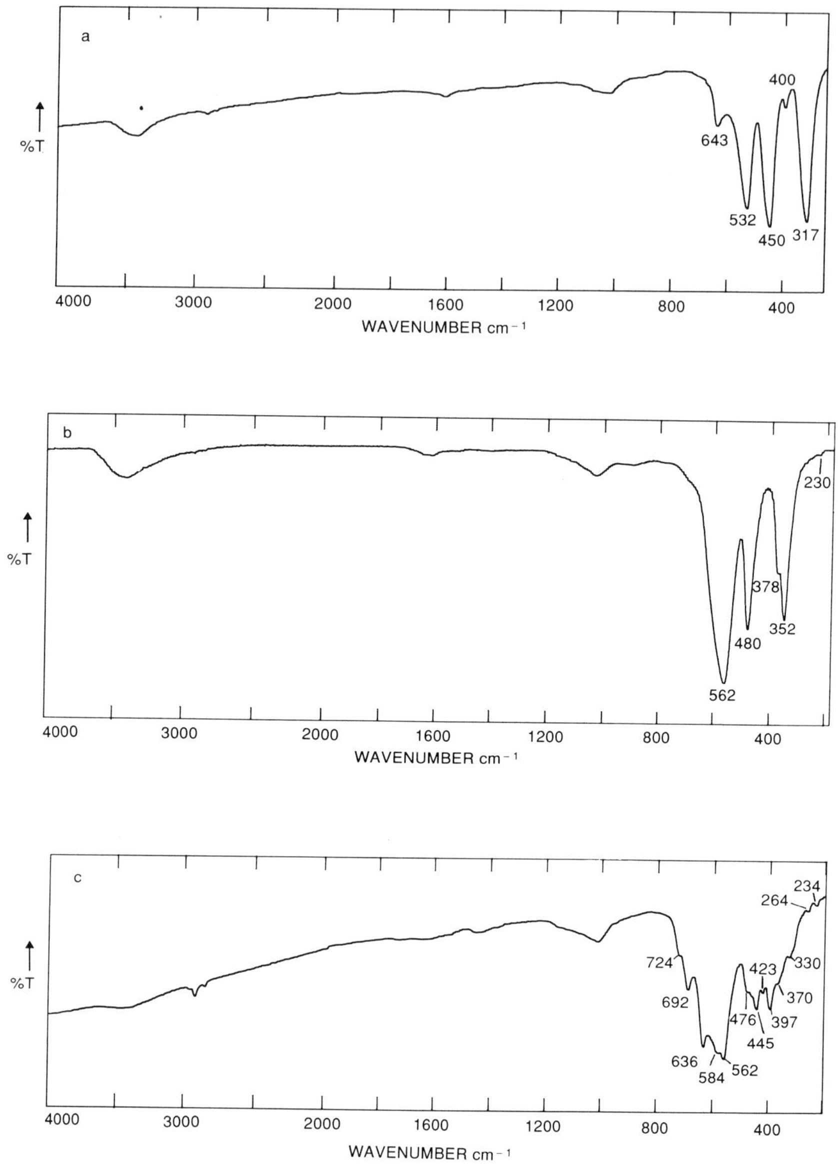

Figure 2.23 Infrared spectra of iron oxides: (a) hematite (kidney ore), (b) hematite (specularite, synthetic), (c) maghemite.

macaulayite (Wilson *et al.*, 1981, 1984). Maghemite, another anhydrous oxide, has a surprisingly sharp pattern of absorption bands (Figure 2.23(c)), with a general resemblance to magnetite, from which it is

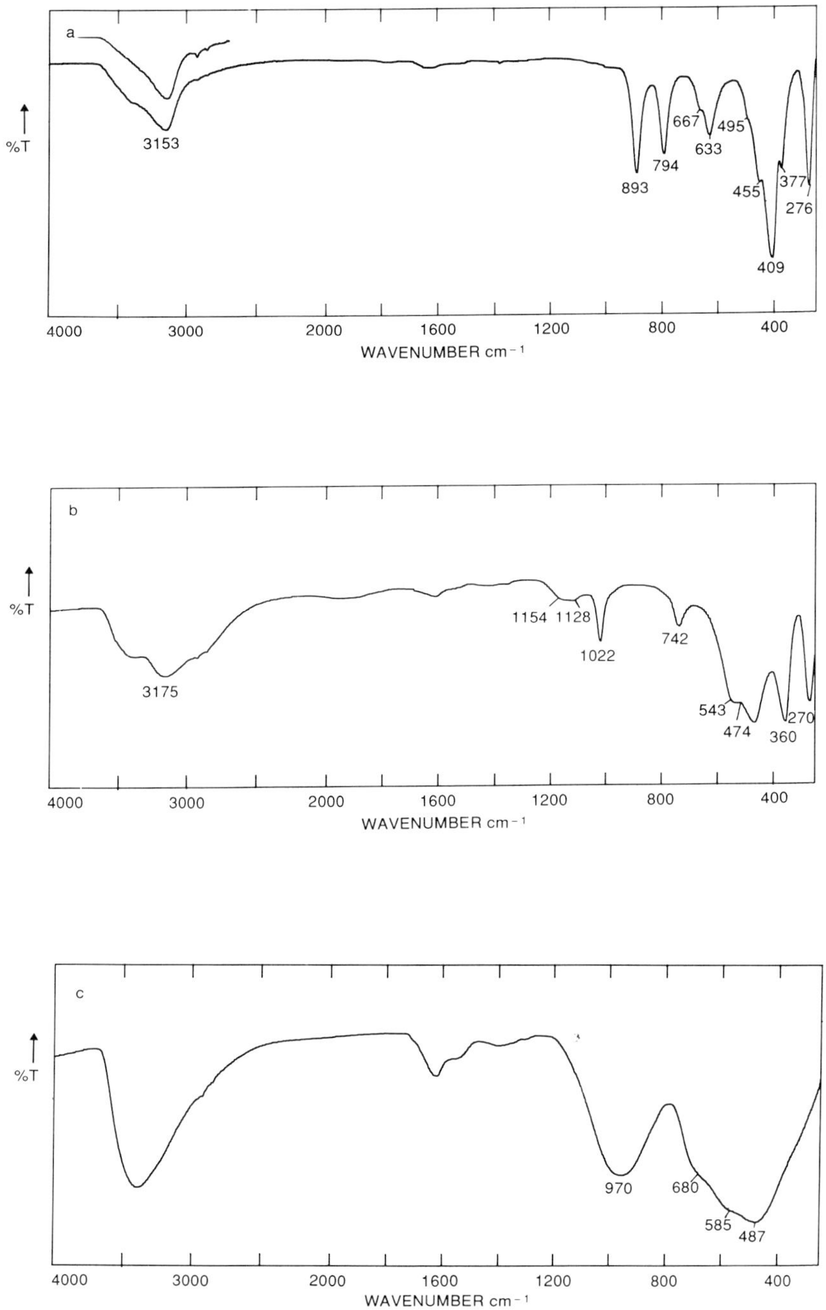

Figure 2.24 Infrared spectra of iron hydroxides: (a) goethite (synthetic), (b) lepidocrocite (synthetic), (c) ferrihydrite (Egmont, New Zealand).

thought to be derived. The IR spectrum of magnetite is broad and diffuse, showing two ill-defined features near 570 and 347 cm^{-1} (Kodama, 1985). All the above oxides generally occur at low abundances in clay materials and may be difficult, if not impossible, to identify by their IR spectra because of overlap by silicates.

The IR spectra of the iron oxyhydroxides goethite and lepidocrocite (Figure 2.24(a), (b)) are easily distinguished. Thus, the strongly hydrogen-bonded OH absorbs at 3153 cm^{-1} in goethite, and at 3175 cm^{-1} in lepidocrocite, indicating a slightly weaker H-bond in the latter. Again, OH deformation bands occur at 893 (in-plane) and 794 cm^{-1} (out-of-plane) for goethite, and 1154, 1022 (in-plane) and 742 cm^{-1} (out-of-plane) for lepidocrocite. The deformation bands and the lower frequency lattice bands are difficult to identify in the spectra of clays rich in layer silicates, but the OH-stretching bands are usually diagnostic of an oxyhydroxide mineral, after removal of adsorbed water by heating. However, it would be difficult to distinguish goethite and lepidocrocite in these circumstances.

Ferrihydrite is a naturally occurring, poorly ordered, hydrated iron oxide, and its IR spectrum (Figure 2.24(c)) is a very simple, broad-banded pattern. Natural specimens are invariably associated with silicate, and from the position of the Si–O band at 970 cm^{-1}, it is present as orthosilicate. Hydroxyl groups are present in the structure (Russell, 1979), which from the band widths is very disordered. Recognition of this phase in a clay would be virtually impossible, except perhaps by difference spectroscopy as described for imogolite.

2.5.2.7 Silica minerals

Quartz is a common constituent of most clay materials, and its spectrum (Figure 2.25(a)) shows several sharp bands which have been used for identification, the 800, 781 cm^{-1} doublet being the most generally useful. In contrast, the spectrum of tridymite (Figure 2.25(b)) shows only broad bands, testifying to its relatively disordered structure. The single broad band at 792 cm^{-1}, with no accompanying absorption down to 600 cm^{-1} has been used for its positive identification in fine-grained siliceous materials (Wilson, Russell and Tait, 1974). The tridymite spectrum closely resembles that of opaline silica. For comparison, the spectrum of cristobalite, the third polymorph of SiO_2 is included (Figure 2.25(c)), the diagnostic bands being at 1196 and 622 cm^{-1}.

2.5.2.8 Carbonates

Calcite is the most common carbonate mineral in clays. Its IR spectrum (Figure 2.26(a)) is easily recognizable, and the bands at 1428, 878 and 714 cm^{-1} are diagnostic. The spectrum of dolomite (Figure 2.26(c)),

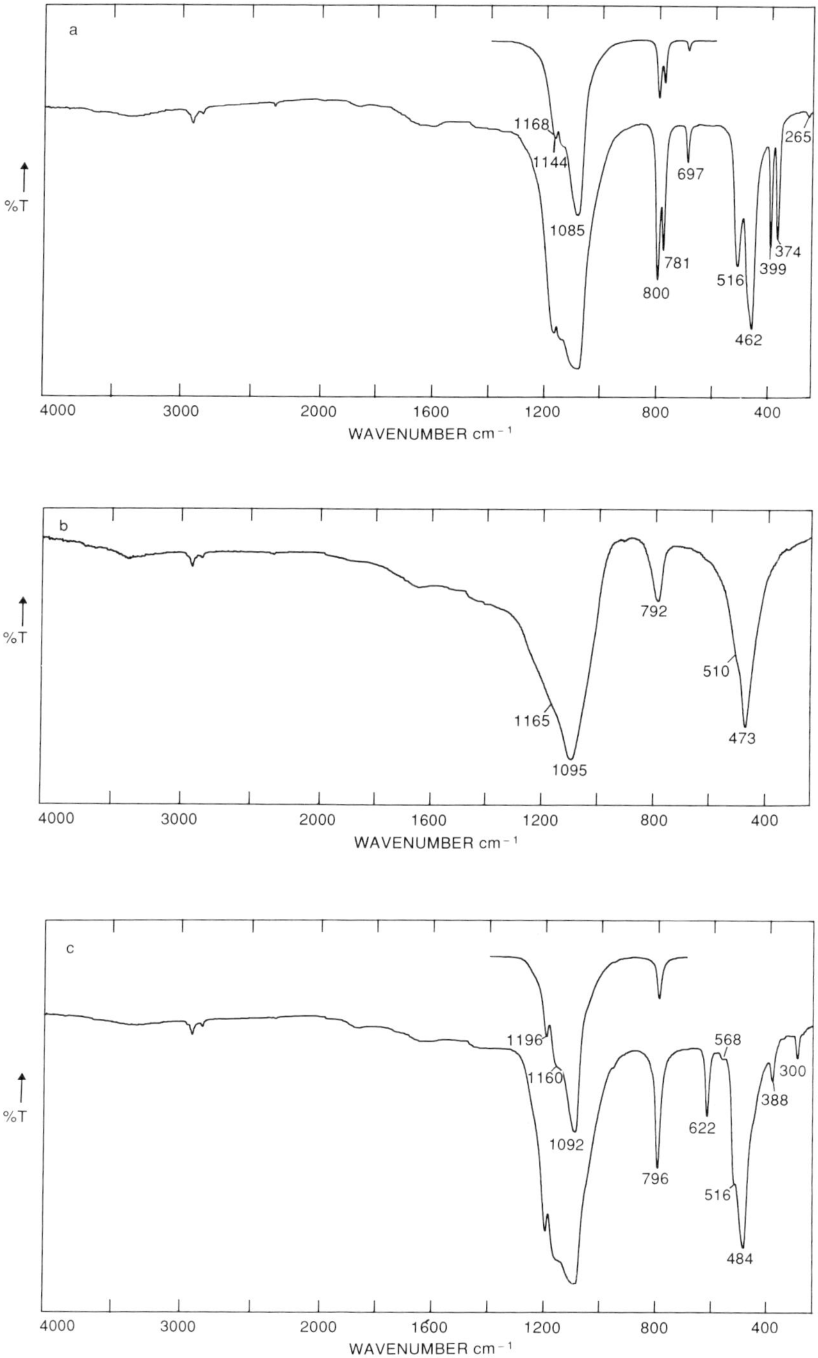

Figure 2.25 Infrared spectra of silicas: (a) quartz, (b) tridymite (Virginia), (c) cristobalite.

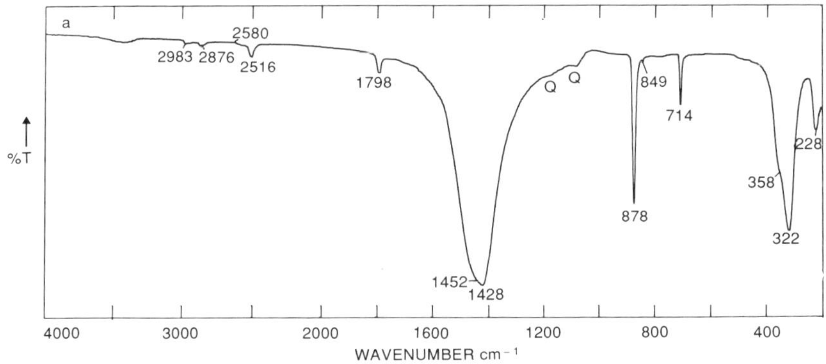
a
2580
2983
2876
2516
1798
Q
Q
849
714
228
358
878
322
1452
1428
%T
4000
3000
2000
1600
1200
800
400
WAVENUMBER cm⁻¹

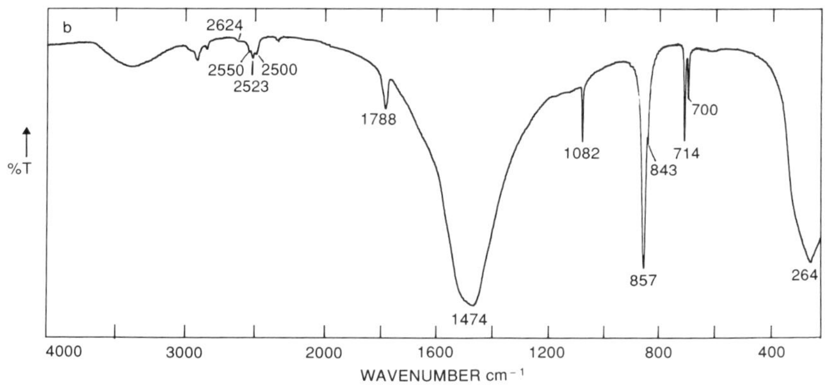
b
2624
2550
2500
2523
1788
1082
700
714
843
857
264
1474
%T
4000
3000
2000
1600
1200
800
400
WAVENUMBER cm⁻¹

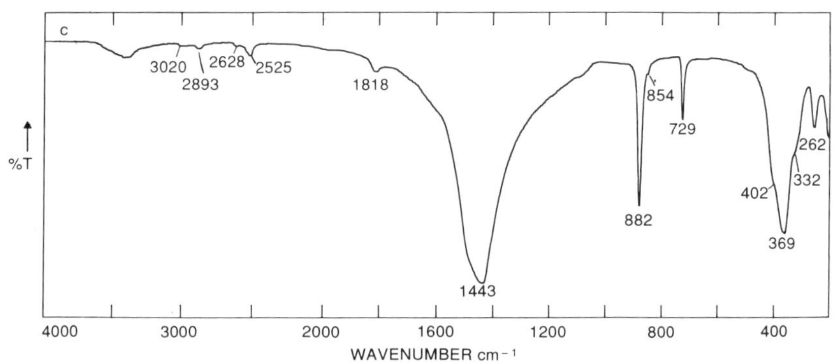
c
3020
2893
2628
2525
1818
854
729
262
332
402
882
369
1443
%T
4000
3000
2000
1600
1200
800
400
WAVENUMBER cm⁻¹

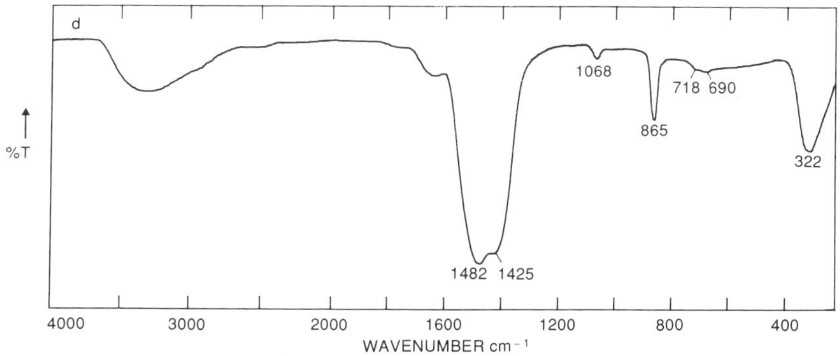

Figure 2.26 Infrared spectra of carbonates: (a) calcite, (b) aragonite (coral), (c) dolomite (Australia), (d) amorphous $CaCO_3$ (synthetic). In (a) Q indicates quartz.

which has a calcite-type structure, shows a shift of all bands to higher frequency. The spectrum of the aragonite form of $CaCO_3$ (Figure 2.26(b)) is distinct from that of calcite, undoubtedly reflecting structural differences. For comparison with the crystalline carbonates, the spectrum of a non-crystalline $CaCO_3$ is included (Figure 2.26(d)), a form recently identified in thin, calcareous weathering crusts (Dupuis *et al.*, 1984).

2.5.2.9 Phosphates

Phosphates usually occur to a minor extent in clay materials. Nevertheless, it is important to recognize the role that IR spectroscopy can play in the identification and characterization of phosphates, a role that could have increasing importance with the advances being made in FTIR microscope investigations of minor components. The literature on the IR spectra of phosphates is extensive, but the spectral collections of Lehr, Brown and Frazier (1967) and Moenke (1962, 1966) and the chapter in Farmer (1974) are to be recommended.

To illustrate the applicability of IR spectroscopy to these minerals, fluor-apatite ($Ca_5[PO_4]_3F$), one of the commonest phosphates in rocks, soils and sediments, has been selected, and its spectrum is compared with those of the hydroxy and carbonato forms (Figure 2.27). In general, and for the major P–O vibrations, all three spectra are alike, as would be expected from their structural and compositional similarity. However, there are several important minor divergences which serve to illustrate the importance of the IR technique. A weak band at 3535 cm^{-1} in the spectrum of apatite (Figure 2.27(a)) arises from a significant proportion of OH groups substituting for F, possibly up to 20% calculated from the absorbancies of the OH bands. Generally, however, the spectrum agrees

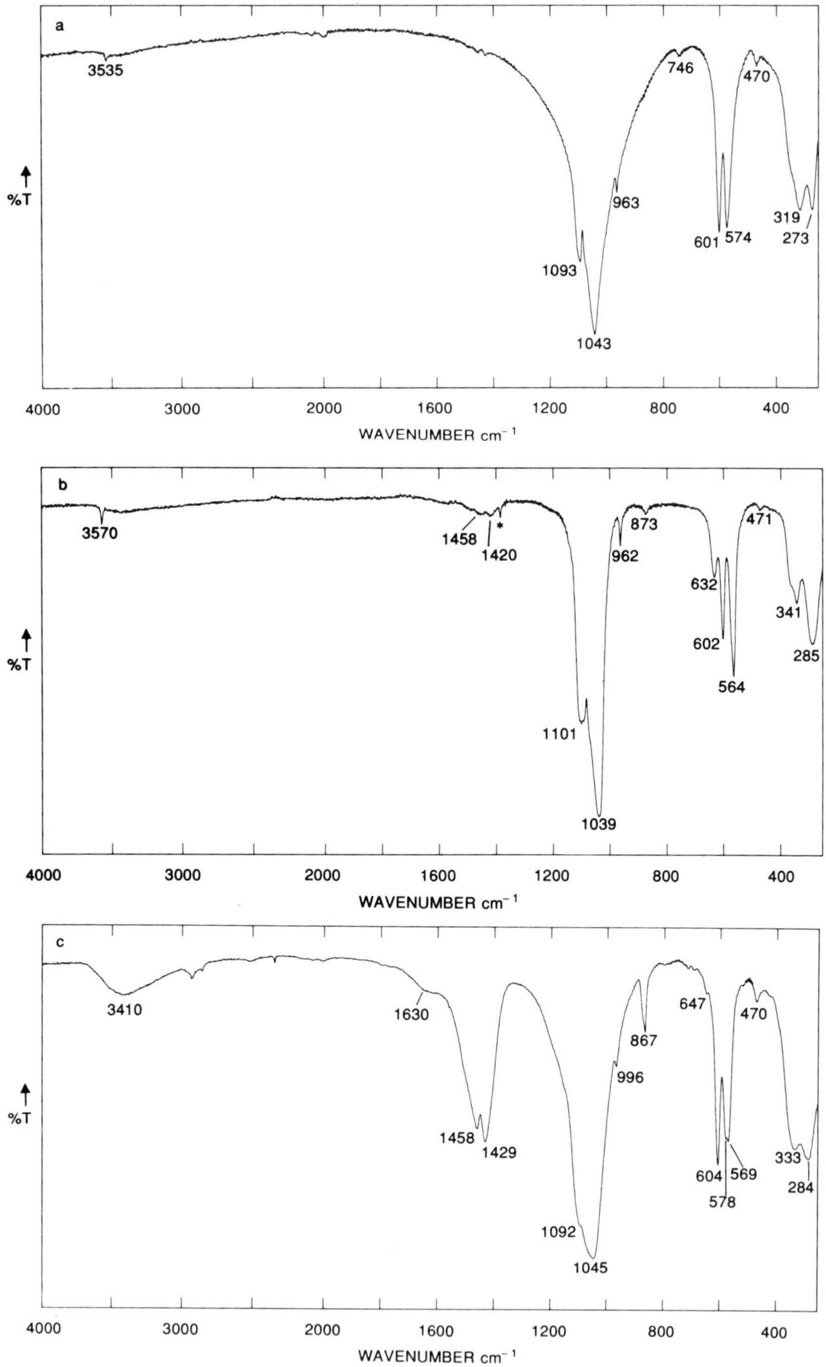

Figure 2.27 Infrared spectra of apatite phosphates: (a) F-apatite, (b) OH-apatite (Kola rock phosphate, Russia), the sharp band indicated by an asterisk (at $1384\,cm^{-1}$) is due to a trace of nitrate (NO_3^-), (c) CO_3-apatite (Gafsa rock phosphate, North Africa). A very weak band at $712\,cm^{-1}$ arises from a trace of calcite, confirmed by XRD. KBr discs heated 16 h at 150°C.

well with those of apatite in the literature (Moenke, 1962; Lehr, Brown and Frazier, 1967). The fluor-apatite structure apparently can tolerate appreciable OH-for-F replacement without significant change in the spectrum.

The spectrum of hydroxy-apatite, $Ca_5(PO_4)_3OH$ (Figure 2.27(b)), shows several new bands compared with that of fluor-apatite. Bands at 3570 and 632 cm^{-1} arise, respectively, from the stretching and bending vibrations of OH groups in the apatite structure. Other weak new bands occurring at 1458, 1420 and 873 cm^{-1} arise from a small amount of carbonate-for-hydroxyl substitution, and would not be expected in the spectrum of a pure hydroxy-apatite. The spectrum of carbonato-apatite, $(Ca_5(PO_4)_3)_2CO_3$ (Figure 2.27(c)), shows well developed and diagnostic carbonate bands at 1458, 1429 and 867 cm^{-1}. It is worth noting that the broad absorption bands at 3410 and 1630 cm^{-1} arise from molecular water adsorbed by the carbonato-apatite and retained even after heating for 16 h at 150°C. These water bands are not observed in the spectra of the other apatite analogues similarly heated, indicating that a relatively stable level of hydration may distinguish carbonato-apatite from the other two. This hydration water was released at 250°C, the spectrum then revealing the virtual absence of absorption of OH groups at 3570–3535 cm^{-1}, signifying that in this carbonato-apatite all the OH groups have been completely replaced, presumably by CO_3.

Small differences in position between the CO_3 bands in Figure 2.27(b) and (c), and between the OH bands in Figure 2.27(a) and (b), are indicative of different bonding regimes encountered by small and large populations of 'foreign' ions in the respective apatite structures.

2.6 INTERPRETING THE IR SPECTRA OF CLAY MATERIALS

This requires the positive identification of as many characteristic absorption bands of individual minerals as possible and will obviously depend on the complexity of the mineral suite present and the degree of overlap of the component spectra. The OH-stretching region is perhaps the most important region of the spectrum, but since some minerals have no hydroxyl groups in their structure, or show only weak absorption bands, the full spectrum is required for the best possible interpretation. High and low concentrations of sample must also be used, so that full weight can be given to weak and strong mineral bands. Finally, because of interference from the absorption bands of adsorbed water, KBr discs of clay samples should be dehydrated by heating at 150°C, in order that underlying absorption bands of mineral OH groups may be unambiguously identified.

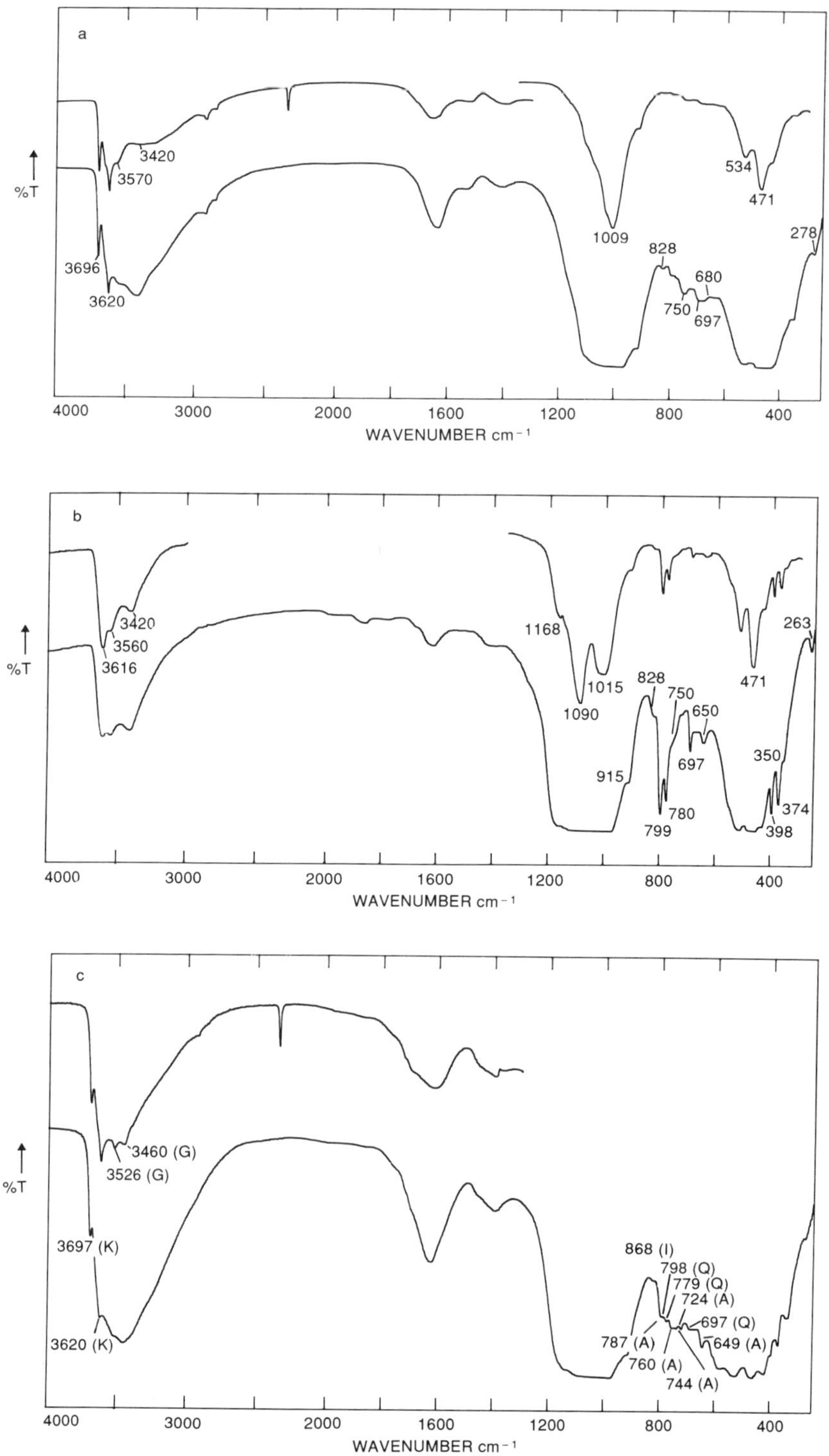
a
%T
3420
3570
3696
3620
1009
534
471
278
828
680
750
697
4000
3000
2000
1600
1200
800
400
WAVENUMBER cm⁻¹
b
%T
3420
3560
3616
1168
1015
1090
828
750
650
697
915
799
780
471
263
350
374
398
4000
3000
2000
1600
1200
800
400
WAVENUMBER cm⁻¹
c
%T
3460 (G)
3526 (G)
3697 (K)
3620 (K)
868 (I)
798 (Q)
779 (Q)
724 (A)
697 (Q)
787 (A)
649 (A)
760 (A)
744 (A)
4000
3000
2000
1600
1200
800
400
WAVENUMBER cm⁻¹

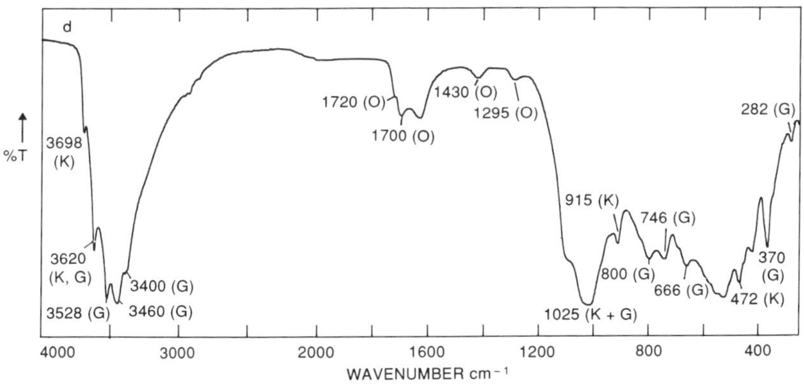

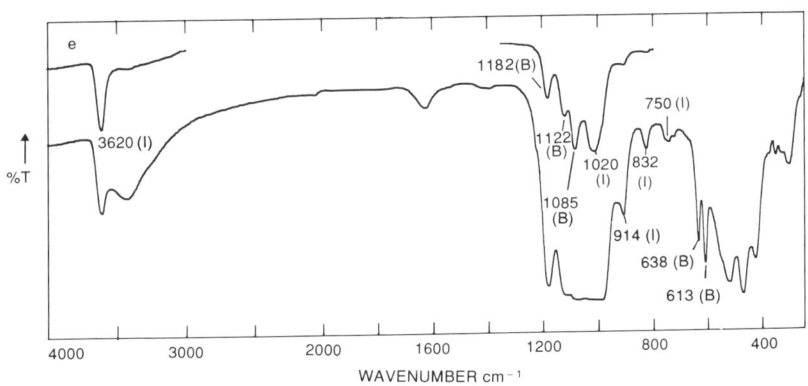

Figure 2.28 (a)–(e) Infrared spectra of soil clays. In (c)–(e) Q, A, G, K, I, O and B indicate quartz, albite, gibbsite, kaolin, illite, complex oxalate and baryte, re-s-pectively. The sharp band near 2340 cm^{-1} arises from molecular CO_2 from thermally decomposed organics.

To illustrate these points several spectra of soil clays have been selected for interpretation, such clays being somewhat more difficult to handle than clays from geological deposits.

(i) Soil clay 1

The spectrum (Figure 2.28(a)), even of the hydrated clay, clearly shows the sharp OH doublet of a kaolin mineral at 3696, 3620 cm^{-1}, with supporting bands at about 800, 700 and 280 cm^{-1} (Figure 2.4). Absorption bands at 828 and 750 cm^{-1} and slight broadening under the 3620 cm^{-1} band indicate the presence of illite (Figure 2.10(b)). After removal of adsorbed water, the clay shows two broad bands at 3570 and 3420 cm^{-1}, which, together with enhancement at 750 cm^{-1} beyond the requirements for illite, a band at 680 cm^{-1} and marked enhancement of the Si–O bands at 1009 and 471 cm^{-1}, is consistent with the presence of an iron-containing chlorite (Figure 2.12(a)). The broad underlying absorption in the OH region, centred on about 3300 cm^{-1}, suggests that an amorphous aluminosilicate (allophane) might also be present. Soil organic matter is responsible for the absorption bands between 1300 and 1800 cm^{-1}.

(ii) Soil clay 2

The spectrum (Figure 2.28(b)) is dominated by the bands of quartz at 1168, 1090, 799, 780, 697, 398 and 263 cm^{-1}. The 3616 cm^{-1} band of a dioctahedral layer silicate is visible from the OH-stretching region after heating. Taken together with bands at 915, 828, 750 and 350 cm^{-1}, the presence of an illite is strongly indicated. The low OH-stretching frequency compared with those in Figure 2.10 suggests that the illite may be ferruginous. The OH-stretching region also reveals the broad doublet of a chlorite at 3560, 3420 cm^{-1}, and this is to some extent supported by the enhanced intensity at 750 cm^{-1}, and the band at 650 cm^{-1}. The latter suggests a ferruginous chlorite, but there is no appreciable Si–O contribution present in the 1000–950 cm^{-1} range (Figure 2.12(a)). The main Si–O in Figure 2.28(b) is, however, quite broad and this may accommodate a low-frequency component. Kaolin minerals are absent, no absorption band being detectable at the characteristic 3700 cm^{-1} position.

(iii) Soil clay 3

Again the sharp kaolin-mineral doublet at 3697, 3620 cm^{-1} is easily recognized from the spectrum (Figure 2.28(c)). In the spectrum of the dehydrated sample, broadening under the 3620 cm^{-1} band and the weak feature at 868 cm^{-1} both suggest the presence of illite, but additional confirmation is difficult because of other overlapping absorption. The well-resolved OH bands at 3526 and 3460 cm^{-1} (a 3620 cm^{-1} component is overlain by the kaolin band) and absorption at *c.* 375 cm^{-1} strongly indicate gibbsite. In addition, sharp, weak features of quartz at 798 and 779 cm^{-1}, and albite at 787, 760, 744, 724 and 649 cm^{-1} can be easily identified. The clay contains significant soil organic matter, and possibly also non-crystalline allophanic material, as judged by broad underlying absorption in the OH-stretching region, even after removal of molecular water.

(iv) Soil clay 4

The spectrum of this clay (Figure 2.28(d)) shows dominance of gibbsite (cf. Figure 2.22(a)), with minor kaolin mineral (3700, 915, 472 cm^{-1}). This clay has been selected because it has been peroxidized and its spectrum exhibits the pattern of absorption bands typical of the adsorbed complex oxalato-aluminate, the oxalate being complexed to structural Al ions (Farmer and Mitchell, 1963) and not removed by washing with water.

(v) Soil clay 5

From the spectrum (Figure 2.22(e)), this soil clay is dominated by illite, which appears to be of the phengite type (cf. Figure 2.10(b)). No other silicate mineral is detectable from the IR spectrum, but the clay is significantly contaminated by barium sulphate, whose absorption bands at 1182, 1122, 1085, 638 and 613 cm^{-1} are very readily identified by comparison with reference spectra in Moenke (1962, 1966), Nyquist and Kagel (1971), and Kodama (1985). This clay is included here because it illustrates one of the most important assets of the IR technique, that of revealing an unexpected component in a sample.

2.7 QUANTITATIVE ANALYSIS

Quantitative determination by IR spectroscopy of clay minerals is sometimes possible if the following conditions are met.

- Particle size of mineral components are all $< 2\,\mu m$, to ensure maximum absorption and minimum scattering by the sample. This may be difficult if the sample contains minerals of widely differing hardness and grindability.
- Good control over weighing and sample transfer during preparation of KBr discs.
- Uniform dispersion and mixing with KBr, and uniform quality of pressed discs.
- One (or more) absorption band in the spectrum is sufficiently isolated from bands of other components to allow easy measurement of absorbance or the peak area.
- The availability of *appropriate* standard minerals of the same particle size. That is, the standard mineral should have the same crystallinity, state of order and similar composition to the component to be determined.
- Reproducibility of technique.

From this, it is obvious that truly quantitative determination of minerals is difficult. The most demanding conditions are the last three listed. For example, when measuring the kaolinite content of a soil clay from the absorbance of the 3700 cm^{-1} band of structural OH groups, estimates can vary by factors of two or three, depending on the crystallinity of the selected standard. Assessment of the crystallinity of the kaolin mineral in a soil clay from the IR spectrum is not easy, because of overlapping absorption bands of other minerals.

Nevertheless, certain minerals can be determined quantitatively, among them quartz (Chester and Green, 1968), quartz and cristobalite (Adamek, Linhova and Ruzek, 1969), quartz and asbestos (Coates, 1977) and biogenic opal (Chester and Elderfield, 1968). Semiquantitative estimates of kaolin minerals (3700 cm^{-1}), carbonates (1400, 870 cm^{-1}), imogolite (348 cm^{-1}), illite (3620, 830 cm^{-1}) and others, are possible with care. Detection levels may be as low as 1–2% for quartz and kaolinite, but for many other minerals they are 5–10%.

2.8 COMPUTER-AIDED METHODS

The ready access to computers, and the development of appropriate in-house software (Adams and Black, 1983) has given the IR spectroscopist, using conventional dispersive spectrometers, the opportunity to tackle the problems associated with the characterization of minor phases in multicomponent mineral assemblages. Two examples presented here illustrate this, and also the way in which other physical procedures are used to enable component differentiation to be achieved.

The first concerns the isolation of the spectrum of a new mineral, macaulayite, present as a minor (< 1 %) phase in a red, kaolinite-rich clay from deeply weathered granite (Wilson *et al.*, 1981). The spectrum of the original clay (Figure 2.29(a)) shows little or no evidence of anything other than a moderately well-ordered kaolinite. However, several passages of the clay through a high-gradient magnetic separator (Russell, Birnie and Fraser, 1984) yielded an iron-rich magnetic fraction with a spectrum (Figure 2.29(b)) showing many new bands in addition to those

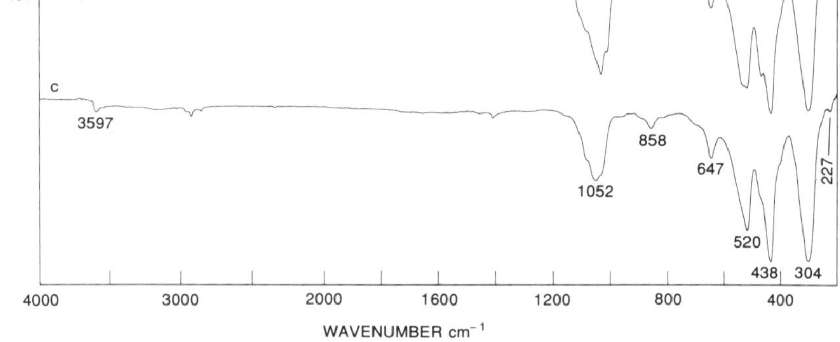

Figure 2.29 Infrared spectra of soil clay from deeply weathered granite (Bennachie, Scotland): (a) < 2 μm soil clay, (b) magnetically concentrated fraction from (a), (c) computer difference spectrum (b) – (a).

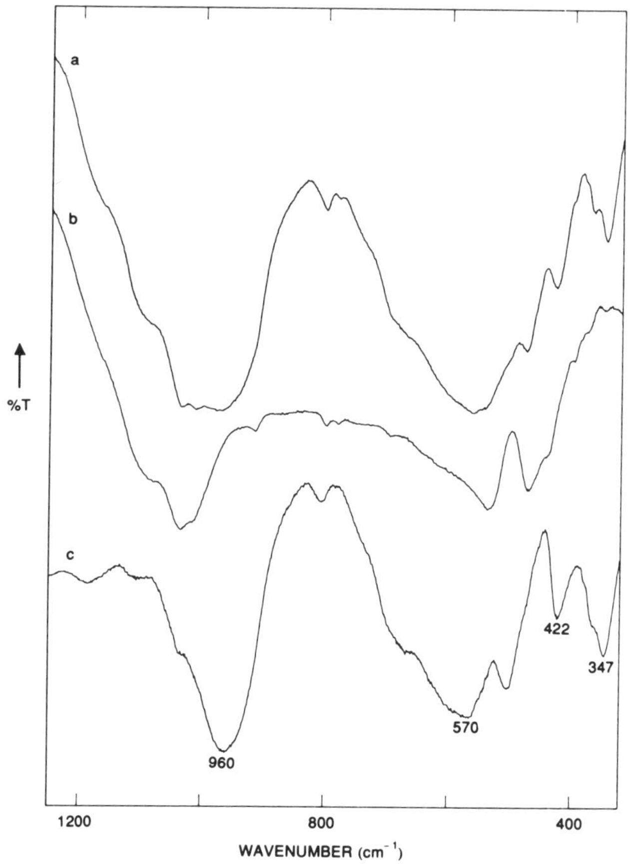

Figure 2.30 Infrared spectra of acid dispersed Australian podzolic soil clay: (a) KBr disc heated 16 h at 150°C, (b) KBr disc heated 16 h at 350°C, (c) computer difference spectrum (a) – (b).

of the kaolinite. Computer subtraction of these two spectra (that of the magnetic concentrate minus a proportion of that of the original clay) removed the kaolinite absorption bands and revealed the spectrum of macaulayite (Figure 2.29(c)), which has a complex hematite–silicate structure (Wilson *et al.*, 1984).

The second example illustrates the confirmation of the presence of the aluminosilicate proto-imogolite allophane (see section 2.5.2.2) in a podzolic soil clay. The spectrum of the soil clay (Figure 2.30(a)) shows weak imogolite type absorption bands near 960, 425 and 348 cm^{-1}. Imogolite structures are destroyed by heating to 350°C and these bands are absent from the spectrum of the clay after this heat treatment (Figure 2.30(b)). The computer difference spectrum (that of the original clay minus that of the thermally decomposed residue) clearly displays most of the bands characteristic of imogolite structures (Figure 2.30(c)), unambiguously confirming the presence of proto-imogolite allophane in the podzolic soil clay (Farmer *et al.*, 1984).

Similar computer-aided procedures can be envisaged for treatments other than the magnetic concentration and thermal decomposition described here, for example, chemical extraction, centrifugation, density gradient separation, indeed any procedure that results in enrichment of some component in the system.

ACKNOWLEDGEMENT

Acknowledgement is given to the Scottish Office Agriculture and Fisheries Department for providing funding for this work.

REFERENCES

Adamek, P., Linhova, R. and Ruzek, J. (1969) Infrared spectroscopy in the study of silica phases. Il *Sborník Vysoké Skoly chemicko-technologické v Praze*, **13**, 259–283.

Adams, M. J. and Black, I. (1983) Application of a microcomputer system in an infrared laboratory. *Journal of Automatic Chemistry*, **5**, 9–13.

Breck, D. W. (1974) *Zeolite Molecular Sieves: Structure Chemistry and Use*, Wiley, New York.

Buckley, H. A., Bevan, J. C., Brown, K. M. *et al.* (1978) Glauconite and celadonite: two separate mineral species. *Mineralogical Magazine*, **42**, 373–382.

Chester, R. and Elderfield, H. (1968) The infrared determination of opal in siliceous deep-sea sediments. *Geochimica et Cosmochimica Acta*, **32**, 1128–1141.

Chester, R. and Green, R. N. (1968) The infrared determination of quartz in sediments and sedimentary rocks. *Chemical Geology*, **3**, 199–213.

Coates, J. P. (1977) IR analysis of toxic dusts. Analysis of collected samples of quartz and asbestos. Part 1. *American Laboratory* (Fairfield, Connecticut, USA), **9**, 105–108, 110–111.

Cradwick, P. D. G., Farmer, V.C., Russell, J. D. *et al.* (1972) Imogolite, a hydrated aluminium silicate of tubular structure. *Nature*, **240**, 187–189.

Dupuis, T., Ducloux, J., Butel, P. and Nahon, D. (1984) Etude par spectrographie infrarouge d'un encroûtement calcaire sous galet. Mise en evidence et modelisation expérimental d'une suite minèrale évolutive à partir de carbonate de calcium amorphe. *Clay Minerals*, **19**, 605–614.

Durig, J. R. (1980) Analytical applications of FTIR to molecular and biological systems. *NATO Advanced Study Institute Series*, D. Reidel, Dordrecht.

Farmer, V. C. (1964) Infrared absorption of hydroxyl groups in kaolinite. *Science*, **145**, 1189–1190.

Farmer, V. C. (1974) *The Infrared Spectra of Minerals*. Mineralogical Society, London.

Farmer, V. C. and Mitchell, B. D. (1963) Occurrence of oxalates in soil clays following hydrogen peroxide treatment. *Soil Science*, **96**, 221–229.

Farmer, V. C. and Russell, J. D. (1964) The infrared spectra of layer silicates. *Spectrochimica Acta*, **20**, 1149–1173.

Farmer, V. C. and Russell, J. D. (1966) Effects of particle size and structure on the vibrational frequencies of layer silicates. *Spectrochimica Acta*, **22**, 389–398.

Farmer, V. C., Fraser, A. R. and Tait, J. M. (1979) Characterization of the chemical structures of natural and synthetic aluminosilicate gels by infrared spectroscopy. *Geochimica et Cosmochimica Acta*, **43**, 1417–1420.

Farmer, V. C., Russell, J. D., Ahlrichs, J. L. and Velde, B. (1967) Vibrations of the hydroxyl group in layer silicates. *Bulletin de la Groupe Française des Argiles*, **19**, 5–10.

Farmer, V. C., Russell, J. D., McHardy, W. J. *et al.* (1971) Evidence of loss of protons and octahedral iron from oxidized biotites and vermiculites. *Mineralogical Magazine*, **38**, 121–137.

Farmer, V. C., Fraser, A. R., Russell, J. D. and Yoshinaga, N. (1977) Recognition of imogolite structures in allophanic clays by infrared spectroscopy. *Clay Minerals*, **12**, 55–57.

Farmer, V. C., Fraser, A. R., Robertson, L. and Sleeman, J. R. (1984) Proto-imogolite allophane in podzol concretions in Australia: possible relationship to aluminous ferralitic (lateritic) cementation. *Journal of Soil Science*, **35**, 333–340.

Ferraro, J. R. (1982) *The Sadtler Infrared Spectra Handbook of Minerals and Clays*. Sadtler/Heyden, London.

Goodman, B. A., Russell, J. D., Fraser, A. R. and Woodhams, F. W. D. (1976) A Mössbauer and IR spectroscopic study of the structure of nontronite. *Clays and Clay Minerals*, **24**, 53–59.

Hadni, A. (1967) *Essentials of Modern Physics Applied to the Study of the Infrared*. Pergamon, Oxford.

Hayashi, H., Otsuka, R. and Imai, N. (1969) Infrared study of sepiolite and palygorskite on heating. *American Mineralogist*, **54**, 1613–1624.

Juo, A. S. R. and White, J. L. (1969) Orientation of the dipole moments of hydroxyl groups in oxidized and unoxidized biotite. *Science*, **165**, 804–805.

Kodama, H. (1985) *Infrared Spectra of Minerals. Reference Guide to Identification and Characterization of Minerals for the Study of Soils*. Technical Bulletin 1985–1E, Research Branch, Agriculture Canada, Ottawa.

Kodama, H. and Oinuma, K. (1963) Identification of kaolin minerals in the presence of chlorite by X-ray diffraction and infrared absorption spectra. *Clays and Clay Minerals*, **11**, 236–249.

Koyama, K. and Takeuchi, Y. (1977) Clinoptilolite: the distribution of potassium atoms and its role in thermal stability. *Zeitschrift fuer Kristallographie, Kristallgeometrie, Kristallphysik, Kristallchemie*, **145**, 216–239.

Langer, K. and Raith, M. (1974). Infrared spectra of Al–Fe(III)-epidotes and zoisites, $Ca_2(Al_{1-p}Fe_p^{3+})Al_2O(OH)[Si_2O_7][SiO_4]$. *American Mineralogist*, **59**, 1249–1258.

Lazarev, A. N. (1974) The dynamics of crystal lattices, in *The Infrared Spectra of Minerals* (ed. V. C. Farmer). Mineralogical Society, London, pp. 69–86.

Ledoux, R. L. and White J. L. (1964) Infrared study of selective deuteration of kaolinite and halloysite at room temperature. *Science*, **145**, 47–49.

Lehr, J. R., Brown, E. H. and Frazier, A. W. (1967). *Crystallographic Properties of Fertilizer Compounds*. Tennessee Valley Authority, Chemical Engineering Bulletin no. 6. National Fertilizer Development Center, Muscle Shoals, AL.

Martin, A. E. (1980) In *Vibrational Spectra and Structures*. Vol. 8 *Infrared Interferometric Spectrometers* (ed. J. R. Durig). Elsevier, Amsterdam.

Mendelovici, E., Yariv, S. and Villalba, R. (1979). Iron-bearing kaolinite in Venezuelan laterites: I. Infrared spectroscopy and chemical dissolution evidence. *Clay Minerals*, **14**, 323–331.

Moenke, H. (1962) *Mineralspektren I*. Akademie-Verlag, Berlin.
Moenke, H. (1966) *Mineralspektren II*. Akademie-Verlag, Berlin.
Nadeau, P. H. (1980) Burial and contact metamorphism in the Mancos Shale. PhD Thesis, Dartmouth College, Hanover, New Hampshire, 200 pp.
Nadeau, P. H. and Bain, D. C. (1986) Composition of some smectites and diagenetic illitic clays and implications for their origin. *Clays and Clay Minerals*, **34**, 455–465.
Nadeau, P. H., Farmer, V. C., McHardy, W. J. and Bain, D. C. (1985) Compositional variations of the Unterrupsroth beidellite. *American Mineralogist*, **70**, 1004–1010.
Neal, M. and Worrall, W. E. (1977). Mineralogy of fireclays: part 1. The crystallinity of kaolinite in fireclays. *Transactions of the British Ceramic Society*, **76**, 57–61.
Nyquist, R. A. and Kagel, O. (1971) *Infrared Spectra of Inorganic Compounds*. Academic Press, London.
Oinuma, K. and Hayashi, H. (1968) Infrared spectra of clay minerals. *Journal of Tokyo University, General Education (Natural Sciences)*, **9**, 57–98.
Prost, R., Dameme, A., Huard, E. *et al.* (1989) Infrared study of structural OH in kaolinite, dickite, nacrite, and poorly crystalline kaolinite at 5 to 600 K. *Clays and Clay Minerals*, **37**, 464–468.
Rendon, J. L. and Serna, C. J. (1981) IR spectra of powder hematite: effects of particle size and shape. *Clay Minerals*, **16**, 375–382.
Rouxhet, P. G., Samudacheata, N., Jacobs, H. and Anton, O. (1977) Attribution of the OH stretching bands of kaolinite. *Clay Minerals*, **12**, 171–180.
Russell, J. D. (1974) Instrumentation and techniques, in *The Infrared Spectra of Minerals* (ed. V. C. Farmer), Mineralogical Society, London.
Russell, J. D. (1979) Infrared spectroscopy of ferrihydrite: evidence for the presence of structural hydroxyl groups. *Clay Minerals*, **14**, 109–114.
Russell, J. D. (1987) Infrared spectroscopy of inorganic compounds, in *Laboratory Methods in Infrared Spectroscopy* (ed. H. Willis), Wiley, New York.
Russell, J. D., Farmer, V. C. and Velde, B. (1970) Replacement of OH by OD in layer silicates and identification of the vibrations of these groups in infrared spectra. *Mineralogical Magazine*, **37**, 869–879.
Russell, J. D., Birnie, A. and Fraser, A. R. (1984). High-gradient magnetic separation (HGMS) in soil clay mineral studies. *Clay Minerals*, **19**, 771–778.
Serna, C. J., Rendon, J. L. and Iglesias, J. E. (1982) Infrared surface modes in corundum-type microcrystalline oxides. *Spectrochimica Acta*, **38A**, 797–802.
Serna, C. J., Vanscoyoc, G. E. and Ahlrichs, J. L. (1977) Hydroxyl groups and water in palygorskite. *American Mineralogist*, **62**, 784–792.
Slonimskaya, M. V., Besson, G., Dainyak, L. G. *et al.* (1986) Interpretation of the IR spectra of celadonites and glauconites in the region of OH-stretching frequencies. *Clay Minerals*, **21**, 377–388.
Tarasevich, Yu. I. (1970) Spectral study of the thermal dehydration of palygorskite. *Dopovidi Akademii Nauk Ukrains'koi RSR, Seriya B*, **32**, 938–942.
Tarte, P. (1965). Experimental study and interpretation of infrared spectra of silicates and germanates. *Memoires de l'Academie Royale de Belgique Classe des Science 8°*, **35**, parts 4a, b.
Theng, B. K. G., Russell, M., Churchman, G. J. and Parfitt, R. L. (1982) Surface properties of allophane, halloysite and imogolite. *Clays and Clay Minerals*, **30**, 143–149.
Tuddenham, W. M. and Lyon, R. J. P. (1959) Relation of infrared spectra and chemical analysis for some chlorites and related minerals. *Analytical Chemistry*, **31**, 377–380.

Turrell, G. (1972) *Infrared and Raman Spectra of Crystals*. Academic Press, London.

Van der Marel, H. W. and Beutelspacher, H. (1976) *Atlas of Infrared Spectroscopy of Clay Minerals and their Admixtures*. Elsevier, Amsterdam.

Vedder, W. (1964) Correlations between infrared spectrum and chemical composition of mica. *American Mineralogist*, **49**, 736–768.

Weiner, S. and Goldberg, P. (1990) On-site Fourier transform–infrared spectrometry at an archaeological excavation. *Spectroscopy*, **5**, 46–50.

White, R. G. (1964) *Handbook of Industrial Infrared Analysis*. Plenum Press, New York.

Wilkins, R. W. T. and Ito, J. (1967) Infrared spectra of some synthetic talcs. *American Mineralogist*, **52**, 1649–1661.

Wilson, M. J., Russell, J. D. and Tait, J. M. (1974) A new interpretation of the structure of disordered α-cristobalite. *Contributions to Mineralogy and Petrology*, **47**, 1–6.

Wilson, M. J., Russell, J. D., Tait, J. M. *et al.* (1981) A swelling hematite/layer silicate complex in weathered granite. *Clay Minerals*, **16**, 261–278.

Wilson, M. J., Russell, J. D., Tait, J. M. *et al.* (1984) Macaulayite, a new mineral from North-East Scotland. *Mineralogical Magazine*, **48**, 127–129.

Yariv, S. H. and Mendelovici, E. (1979) The effect of degree of crystallinity on the infrared spectrum of hematite. *Applied Spectroscopy*, **33**, 410–411.

CHAPTER 3

Mössbauer spectroscopy

B. A. Goodman

3.1 INTRODUCTION

In Mössbauer spectroscopy transitions are observed between the energy levels in an atomic nucleus in the gamma-ray region of the electromagnetic spectrum. When a gamma-ray is emitted from a nucleus in a free state, such as a gas, that nucleus recoils in the same way as a gun when a bullet is fired. The energy of recoil, E_R, is related to the mass of the nucleus, m, and to the velocity, v, by the relationship

$$E_R = \frac{mv^2}{2}.$$

If, however, the nucleus is not in a free state but is part of a crystalline structure, such as a mineral, then the recoil energy is absorbed by many atoms, to such an extent that a fraction, f, of the emissions is recoil-free. A similar process occurs on absorption. It is the study of the recoil-free absorption of those gamma-rays that were emitted without recoil that is the basis of Mössbauer spectroscopy.

The energy levels of the nuclear transitions are precisely defined and their changes with variations in the environment of the nucleus are very small. It is thus necessary to use radiation that has an exceedingly small spread of energies in order to observe the spectral structure from interactions between nuclei and their surrounding electrons. In Mössbauer spectroscopy this is achieved by modulating the energy of the gamma-rays by simple Döppler motion, whereby the radiation source is moved relative to the absorbing specimen. Only small velocities are required and ± 1 cm/s covers the range of energies required for ^{57}Fe Mössbauer spectroscopy.

There are several factors that limit the numbers of isotopes that exhibit the Mössbauer effect. Firstly, it is necessary for a source to exist that

Clay Mineralogy: Spectroscopic and Chemical Determinative Methods.
Edited by M. J. Wilson. Published in 1994 by Chapman & Hall, London.
ISBN 0 412 53380 4.

decays to produce an isotope of the nucleus under investigation in an excited state. Thus all low atomic weight < 19 nuclei are excluded. Secondly, the isotope must have the energy of its first excited state low enough for the recoil energy to be lower than the lattice phonon energy (an absolute limit here is about 140 keV but *c*. 25 keV is the limit for ambient temperature operation) and to have a half-life that is long enough to produce a linewidth that is smaller than the differences in magnetic dipole and electric quadrupole effects that are produced by changes in the chemical environment of the nucleus. Furthermore, for routine operation the half-life of the source should be at least several days and preferably months or years. For the study of clay minerals there is the additional requirement that the natural abundance of the isotope should be high enough for the Mössbauer effect to be readily detectable (i.e. > 0.01 % of the total material). After taking into account all of these factors the only isotope suitable for routine investigations in clay minerals is ^{57}Fe, which has a natural abundance of *c*. 2%. Consequently, the remainder of this chapter will be concerned solely with this isotope.

The nuclear properties of ^{57}Fe are such that its first excited state lies *c*. 14.4 keV above its ground state and has a half-life of *c*. 100 ns. It is populated as a result of the radioactive decay by electron capture of the ^{57}Co nucleus, which has a half-life of *c*. 270 days (Figure 3.1). The ground and first excited states have spins of $\frac{1}{2}$ and $\frac{3}{2}$, respectively, and interactions between the nucleus and its electronic environment can cause the energy states to shift or split. The various interactions and other factors that determine the nature of a Mössbauer spectrum are considered in the following section.

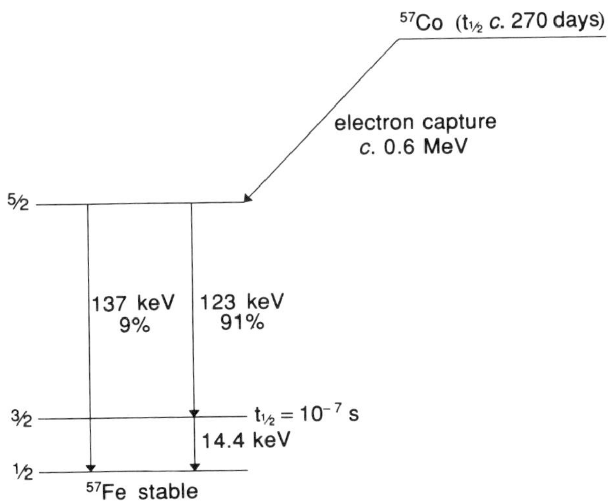

Figure 3.1 Simplified decay scheme for the ^{57}Co nucleus. $t_{\frac{1}{2}}$ = half-life.

3.2 PRINCIPLES OF MÖSSBAUER SPECTROSCOPY

3.2.1 Electric monopole interaction (isomer shift)

The magnitude of the electric monopole interaction between the nucleus and its surrounding electrons is determined by the electron density at the nucleus (Figure 3.2). A parameter called the isomer shift, δ, is used to describe the shift in the centre of the spectrum between the sample being investigated and a known standard, usually metallic iron. The isomer shift is given by the equation

$$\delta = \left(\frac{2\pi}{5}\right) Ze^2 \{ |\Psi_{(0)}|_a^2 - |\Psi_{(0)}|_r^2 \} (R_e^2 - R_g^2),$$

where Z is the atomic number, e is the electronic charge, $|\Psi_{(0)}|_a^2$ and $|\Psi_{(0)}|_r^2$ are the electron densities at the nucleus of the sample and reference respectively, and R_e and R_g are the radii of the nuclear excited and ground states respectively.

The predominant source of the isomer shift in oxide and aluminosilicate minerals is a process described as core polarization. This occurs because the radial distributions of atomic orbitals are such that electrons in 3d orbitals spend part of their time closer to the nucleus than electrons in core s-orbitals. When this occurs the attractive Coulomb potential between the nucleus and the s-electrons is decreased. Therefore, any change in the number of d-electrons changes the magnitude of $|\Psi_{(0)}|^2$. The resulting change in δ is positive for an increase in d-electron density, because the term $(R_e^2 - R_g^2)$ is negative for ^{57}Fe.

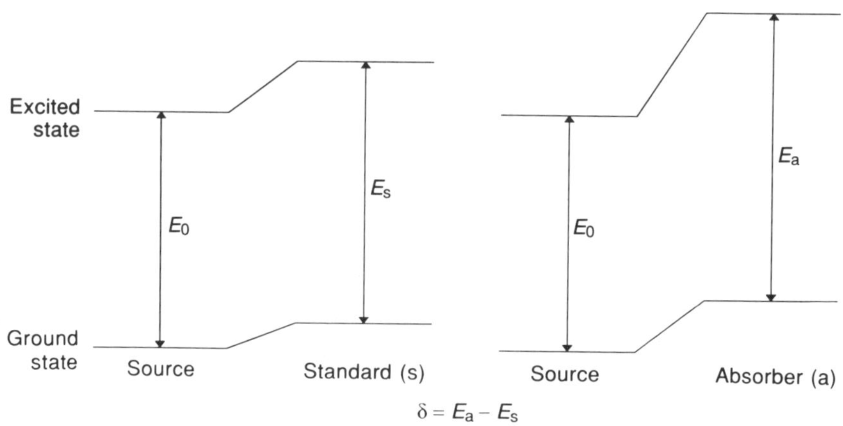

Figure 3.2 Diagrammatic representation of the electric monopole interaction, which yields the isomer shift, δ.

The d-electron density on the iron is determined primarily by the oxidation state (Fe^{2+} has six 3d electrons whereas Fe^{3+} has five), but it is also influenced by the degree of covalency of the molecular orbitals in which the d-electrons are involved. Generally, the degree of covalency increases with decrease in coordination number if the liganded group remains unaltered. Thus δ decreases with increasing oxidation state and increases to a lesser extent with increasing coordination number.

3.2.2 Electric quadrupole interaction

If a nucleus has a non-uniform charge density, which is the case for nuclear spin, $I > \frac{1}{2}$, the energy levels are split by interaction with an electric field gradient, the electric quadrupole interaction (Figure 3.3). The magnitude of the resulting spectral splitting is known as the quadrupole splitting, Δ, and is determined by the distributions of valence electrons and charges in the surrounding lattice.

In Fe^{3+} the high spin ion has one electron in each of the 3d orbitals. The contribution to the electric field gradient from the valence electrons, q_{val}, is, therefore, zero and the quadrupole splitting is to a first approximation determined by the electric field gradient resulting from the crystal lattice, q_{latt}. The high spin Fe^{2+} ion has one electron over a half-filled set of 3d orbitals. For the ground state there is a large valence electron contribution to the quadrupole interaction. However, excited electronic states are often of only slightly higher energy and may produce a

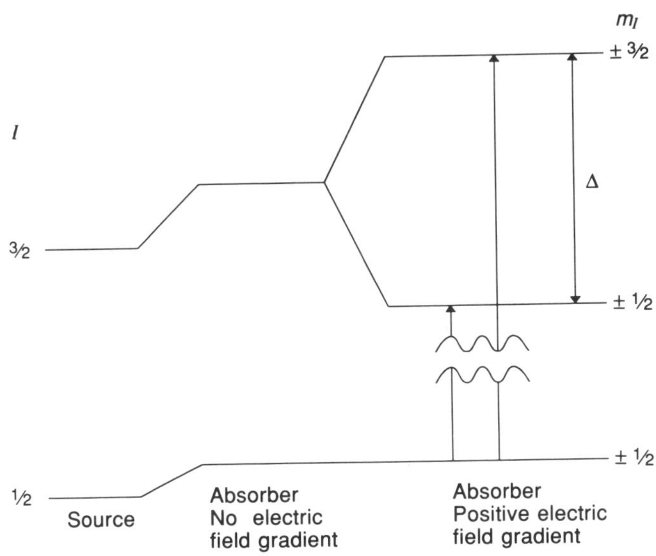

Figure 3.3 Diagrammatic representation of the electric quadrupole interaction, which gives the quadrupole splitting, Δ.

contribution to the electric field gradient of opposite sign to that of the ground state. It is common, therefore, for the Fe^{2+} quadrupole splitting to decrease with increasing temperature as a result of increasing population of excited electronic states. Also, with Fe^{2+} the q_{val} and q_{latt} terms are generally of opposite sign, with the result that the most distorted environments have the smallest quadrupole splittings.

The intensities of the two peaks in a quadrupole split spectrum are equal if the sample is randomly oriented. However, the intensity does vary with the angle between the principal axis of the electric field gradient and the direction of the gamma-ray beam (Table 3.1).

Table 3.1 Orientation dependence of the peaks in a quadrupole split spectrum

Transition	*Relative intensity*[a]
$\pm\frac{1}{2} \rightarrow \pm\frac{3}{2}$	$1 + \cos^2\theta$
$\pm\frac{1}{2} \rightarrow \pm\frac{1}{2}$	$\frac{5}{3} - \cos^2\theta$

[a] θ is the angle between the principal axis of the electric field gradient and the direction of the gamma-ray beam.

3.2.3 Magnetic hyperfine interaction

The coupling of the nuclear magnetic moment with a magnetic field causes a complete splitting of both ground and excited state energy levels (Figure 3.4). The separations are given by $g\mu_n B$, where B is the magnitude of the magnetic field, μ_n the nuclear magneton (0.50505×10^{-27} erg/T), and g the nuclear gyromagnetic ratio, which is different for

Table 3.2 Relative energies and intensities of the peaks in a magnetically hyperfine split spectrum

Transition	*Relative energy*[a]	*Relative intensity*[b]
$-\frac{1}{2} \rightarrow -\frac{3}{2}$	$-\frac{1}{2}\mu_n B(3g_e + g_0)$	$\frac{9}{4}(1 + \cos^2\theta)$
$-\frac{1}{2} \rightarrow -\frac{1}{2}$	$-\frac{1}{2}\mu_n B(g_e + g_0)$	$3\sin^2\theta$
$-\frac{1}{2} \rightarrow +\frac{1}{2}$	$\frac{1}{2}\mu_n B(g_e - g_0)$	$\frac{3}{4}(1 + \cos^2\theta)$
$+\frac{1}{2} \rightarrow -\frac{1}{2}$	$-\frac{1}{2}\mu_n B(g_e - g_0)$	$\frac{3}{4}(1 + \cos^2\theta)$
$+\frac{1}{2} \rightarrow +\frac{1}{2}$	$\frac{1}{2}\mu_n B(g_e + g_0)$	$3\sin^2\theta$
$+\frac{1}{2} \rightarrow +\frac{3}{2}$	$\frac{1}{2}\mu_n B(3g_e + g_0)$	$\frac{9}{4}(1 + \cos^2\theta)$

[a] g_e and g_0 are the gyromagnetic ratios for the excited and ground states respectively.
[b] θ is the angle between the magnetic field and the gamma-ray beam.

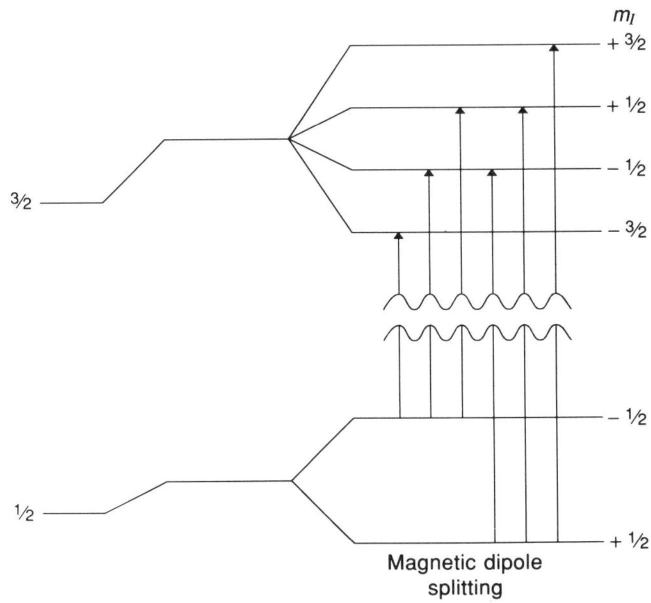

Figure 3.4 Diagrammatic representation of the magnetic dipolar interaction and the allowed transitions between the ground and first excited state of ^{57}Fe.

the ground and excited states. The selection rule for a transition to be allowed is that $m_I = \pm 1$ or 0, with the result that six peaks are observed in a spectrum. Their relative energies and intensities are given in Table 3.2.

For a powder specimen, where the principal axis of the magnetic hyperfine coupling is randomly oriented, the peak intensities vary in the ratio 3 : 2 : 1 : 1 : 2 : 3 (e.g. Figure 3.6(h)). However, if there is some preferential orientation, as may occur for example with ferrimagnetic particles in an external magnetic field, then the intensities of peaks 2 and 5 may be either increased or decreased depending on the angle between the magnetic axis and the gamma-ray beam.

In magnetically ordered materials the magnitude of the magnetic hyperfine structure is not constant, but increases with decreasing temperature below the magnetic ordering temperature (Curie temperature for ferro- or ferrimagnetic materials, Néel temperature for antiferromagnetic materials). This takes the form of a Brillouin function (Figure 3.5) and it can be seen that the characteristic field, B_{hf}, is only approached at very low temperatures.

It should be noted that the appearance of magnetic hyperfine structure in a Mössbauer spectrum does not necessarily require that the specimen is magnetically ordered. The motion of electrons in paramagnetic materials is able to generate a magnetic field at the nucleus. Usually the rate of transition between the electron spin states is fast and hyperfine structure is not observed (see below). However, if this rate of relaxation is slowed, e.g. by the

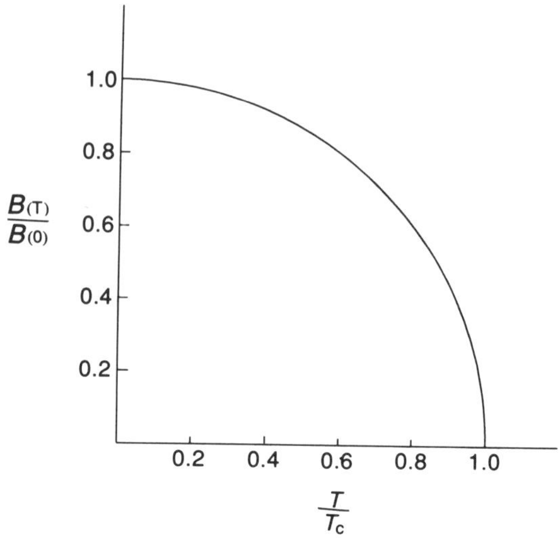

Figure 3.5 Dependence of the ratio of the internal magnetic hyperfine field ($B_{(T)}$) to its saturation value ($B_{(0)}$) in magnetically ordered materials on the ratio of the temperature of observation (T) to the ordering temperature (T_c).

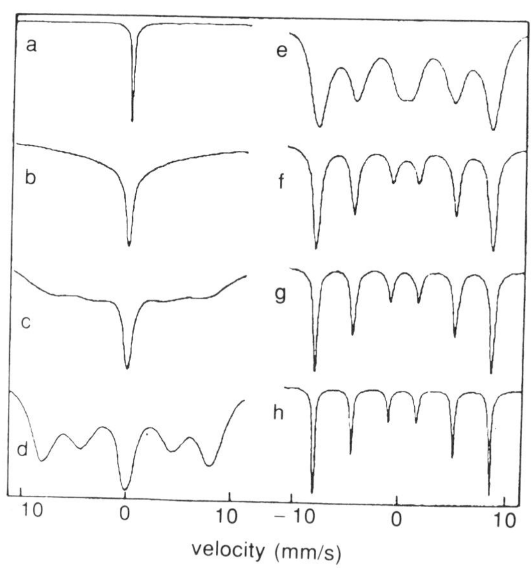

Figure 3.6 Variation of shape of magnetic hyperfine structure with relaxation time, t. (a) 10^{-12} s, (b) 10^{-9} s, (c) 2.5×10^{-9} s, (d) 5×10^{-9}s, (e) 7.5×10^{-9} s, (f) 2.5×10^{-8} s, (g) 7.5×10^{-8} s, (h) 10^{-6} s. (Adapted from Wickman, 1966.)

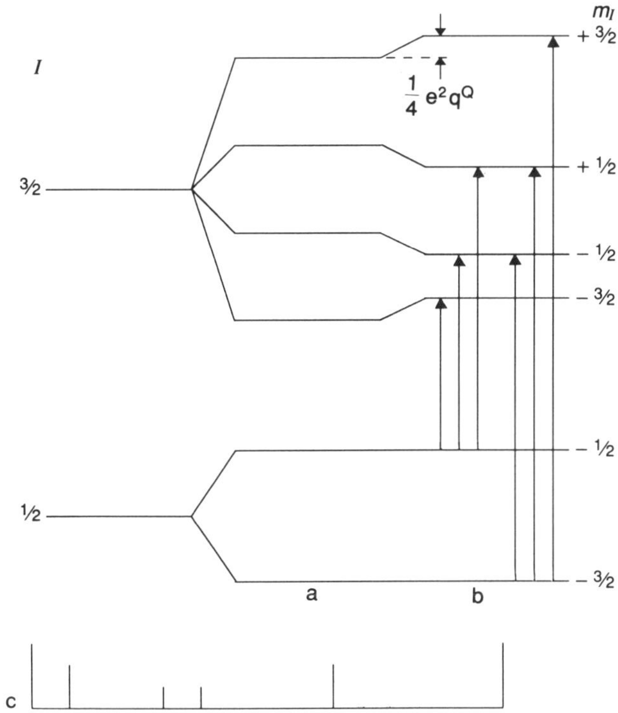

Figure 3.7 Diagrammatic representation of (a) simple magnetic dipolar, (b) combined magnetic dipolar and electric quadrupolar interactions, and (c) the positions of the spectral peaks that arise from (b).

application of an external magnetic field, or by decreasing the temperature of magnetically dilute samples (Wickman, 1966), then magnetic hyperfine structure may be seen in paramagnetic materials (Figure 3.6).

When there are combined magnetic hyperfine and electric quadrupole interactions it is not possible to determine the absolute magnitudes of both from a single spectrum unless their principal axes are colinear. With Fe^{3+} the magnetic term usually dominates and the quadrupolar term produces a perturbation in the relative energies of the transitions (Figure 3.7).

3.2.4 Peak shapes and intensities

For thin specimens Mössbauer peaks are Lorentzian in shape, but with thick absorbers the line shape is modified by saturation effects. However, in order to obtain a reasonable signal-to-noise ratio, fairly thick absorbers are generally used and the assumption of Lorentzian shape in the analysis of partially resolved spectra may lead to considerable inaccuracies in the determination of the positions, widths and intensities of the component peaks, especially where there is a high degree of overlap (see e.g. Longworth, 1984a). Ways of overcoming this problem are discussed in section 3.3.4. Furthermore, with components undergoing relaxation processes (i.e. hyperfine parameters fluctuating at a rate comparable to the time of the transition – *c.* 10^{-7} to 10^{-9} s) this line-shape approximation breaks down completely. An example of the dependence of the shape of magnetic hyperfine structure on relaxation time is given in Figure 3.6.

This behaviour often results from the collective reorientation of the magnetic moment direction in very small particles of magnetically ordered materials, a process that is very important in clay minerals, where such small particles are common.

The area under an absorption peak is proportional to the effective thickness, t, of the absorber,

$$t = nf_a\sigma_0,$$

where n is the number of atoms of the Mössbauer isotope per unit area, f_a is the recoil-free fraction for the absorber, and σ_0 is the absorption cross-section. In the thin-absorber approximation the absorption area, A, is given by

$$A = \left(\frac{\pi}{2}\right) f_s \Gamma t,$$

where f_s is the recoil-free fraction of the source and Γ the natural linewidth of the transition.

For moderately thick absorbers (< 0.1 mg $^{57}Fe/cm^2$) the Lorentzian shape is a reasonable approximation to the experimental peak shape, and the experimental linewidth, Γ_{ex}, can be expressed (Bancroft, 1973) as

$$\Gamma_{ex} = \Gamma_a + \Gamma_s + 0.27\Gamma t,$$

where Γ_a and Γ_s are the widths for a thin absorber and the source, respectively.

Experimental linewidths in clay mineral specimens can be broadened by a number of factors in addition to sample thickness, the most important being the existence of inhomogeneities in the coordination environment of the iron atoms. This results in the spectral peaks each being spread over a range of values, instead of being centred on a unique value. These individual peaks are not generally resolved and the result is inhomogeneously broadened peak shapes.

The magnitudes of the recoil-free fraction, f, can vary appreciably from one sample to another, and decrease rapidly at elevated temperatures as a result of increased vibrational motion of atoms in the crystal lattices. Therefore, differences in f factors between different structures are at a minimum at low temperatures.

3.3 INSTRUMENTATION AND EXPERIMENTAL TECHNIQUES

3.3.1 The Mössbauer spectrometer

A Mössbauer spectrometer (Figure 3.8) consists of a drive unit which moves the source to modulate its energy, a gamma-ray detector, a data storage system along with various amplifiers, and some form of output device.

3.3.1.1 Velocity modulation

The drive unit usually consists of a linear velocity transducer, which operates like a moving coil loudspeaker, consisting of a driving coil, an electromagnetically isolated pick-up coil, and a function generator which controls the motion of the drive. Sources may be held at one or both ends of the transducer.

Two different types of function generator are in use, namely constant velocity or velocity sweep devices. In the former the source is moved towards or away from the absorber at a constant velocity for a fixed period of time, during which the amount of transmitted radiation is counted. This procedure is then repeated for a succession of different velocities. With a velocity sweep system the spectrum is scanned by varying the velocity of the source during a single sweep, usually at constant acceleration. In a typical system the data-logging device, usually a microcomputer, and the waveform generator are synchronized so that the gamma-rays corresponding to particular velocities are always

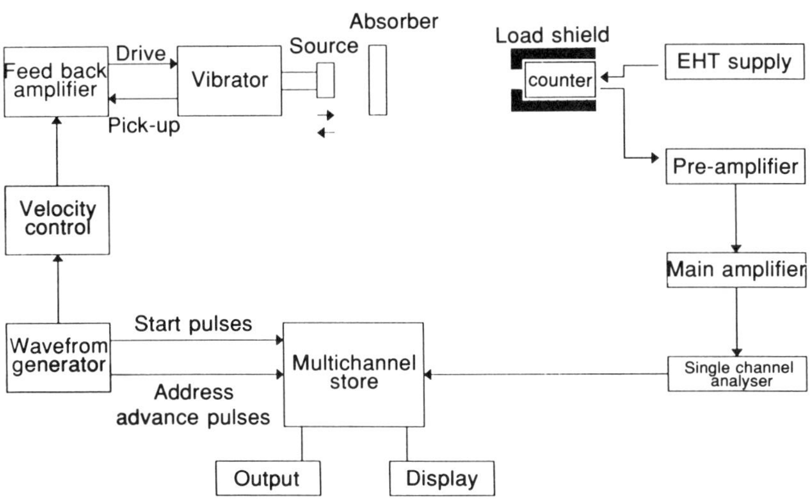

Figure 3.8 Block diagram of a Mössbauer spectrometer.

fed into the same channels, each of which corresponds to a fixed velocity increment.

Devices have been produced which allow a pre-selected velocity range to be swept normally while unrequired velocities are swept quickly. By this means maximum resolution can be achieved for a particular region of interest in a reasonable time. This is sometimes of value when dealing with complex peaks from overlapping components. It should be pointed out that constant velocity systems are of particular value in this respect because a region of the spectrum well-removed from zero velocity can be scanned with high statistical accuracy in a short time.

3.3.1.2 Sources

For ^{57}Co the radioactive isotope is usually bound in an environment that produces a single peak emission with as high a recoil-free fraction as possible and with linewidth close to the natural value. Several alloys satisfy these criteria and are produced commercially by either evaporating or electroplating the ^{57}Co on to a metal foil, then heating it *in vacuo* and finally mounting it in a suitable holder. Two of the most popular alloying metals are Ir and Pd, with Ir being the choice for experiments where the source is cooled because it remains as a single line at low temperatures.

3.3.1.3 Detection systems

The nature of the detection system to be used is determined by the nature of the experiment being performed. In all cases though it must perform the function of detecting the selected photons as efficiently as possible, while at the same time excluding any other radiation. Typically, it consists of a radiation counter, pre-amplifier, amplifier, single channel analyser and multichannel analyser. In transmission geometry, which is the most common set-up, gas-filled proportional counters are normally used as gamma-ray counters. These are filled with one of the heavier inert gases, e.g. Ar, Kr or Xe with nitrogen or methane as the quenching gas, and provide good resolution of the ^{57}Fe 14.4 keV gamma-ray with a high signal-to-noise ratio. For experiments in scattering geometry the absorber is often incorporated within the detector, which should be able to detect selectively re-emitted gamma-rays, conversion electrons, Auger electrons or X-rays (Table 3.3). It is possible, therefore, with this type of arrangement to produce depth-resolved Mössbauer spectra, although sensitivity is usually appreciably lower than for measurements in transmission geometry with a properly prepared absorber.

Table 3.3 Energies, intensities and ranges of photons emitted from ^{57}Fe in its first excited state

Emitted radiation	*Energy (keV)*	*Intensity*[a]	*Lower range*[b]	*Upper range*[c]
Gamma-ray	14.4	0.10		
K X-rays	6.4	0.28		
L X-rays	0.7	0.002		
K conversion electrons	7.3	0.79	10	400
L conversion electrons	13.6	0.08	20	1300
M conversion electrons	14.3	0.01	20	1500
KLL Auger electrons	5.5	0.63	7	200
LMM Auger electrons	0.53	0.6	1	2

[a] Relative intensities per decay of excited ^{57}Fe.
[b] Depth of the mean free path without energy loss (in mm).
[c] Depth to stop all electrons (in mm).

3.3.2 Drive velocity calibration

The usual method for calibration of the drive system is through the use of a standard absorber, such as iron metal in the form of a thin foil, which also provides a standard reference point for isomer shifts because different sources emit at slightly different energies. The spectrum of metallic iron consists of six peaks, whose positions are known with a high degree of accuracy. By comparing peak channels with the known energies of these peaks, the channel velocity increment of the spectrometer can be calculated. Also, because there are six points, any deviations from linearity of a velocity sweep waveform can be detected.

Absolute calibration of a velocity sweep drive system can be performed with a laser interferometer. This is performed by splitting the laser beam into two parts, one part being reflected by a prism attached to the beam splitter and the other by a prism attached to the driving tube of the velocity transducer. A displacement of this latter prism generates a fringe pattern which is detected by a photodiode. The fringe counts are transformed into pulses, each of which corresponds to a displacement of one-half of a wavelength of the laser light (i.e. 3.164×10^{-4} mm for a He–Ne laser).

3.3.3 Sample preparation

For experiments in transmission geometry routine sample preparation is straightforward and simply requires that the specimen is powdered and inserted between the source and detector in the form of a disc with a known thickness of ^{57}Fe. Control of thickness is important, because if it is too low there will be a poor signal-to-noise ratio, whereas if it is too high there will be saturation and peak distortion. Depending on the number of peaks in the spectrum and the temperature of operation, the optimum amounts of ^{57}Fe are in the range 3–10 mg/cm^2. Absorbers also need to be uniform because any gross local variations in thickness will lead to either saturation or high background counts on a local scale and a consequent decrease in spectral quality. With dense iron-rich specimens it is sometimes helpful to mix the specimen with an inert matrix of low atomic number elements, such as polyethylene powder, alumina or icing sugar, in order to prepare an absorber of uniform thickness.

With platy samples, such as sheet silicates, it may be difficult to prepare absorbers in which the individual particles are completely randomly oriented because there will be some preference for alignment of the sheets with the face of the holder. Such texture effects can cause the ratios of the peaks of quadrupole split spectra to deviate from unity (see Table 3.1). If such an effect is not recognized it can lead to errors in the spectral analysis. Texture effects can be minimized by mixing specimens with an excess of an inert matrix, as mentioned above, or by orienting the sample holder at the 'magic angle' of 54.7° to the gamma-ray beam. (Note though that the path length of the gamma-ray beam through the absorber is now twice that for an absorber aligned perpendicular to the beam and its thickness should be reduced accordingly.) Excessive grinding of a specimen to reduce texture effects is not recommended, because this often results in some chemical alteration as a result of the increased temperatures and/or pressures that are generated.

Absorber holders need to be of a material with low absorption capacity for the gamma-rays and also cheap to produce. Thus organic polymers, such as polymethylmethacrylate or polyethylene, are ideal for this purpose.

For scattering experiments there is very little absorber preparation required because the specimen does not have to be powdered and there are no limitations on the thickness of the absorber.

3.3.4 Analysis of spectra

Analysis of Mössbauer spectra is invariably carried out by computer and a number of programs are now readily available. The most common approach (Longworth, 1984a) is to use a least-squares minimization fitting routine, in which estimates of spectral parameters are systematically refined by an iterative process. It is usual to assume that the thin-absorber approximation is valid and that the peaks are Lorentzian in shape, although it is a straightforward matter to use any other line-shape function if there is a justifiable reason for doing so. For randomly oriented absorbers the number of variables required to define each peak (position, width and height or area) may be decreased for quadrupole or magnetically hyperfine split spectra. In a quadrupole doublet the two components have equal areas and widths, whereas with a magnetic hyperfine spectrum the areas are in the ratio 3 : 2 : 1 : 1 : 2 : 3 for a randomly oriented powder and the positions of all six lines are not all independent, i.e. they are defined by three independent variables, the centre, the magnetic hyperfine term and the electric field gradient term. For complex spectra with two or more overlapping components it is usually necessary to use all justifiable constraints in order to achieve a converging fit with the computer.

When the thin-absorber approximation is not valid, spectra suffer from selective saturation effects, so that the area under a peak is no longer proportional to the concentration of iron atoms. Various numerical methods have been developed to compensate for this. For example, Meads *et al.* (1972) generated correction curves by comparing spectral fits obtained with the thin-absorber approximation with simulations obtained by numerical integration of the transmission integral. Others have attempted to use the transmission integral itself in fitting routines, but this tends to be very costly in terms of computation time. This was overcome neatly by Cranshaw (1974), who replaced the integral by arrays calculated from the source profile and absorber response. An alternative approach was developed by Dibar de Ure (1973), who used fast Fourier transform techniques to remove the thickness distortion from a spectrum, which could then be fitted legitimately by a sum of Lorentzian-shaped peaks as a thin absorber.

In many microcrystalline and poorly ordered materials there are pronounced surface effects and high concentrations of lattice defects, each of which affects the Mössbauer parameters of associated Fe atoms. In such materials the spectra are characterized by a distribution of electric field gradients, and magnetic hyperfine fields if there is magnetic ordering. The result is spectral broadening. The situation of magnetic hyperfine field distributions is most easily recognized because it usually results in asymmetric peak broadening, which is most pronounced on the outermost lines. Fitting such envelopes is more complicated than is the case with simple Lorentzian peaks, but procedures have been developed based on Fourier analysis (Window, 1971) or stepwise field variations (Wivel and Morup, 1981; Murad, 1982a). Generally it is not possible to fit Mössbauer spectra to electric field gradient distributions without some information or assumptions concerning the nature of the distribution function. The approach that has been adopted here has been to calculate the shape of the electric field gradient distribution function for different distributions of ions in the mineral lattice. For example, Goodman (1976) has used a Monte-Carlo procedure to calculate the electric field gradient envelope for a random distribution of ions in the octahedral sites of a layer silicate. More recently, Dainyak, Drits and Heifits (1992) have used an analysis of the IR spectra to determine the distribution of different octahedral cations around the OH groups and have then used these results in fitting the Mössbauer spectra.

In some situations, where spectra result from the presence of several components, it may be possible to subtract the contributions of known components to reveal the spectra from other species, which can then be analysed by normal least-squares minimization procedures. (It should be noted that it is usually necessary to subtract a simulation rather than actual experimental data points in order to avoid increasing noise levels.)

3.4 MÖSSBAUER SPECTROSCOPY OF OXIDE MINERALS

3.4.1 Introduction

Iron oxide minerals are of widespread occurrence in the Earth's crust and are often formed as clay-sized particles as a result of mineral weathering reactions. The principal oxide and oxyhydroxide species are listed in Table 3.4, along with details of their magnetic properties. All of the minerals order magnetically, but only magnetite and maghemite, which have spinel-type structures, are strongly ferrimagnetic. The others are either antiferromagnetic or very weakly ferrimagnetic, although the arrangement of spins in the poorly crystalline mineral, ferrihydrite, has been described as sperimagnetic (Coey and Readman, 1973). The Mössbauer spectra of iron oxides have been reviewed by Bowen and Weed (1984), Murad and Johnston (1987) and Murad (1988) and the spectroscopic parameters for well-defined mineral specimens are summarized in Table 3.5. It has been suggested recently (Pollard *et al.*, 1992) that the use of large applied magnetic fields in low temperature Mössbauer spectra is the most effective approach for discrimination between the

Tabel 3.4 Structural and magnetic properties of the major oxide and oxyhydroxide species of iron

Mineral	*Formula*	*Magnetic properties*[a]	*Ordering temperature (K)*
Hematite	α-Fe_2O_3	Weak F (> 260 K) A (< 260 K)	955
Maghemite	γ-Fe_2O_3[b]	F	*c.* 750
Magnetite	Fe_3O_4	F	*c.* 850
Goethite	α-FeOOH	A	393
Lepidocrocite	γ-FeOOH	A	77
Akaganeite	β-FeOOH	A	295
Ferrihydrite	$Fe_5HO_8 \cdot 4H_2O$[b]		
Feroxyhite	δ'-FeOOH	F	*c.* 440

[a] F = ferrimagnetic; A = antiferromagnetic.
[b] Defective structures.

Table 3.5 Mössbauer spectroscopic parameters for the major oxides and oxyhydroxides of iron

Minerals	T (K)	B_{hf} (T)	*Isomer shift*[a] *(mm/s)*	*Quadrupole splitting (mm/s)*
Hematite	297	51.7	0.37	−0.20
	4.2	54.2	0.49	0.41
Maghemite	297	49.9	0.32	*c.* 0
Magnetite	297	49.0	0.26	*c.* 0
		46.0	0.67	*c.* 0
Goethite	297	38.2	0.37	−0.26
	4.2	50.6	0.48	−0.24
Lepidocrocite	297		0.37	0.53
	4.2	45.8	0.47	*c.* 0
Akaganeite	297		0.38	0.55
			0.37	0.95
	4.2	48.9	0.50	*c.* 0
		47.8	0.48	−0.24
		47.3	0.49	−0.81
Ferrihydrite	292		0.35	*c.* 0.75
	4.2	46–50	0.49	*c.* 0
Feroxyhite	4.2	53.0	0.48	0.17
		50.8	0.48	0.07

[a] Relative to iron metal.

various iron oxyhydroxide species, because it minimizes particle size effects. However, it is a costly procedure and is not likely to be used extensively, except for problems that cannot be resolved with measurements at higher temperatures.

A common feature of many clay materials is a combination of small particle size and extensive isomorphous substitutions, both of which strongly affect the spectral characteristics. These effects will be included in the discussions of the major crystalline phases in the following paragraphs. Poorly crystalline structures and other oxide minerals will be dealt with separately at the end of this section.

3.4.2 Hematite (α-Fe_2O_3)

The structure of hematite is based on a hexagonal close-packed arrangement of oxygen atoms with Fe^{3+} ions occupying two-thirds of the interstitial sites in an ordered arrangement. Thus each of the Fe atoms is coordinated to six oxygens in a distorted octahedral arrangement. At low temperatures well-crystalline specimens are antiferromagnetic with the magnetic moments aligned at 7° to the [001] axis. On raising the temperature a 'spin-flip' transition occurs and the magnetic moments reorientate so that they lie in the (001) plane. This transition, which is known as the

Morin transition, occurs at *c*. 260 K. Above this temperature the magnetic moments are no longer exactly antiparallel and the mineral is weakly ferrimagnetic. The principal axis of the electric field gradient is parallel to the [001] axis and is not affected by the Morin transition. Consequently the quadrupole term in Mössbauer spectra undergoes a change in both magnitude and sign as a result of this transition (*c*. + 0.4 mm/s below and – 0.2 mm/s above). The sign of this quadrupole term is the most convenient way of distinguishing whether the weakly ferrimagnetic or the antiferromagnetic phase is responsible for a Mössbauer spectrum (Table 3.5). The saturation hyperfine fields also change slightly as a result of the Morin transition.

With microcrystalline and/or isomorphously substituted specimens the temperature of the Morin transition is decreased and in some cases the weakly ferrimagnetic phase may exist to very low temperatures. Sometimes, with specimens which exhibit a range of particle sizes, both weakly ferrimagnetic and antiferromagnetic phases may coexist over a range of temperatures. The amounts of these components can be determined readily from the Mössbauer spectra because of the large separation of the sextets at the second and sixth peaks (Fig. 3.9).

Both particle size and level of Al substitution, Al_S, affect the magnitude of the magnetic hyperfine field. These have been studied in some detail by Murad and Schwertmann (1986), who derived the following empirical relationship for room temperature spectra

$$B_{hf} = 51.72 - 7.6Al_S - 32/MCD_{001}$$

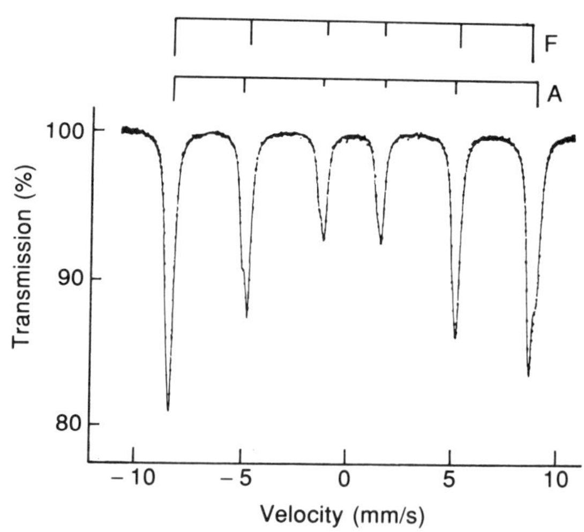

Figure 3.9 Mössbauer spectrum at 4.2 K of a synthetic hematite containing 4.3% Al showing individual contributions from the antiferromagnetic (A) and ferrimagnetic (F) phases. (Reproduced from Murad, 1988.)

where MCD_{001} is the mean crystallite dimension in nanometres along the [001] direction, as estimated by XRD. Thus the level of Al substitution in hematite specimens can be determined from Mössbauer spectra only if the crystal dimensions are determined independently.

3.4.3 Maghemite (γ-Fe_2O_3) and magnetite (Fe_3O_4)

Both magnetite and maghemite have spinel structures, which contain unit cells of 32 close packed oxygen atoms in which 8 tetrahedral (A) and 16 octahedral (B) sites are available for occupancy by cations. In magnetite the Fe^{2+} ions occupy half of the B sites and Fe^{3+} fill the A sites and the remaining B sites. Maghemite is non-stoichiometric and has an average of $\frac{1}{9}$ of the cation sites vacant. Both minerals are ferrimagnetic in character at room temperature and below. The spin arrangement on the A and B sites is antiparallel, but the greater abundance of the B sites means that both minerals have strong magnetic properties.

In magnetite there is rapid exchange between the Fe^{2+} and Fe^{3+} ions (electron hopping) on the B sites between the Curie temperature of *c.* 850 K and the Vervey temperature of 120 K. (The Vervey temperature is the low temperature limit for electron hopping.) In this temperature range the Mössbauer spectra consist of two components (Figure 3.10)

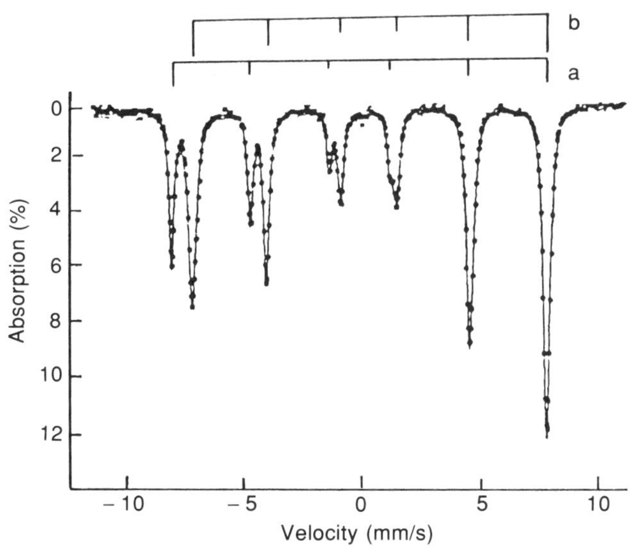

Figure 3.10 Mössbauer spectrum of magnetite at room temperature; sextet (a) corresponds to Fe^{3+} in tetrahedral coordination and sextet (b) corresponds to a mixture of equal amounts of Fe^{2+} and Fe^{3+} undergoing rapid electron exchange (i.e. $Fe^{2.5+}$) in octahedral coordination.

corresponding to Fe in the A and B sites, the latter having spectral parameters intermediate between those of Fe^{2+} and Fe^{3+}.

The peaks from the B-site ions are significantly broader than those from the A site (typically *c.* 0.48 mm/s compared with *c.* 0.30 mm/s). This has been interpreted as resulting from either the existence of more than one inequivalent B site (Häggström *et al.*, 1978) or the relaxation effects from the electron hopping process (Kündig and Hargrove, 1969; Sawatzky, Coey and Morrish, 1969). Below the Vervey transition temperature, electron hopping is inhibited and separate spectral components are observed for each ion and crystallographic site. Complex spectra are obtained because of the presence of a number of inequivalent B sites and fitting such spectra requires at least five components (Hargrove and Kündig, 1970).

Aluminium substitution in magnetites does not occur readily, but Schwertmann and Murad (1990) have reported an extensive study of specimens synthesized at ambient temperatures. Although increasing amounts of aluminium in the reaction media favour the formation of goethite, magnetite specimens with limited Al-for-Fe substitutions were produced. Mössbauer spectra showed that this Al substituted preferentially in the tetrahedral sites. Increasing Al substitution resulted in decreasing crystal size, demonstrating that the structure had limited capacity to incorporate aluminium.

Spectra of pedogenic magnetite may be influenced by isomorphous substitutions and small-particle-size effects. In the absence of magnetic splitting, spectra consist of a single broad peak, which can be resolved into a doublet from B sites and a single peak from A sites (Häggström *et al.*, 1978).

The peaks in the Mössbauer spectra of maghemite tend to be rather broad and individual sextets are not well resolved. At room temperature well-crystalline specimens give spectra that are fitted to a single sextet with a magnetic hyperfine field *c.* 50 T and a small quadrupolar component. In the presence of an external magnetic field this splits into two components (Coey, Morrish and Sawatsky, 1971) with hyperfine fields of 49.9 and 50.5 T. Determination of isomer shifts was apparently not accurate enough in that work for the octahedral and tetrahedral ions to be distinguished. At low temperatures the spectra show some sign of resolution in the absence of an applied magnetic field, and Murad (1988) has reported a spectrum containing at least two overlapping sextets with similar hyperfine fields (52.6 T) and isomer shifts, but different quadrupole splittings. From these various reports it would appear to be difficult to use Mössbauer spectroscopy to determine the distribution of vacancies in maghemite, although in principle it should be a straightforward procedure.

3.4.4 Goethite (α-FeOOH)

Goethite has a structure that consists of a hexagonal close-packed array of oxygen atoms and hydroxyl groups that are stacked in an alternating sequence along the [001] axis, with Fe^{3+} ions occupying double rows of octahedral sites. Its Néel temperature of *c.* 393 K means that there is a large difference in the hyperfine field at room temperature (38.2 T) and 77 K (50.4 T) (Figure 3.11). Thus a room temperature spectrum is, in principle, a good method for the identification of goethite. However, because goethite is isostructural with diaspore, natural goethites often

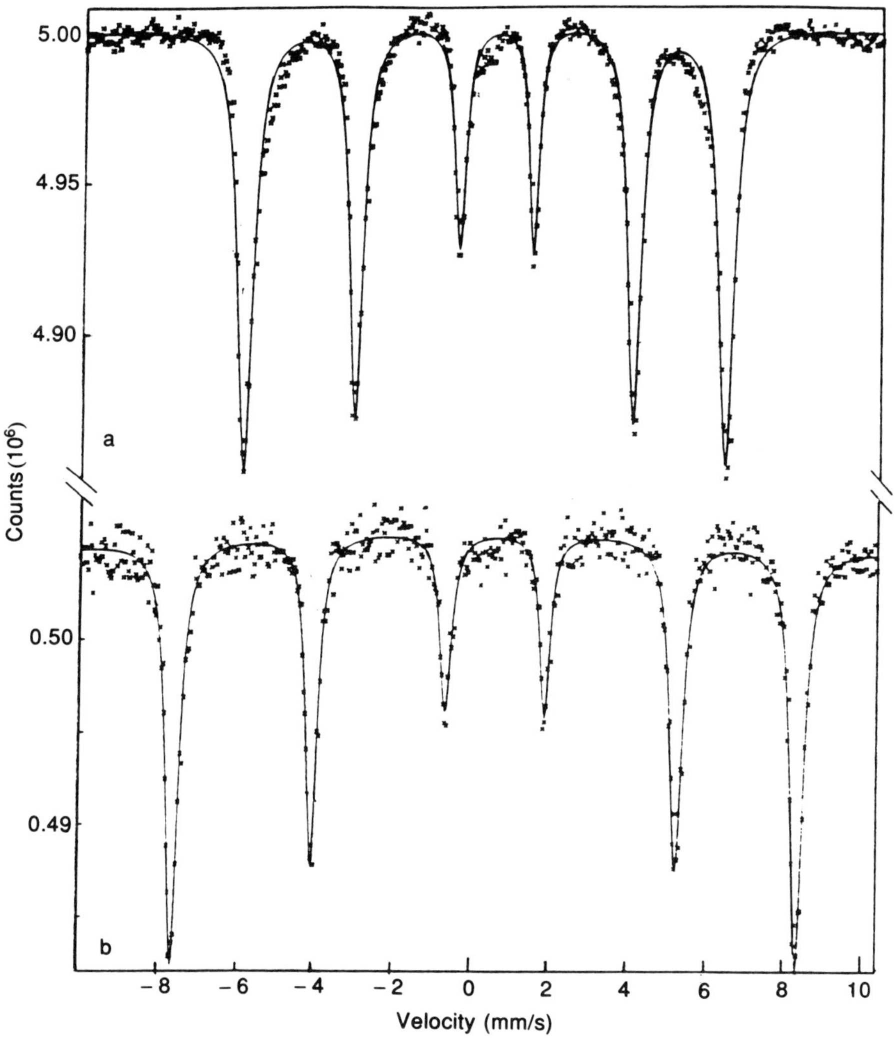

Figure 3.11 Mössbauer spectra of a natural goethite specimen at (a) room temperature and (b) 77 K. Note the poor fit to the outermost lines of spectrum (a) is the result of relaxation effects.

accumulate significant quantities ($\leqslant$ 33 atom%) of Al in their structures. These substitutions result in a lowering of the Néel temperature, and consequently magnetic hyperfine structure may not always be observable at room temperatures. Indeed with high levels of Al substitution and poor crystallinity, goethite may sometimes produce a superparamagnetic doublet at 77 K (Goodman and Lewis, 1981). Several authors have attempted to relate the Mössbauer parameters to levels of isomorphous substitutions and sample crystallinity, and as was discussed in section 3.4.2 for hematite it appears that both phenomena produce similar effects in Mössbauer spectra. Murad and Schwertmann (1983) have made a comprehensive study of these effects and have derived the empirical relationship

$$B_{hf}(4.2\,K) = 50.65 - 4.2Al - 8.7/MCD_{111},$$

where MCD_{111} represents the mean crystallite dimension in nanometres in the [111] direction, as determined by analysis of XRD data.

In contrast to the behaviour with aluminium, when synthetic goethites were prepared with varying amounts of silicon and phosphorus these ions were not incorporated into the goethite structure (Quin *et al.*, 1988). Mössbauer spectra at 4.2 K demonstrated that the goethite ordering temperature decreased with increasing Si and P, but the hyperfine parameters were typical of goethite. Thus the effect of these substituents is to adsorb on to the crystal-growth sites and to prevent uniform crystal growth.

3.4.5 Lepidocrocite (γ-FeOOH)

The coordination of the Fe in lepidocrocite is similar to that in goethite except that the octahedral groupings are linked to form layers that are held together by hydrogen bonds from the hydroxyl groups. The magnetic exchange interactions between Fe^{3+} ions in this structure are considerably weaker than in goethite and as a consequence the Néel temperature is lower (at *c.* 77 K). Thus Mössbauer spectra consist of a doublet at temperatures $\geqslant$ 77 K and magnetic hyperfine structure is observed only at low temperatures. Also, for pure specimens, it is smaller than those of the other oxides and hydroxides (Table 3.5).

3.4.6 Other iron oxide minerals

This section will cover briefly the Mössbauer spectroscopic characteristics of the rare oxyhydroxide mineral, akaganéite, and the poorly defined phases, ferrihydrite and feroxyhite.

3.4.6.1 Akaganéite (β-FeOOH)

Akaganéite has an open structure similar to that of hollandite the manganese oxide mineral. Its Mössbauer spectra show the presence of more than one type of iron atom (e.g. Childs *et al.*, 1980). Room-temperature spectra indicate the presence of at least two inequivalent Fe^{3+} doublets, whereas low-temperature spectra show at least three distinct magnetically ordered components.

3.4.6.2 Ferrihydrite

The structure of ferrihydrite is generally considered to consist of octahedrally coordinated Fe^{3+} ions in hexagonally close-packed layers of O, OH and OH_2 ions. Thus it may be considered to be a poorly ordered hydrous oxide. At room temperature its Mössbauer spectrum consists of a broad doublet which requires a minimum of two quadrupole components for a satisfactory computer fit. Parameters are similar to those of akaganéite and on occasions ferrihydrite samples have been wrongly identified as akaganéite. At 4.2 K sextet spectra are produced with broad peaks, whose widths are strongly influenced by sample crystallinity. It is common to use magnetic field distribution functions to describe the details of these spectra. Maxima in B_{hf} distributions have been reported to vary between 50 T for well-crystalline specimens and 46.5 T for poorly defined samples (Murad, 1982b). Behaviour of spectra at 77 K is varied; poorly crystalline specimens produce spectra that resemble those obtained at room temperature, whereas better defined specimens produce sextet structures.

Aluminium substitution in synthetic ferrihydrites has been shown by Chadwick *et al.* (1986) to result in a lowering of the magnetic ordering temperature consistent with the behaviour of other iron oxide minerals. Isomorphous substitutions also appear to influence the magnetic properties of natural ferrihydrites (Childs *et al.*, 1984; Murad and Schwertmann, 1988). In an extensive study of three natural ferrihydrites Murad *et al.* (1988) have demonstrated that the appearance of the magnetic sextets occurred over an extensive temperature range, in which they gradually replaced the superparamagnetic doublets with decreasing temperature. Blocking temperatures were reported to range from 115 to 28 K and average particle dimensions, calculated from the relative areas of the doublet and sextet components, between 5 and 3 nm.

In a detailed computer analysis of the ambient temperature Mössbauer spectrum of a well-defined (six-line XRD pattern) ferrihydrite, Cardile (1988) has shown that it is not possible to distinguish between a situation where there are three components with octahedral coordination and that where there are one tetrahedral and two octahedral components, as proposed by Eggleton and Fitzpatrick (1988). This difficulty in using Mössbauer spectroscopy to distinguish between models with such large differences in tetrahedral/octahedral ratios is probably the consequence of extreme disorder, which results in an extensive (and possibly nonuniform) distribution of environments in the ferrihydrite structure.

3.4.6.3 *Feroxyhite (δ′-FeOOH)*

Feroxyhite is thought to have a disordered CaI_2 structure in which the Fe^{3+} ions are randomly distributed over the octahedral sites. The related δ-FeOOH, which does not occur naturally, is ferrimagnetic with an ordering temperature of *c.* 440 K and a hyperfine field of *c.* 38 T at room temperature (Dézsi *et al.*, 1967). Usually this structure is not observed at room temperature because samples tend to be poorly crystalline, although it can be seen if an external magnetic field is applied. At low temperatures both δ-FeOOH and feroxyhite give spectra with two sextet components, which probably correspond to two different structural sites (Persoons, Chambere and DeGrave, 1986; Murad and Johnston, 1987).

3.4.7 Characterization of green rusts

Green rusts are unstable hydroxides that contain a mixture of Fe^{2+} and Fe^{3+} ions along with anions such as chloride, sulphate or carbonate. Mössbauer spectroscopy is a powerful method for characterizing the forms of iron in these various phases and furthermore its application is not limited by the instabilities of the minerals.

Hydroxycarbonate green rusts have been studied by Murad and Taylor (1984), who found that their Mössbauer spectra in the paramagnetic state could be fitted to two Fe^{2+} and two Fe^{3+} doublets. This result was rather more complex than was envisaged for a structure based on that of pyroaurite (Taylor, 1980), in which the irons are surrounded by an octahedral arrangement of oxygens in brucite-like sheets that are separated by interlayers containing H_2O and CO_3^{2-}. It was interpreted in terms of a basic structure that contains an ordered distribution of Fe^{2+} and Fe^{3+} in the ratio 3 : 1 (Figure 3.12), which should produce a single component for each type of ion. Variations from this composition will, however, generate sites with electric field gradients different from those in the basic structure, with the result that two doublets are generated for each type of ion. A similar model was also used to explain the Mössbauer spectra of iron/aluminium hydroxycarbonates.

The preparation and Mössbauer spectra of green rusts from sulphate solutions have been described by Cuttler *et al.* (1990). These authors concentrated on measurements at 77 and 4 K. They found that at 77 K the spectra could be fitted satisfactorily to two components, one Fe^{2+} and one Fe^{3+}. The quadrupole splitting of the Fe^{2+} at 2.93 mm/s was similar to that reported by Murad and Taylor (1984) for the outermost component of their iron hydroxycarbonate green rusts. At 4.2 K both Fe^{2+} and Fe^{3+} components exhibited magnetic hyperfine structure, with values for B_{hf} of 12.4 and 50.4 T, respectively.

Bender Koch and Morup (1991) have reported Mössbauer spectra of a natural ochre sample in which they conclude that about 10% of the iron

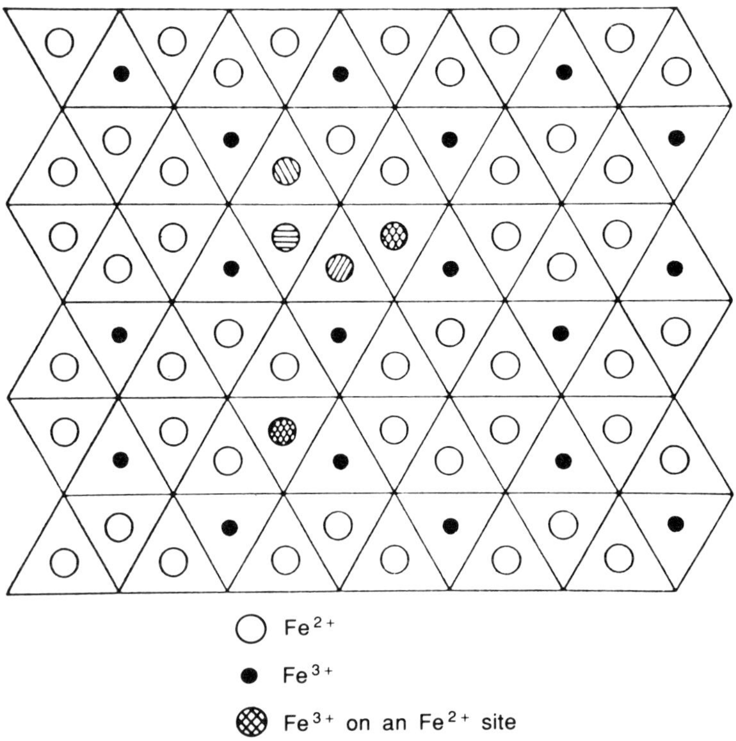

Figure 3.12 Model of the arrangement of ions in the brucite sheet of iron hydroxycarbonate 'green rust'. Hatched circles show the different orientations of Fe^{2+} relative to a Fe^{3+} formed by oxidation of Fe^{2+}. (Reproduced from Murad and Taylor, 1984.)

is in the form of a green rust. This conclusion was based on observations of the behaviour of a minor ferrous component, which showed sensitivity to oxidation by atmospheric oxygen and produced partial magnetic hyperfine structure at 12 K. Because XRD revealed the presence of only calcite and iron oxides, the assignment of the Fe^{2+} to a layer silicate could be excluded.

3.4.8 Iron-substituted aluminium oxides

When iron is substituted at low levels in oxide lattices, paramagnetic hyperfine structure may be observed at low temperatures (i.e. when the electronic relaxation time is long compared to the the time-scale of the Mössbauer transition – *c.* 10^{-8} s). The general appearance of such spectra is similar to that of magnetically ordered species, but the magnitude of the hyperfine field is typically around 55 T, if the $S = \frac{5}{2}$ state is occupied. This is greater than for any magnetically ordered oxide, with the possible

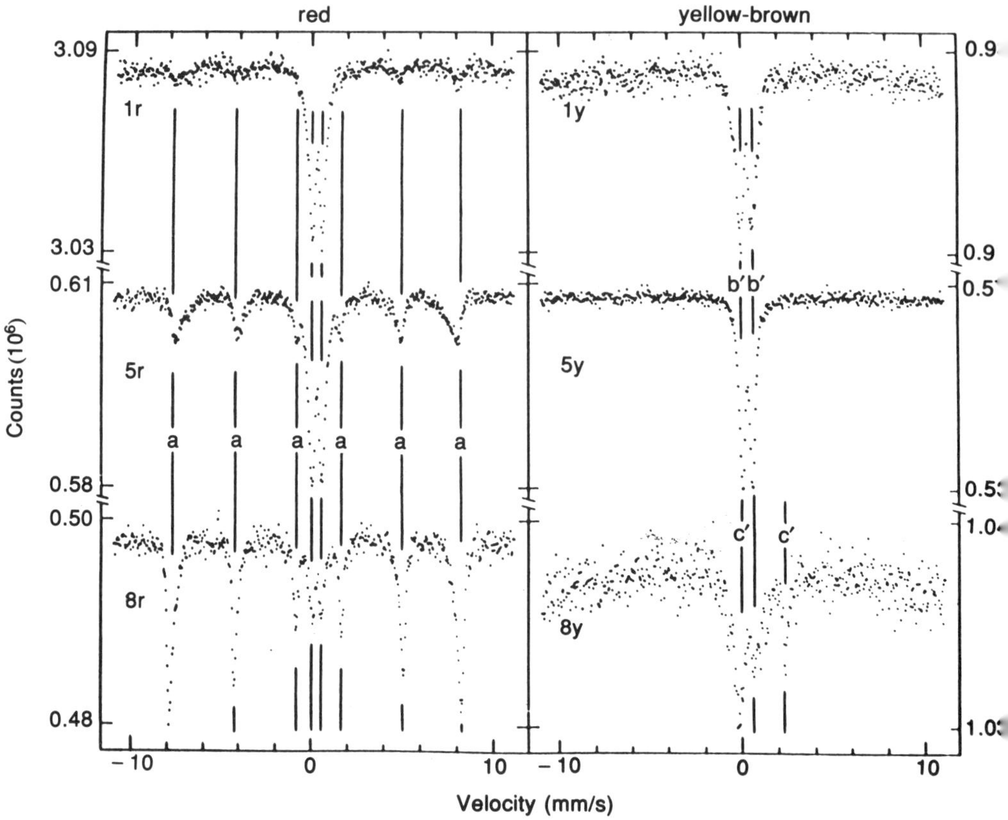

Figure 3.13 Mössbauer spectra at 300 K of three pairs of red (r) and yellow-brown (y) soil samples from New Zealand. Note that the main differences relate to the presence in the red and absence from the yellow-brown of sextet (a) from hematite. (Reproduced from Childs, Goodman and Churchman, 1978.)

exception of hematite below its Morin transition, but as paramagnetic hyperfine structure is usually observed only at low temperatures there is no problem in distinguishing it from a hematitic phase. Also, in slowly relaxing paramagnetic materials there is a strong dependence of the energy levels on the magnitude of any external magnetic field (Chapter 5), and Wickman (1966) has shown how the spectra of ^{57}Fe in α-Al_2O_3 in external magnetic fields change dramatically at the positions of cross-over of energy states.

The electron spin lattice relaxation time increases with increasing temperature and consequently the paramagnetic hyperfine structure collapses with increasing temperature; spectra are then determined by the electric monopole and quadrupole interactions.

3.4.9 Characterization of iron oxides in soils

This represents one of the more valuable uses of Mössbauer spectroscopy. It has a distinct advantage over other methods in that no chemical pretreatments are required, thus eliminating the potential formation of artefacts in the laboratory. Furthermore, the samples are not destroyed during the analysis. As examples of practical applications this section will briefly review two papers in which Mössbauer spectroscopy has been used as the principal technique for characterizing the various forms of iron in soil specimens.

3.4.9.1 Identification of red and yellow-brown mottles

This section describes the identification of the secondary iron oxides in some red and yellow-brown New Zealand soil specimens that were closely associated within the same profiles (Childs, Goodman and Churchman, 1978). The room temperature spectra of three pairs of samples are presented in Figure 3.13. In each case the red sample had a sextet component with parameters characteristic of hematite, and this was absent from the corresponding yellow-brown sample. At 77 K (Figure 3.14) each sample exhibited an additional sextet component with parameters consistent with goethite. Thus the main distinction between the two types of soil sample is the presence or absence of hematite, there being little differences in other components between pairs of samples.

3.4.9.2 Detailed characterization of magnetic components in soils

By performing magnetic separations it is possible to physically concentrate any ferri- and strongly paramagnetic components in soil samples. This section describes the work of Longworth *et al.*, (1979) on the characterization of such specimens by Mössbauer spectroscopy. Room-temperature spectra of whole soils and magnetically separated fractions are presented in Figures 3.15 and 3.16. Samples TB3, TB7 and Caldy Hill came from soils that had been subjected to recent burning and it can be seen that the magnetic separation technique represents a very effective way of increasing the spectral intensity. The Annecy sample also had a history of burning and its sextet component was also enhanced by magnetic separation, but sample TB1 from a stream bed in an unburnt soil catchment was not greatly altered. The parameters for the sextet components in Figure 3.16 are consistent with the presence of both hematite and magnetite.

It is difficult to effectively deconvolute the spectra obtained at 4.2 K (Figure 3.17), because of the large amount of overlap between the spectra of the various magnetic components. However, by applying large external magnetic fields to the samples, it was possible to increase significantly the amount of information that could be obtained from these

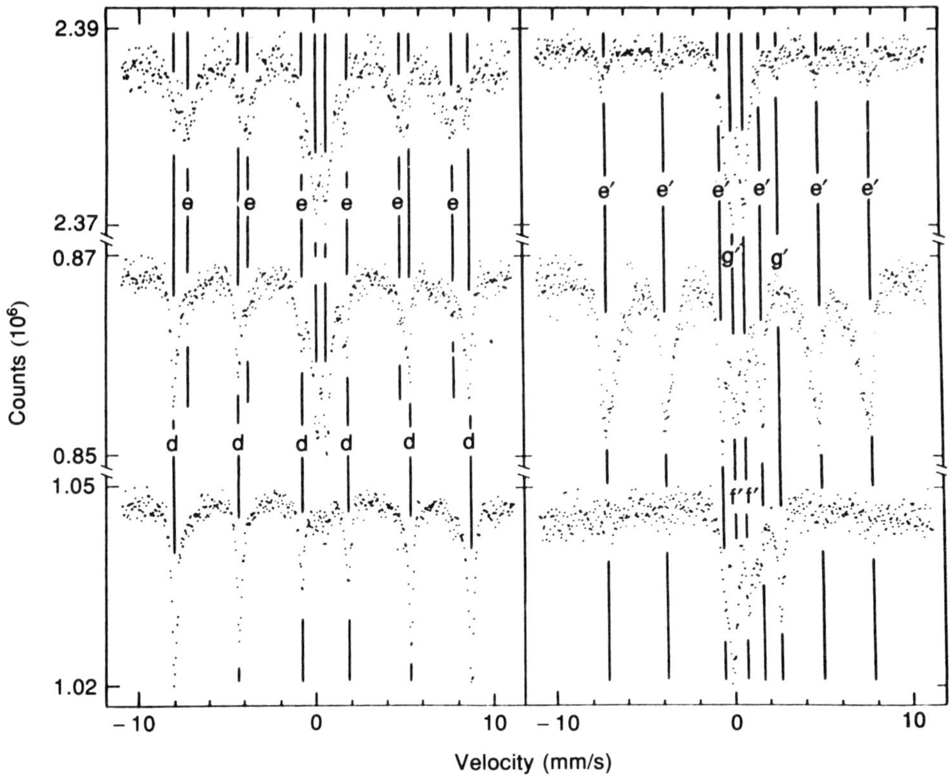

Figure 3.14 Mössbauer spectra at 77 K of the red (r) and yellow-brown (y) soil samples of Figure 3.13. Note that at this temperature the two sets of sample are again distinguished by the presence or absence of the hematite component (d). (Reproduced from Childs, Goodman and Churchman, 1978.)

low-temperature spectra (Figure 3.18). For an external field applied parallel to the direction of the gamma-ray beam, the intensity ratios of the peaks from ferrimagnets, such as magnetite and maghemite, are changed from 3 : 2 : 1 : 1 : 2 : 3 to 3 : 0 : 1 : 1 : 0: 3, whereas those from antiferromagnets, such as hematite and goethite, are unaltered. In interpreting the spectra of Figure 3.18 it should be borne in mind that for microcrystalline ferrimagnets alignment of the spins in the external magnetic field may be incomplete and hence the intensities of their second and fifth peaks may not be reduced to zero. Nevertheless, in samples TB3 and TB7, where magnetite is the major magnetic component, there is an appreciable reduction in the intensities of these peaks. It can also be seen that the spectrum of the component in TB1, which was superparamagnetic at room temperature, has the appearance of an antiferromagnetic material.

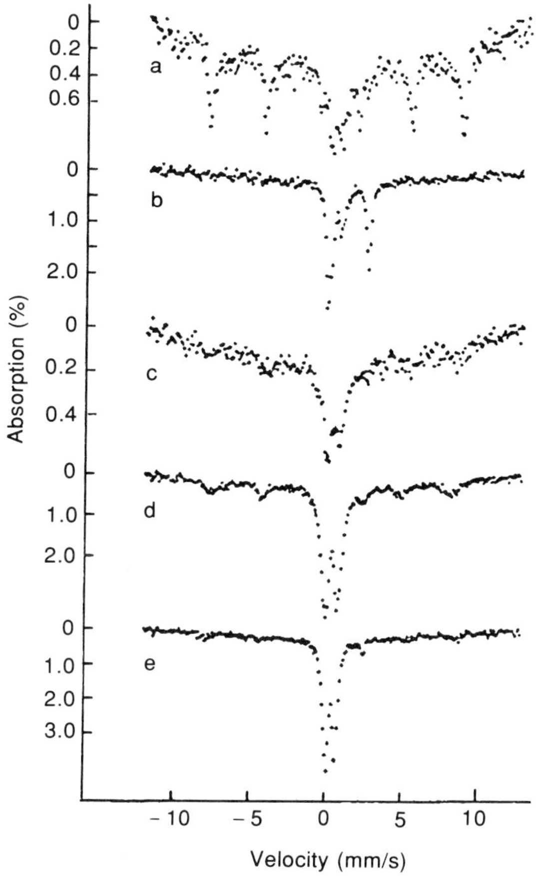

Figure 3.15 Mössbauer spectra at 300 K of some whole soil samples; (a) Caldy Hill, (b) TB1, (c) TB3, (d) TB7, and (e) Annecy. (Reproduced from Longworth *et al.*, 1979.)

Its hyperfine field at 4.2 K was 48.3 T, suggesting that it is probably goethite, although Longworth *et al.* (1979) did not rule out the possibility of microcrystalline hematite.

This work shows that differentiation between hematite and maghemite, which have similar Mössbauer parameters, is possible in the presence of a magnetic field. The two hyperfine fields in magnetite and their appreciably lower values allow it to be distinguished from both of these oxides in room-temperature spectra, although of course an applied magnetic field at low temperatures could be used to confirm its distinction from hematite.

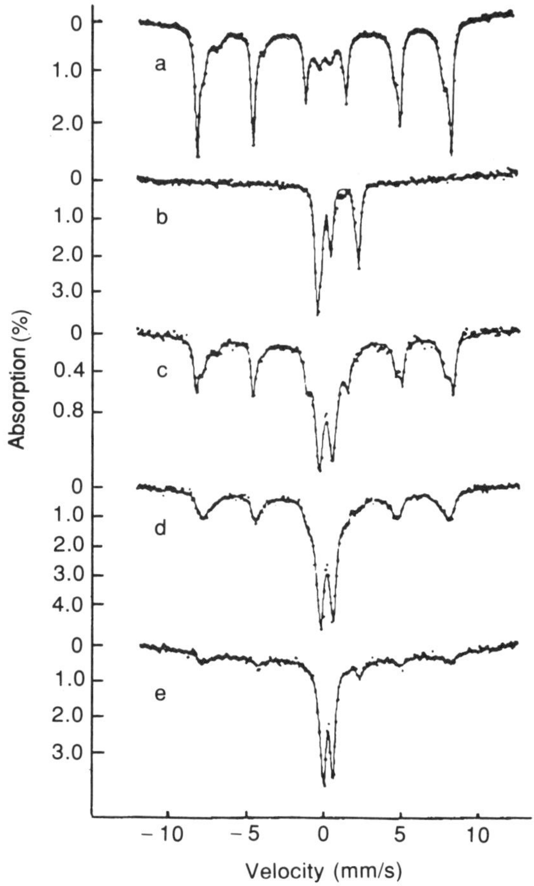

Figure 3.16 Mössbauer spectra at 300 K of magnetic separates of the soil samples used for Figure 3.15. (Reproduced from Longworth *et al.*, 1979.)

3.5 MÖSSBAUER SPECTROSCOPY OF ALUMINOSILICATE MINERALS

3.5.1 Introduction

This section will concentrate on the various layer-silicate minerals, which collectively account for most of the natural clay-sized aluminosilicate minerals. Other types of aluminosilicates will be considered briefly at the end of this section.

The existence of widespread isomorphous substitutions in the idealized structures is a general phenomenon in most aluminosilicate minerals. Consequently, different specimens of a particular type of mineral generally have different chemical compositions. Thus spectra may vary from sample to sample and even a single specimen may produce a spectrum that is composed of a number of similar, but not identical components. The correct interpretation of such spectra presents considerable difficulties and there is often not a unique fit to the data. Oxidation states can be determined from the magnitudes of the isomer shifts, and these can in addition sometimes be used to estimate coordination numbers. Typical values for Fe^{2+} and Fe^{3+} bound to oxygen in different coordination environments are illustrated in Figure 3.19.

Far more difficult is the assignment of spectral components to structural sites that have the same coordination number, because of the

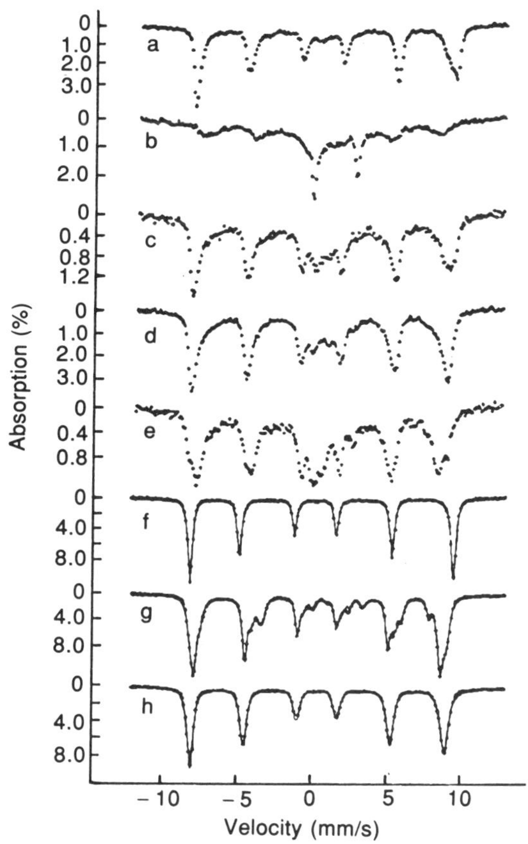

Figure 3.17 Mössbauer spectra at 4.2 K of the magnetically separated soil samples from Figure 3.16 (a–e) and hematite (f), magnetite (g), and maghemite (h) standard samples. (Reproduced from Longworth *et al.*, 1979.)

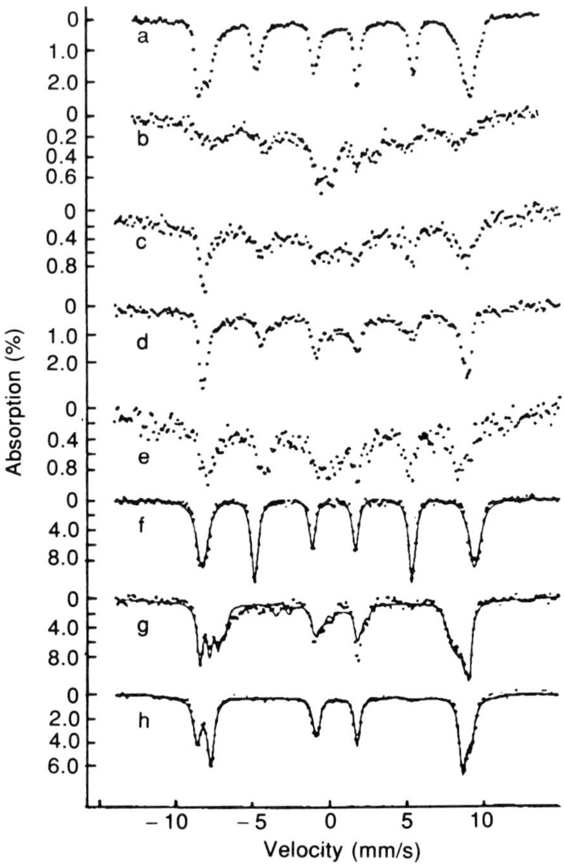

Figure 3.18 Mössbauer spectra at 4.2 K of the samples in Figure 3.17 in a magnetic field of 3 T applied parallel to the gamma-ray beam. (Reproduced from Longworth *et al.*, 1979.)

problems involved in determining the nature of the distribution functions for the quadrupole splittings, as was described in section 3.3.4.

A further complication that exists in mineral analysis is the assumption that is usually made that the recoil-free fraction is the same at each non-equivalent site. However, since the recoil-free fraction is proportional to the mean square vibrational displacement of the Mössbauer nucleus, which in turn is related to both coordination number and valence, there may be some variations in recoil-free fractions at different sites within the same structure. There is certainly the likelihood of significant variations in recoil-free fractions for different components in mineral mixtures.

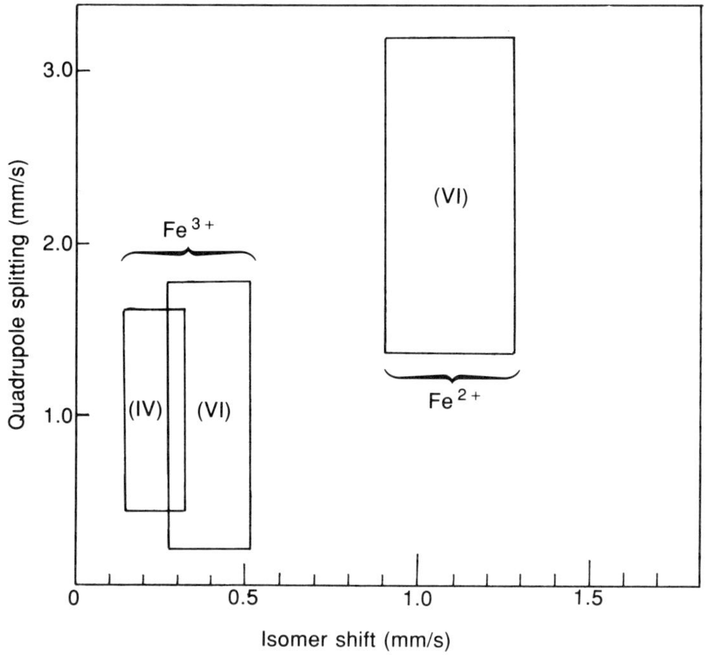

Figure 3.19 Ranges of values for isomer shift (relative to Fe metal) and quadrupole splitting for Fe^{2+} and Fe^{3+} in silicate minerals. (Adapted from Coey, 1980.)

The crystal structures of all layer silicates are based upon hexagonal networks of linked silicon–oxygen tetrahedra. The 1 : 1 family is formed when each tetrahedral sheet is joined to an $X(O, OH)_6$ octahedron, where X = Al, Fe or Mg and the O atoms are those at the apices of the tetrahedra. This represents a basic layer, and successive layers are held together by the relatively weak van der Waals' forces or hydrogen bonds. In the 2 : 1 family the basic layer consists of an octahedral sheet sandwiched between two tetrahedral sheets. In both types of structure there is the possibility that either all of the octahedral sites are filled, which is called trioctahedral, or that one-third is empty, when the structure is called dioctahedral.

3.5.2 1 : 1 layer silicates

3.5.2.1 Dioctahedral minerals

Kaolinite with an idealized chemical composition of $Al_2Si_2O_5(OH)_4$ is the dioctahedral end-member. It may have some substitution of Fe for Al, but in natural samples analysis of Mössbauer spectra is often complicated by the tendency for iron oxide phases, such as hematite or goethite, to associate with kaolinite surfaces. These can be distinguished from the structural Fe at room temperature only if they are well crystalline and produce magnetic hyperfine structure, because the spectrum for poorly crystalline goethite is very similar to that from the structural Fe in kaolinite. Distinction between Fe in the structure and oxide impurity phases is made more readily at low temperatures, at which the oxide phases order magnetically and show their characteristic sextet structures.

The levels of isomorphous substitutions of Fe for Al in kaolinites are usually small (substitution of Fe for Si is not thought to occur to any measurable extent). Consequently, paramagnetic hyperfine structure might be expected to occur as a result of slow relaxation rates. The sextet structures that result can be distinguished from any oxide phases by careful measurement of B_{hf} values. These are typically *c.* 55.5 T for the structural Fe^{3+} (Fysh, Cashion and Clark, 1983), which is significantly higher than for any oxide phase (Table 3.5).

3.5.2.2 *Trioctahedral minerals*

With the trioctahedral minerals iron is usually present as Fe^{2+} when there is no substitution of trivalent ions for Si. Should this be the case, significant quantities of Fe^{3+} may be found in both the octahedral and tetrahedral sheets.

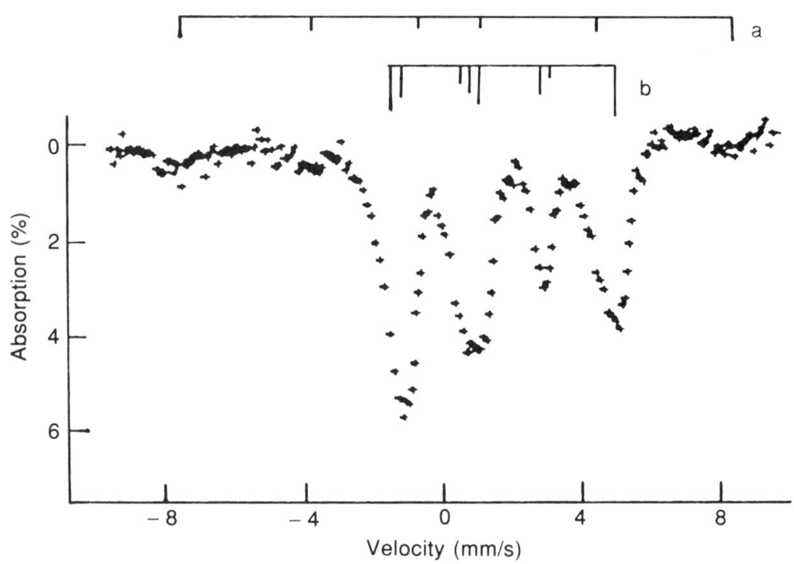

Figure 3.20 Mössbauer spectrum of greenalite at 4.2 K showing magnetic structure from Fe^{3+} (a) and Fe^{2+} (b). (Reproduced from Ballet and Coey, 1978.)

The iron-rich mineral greenalite has an idealized formula of $Fe_3^{2+}Si_2O_5(OH)_4$. Its Mössbauer spectrum at low temperatures shows magnetic ordering (Figure 3.20); Ballet and Coey (1978) have reported values of 15.8 T for B_{hf} and −2.96 mm/s for the quadrupole splitting, whose principal axis is perpendicular to that of the internal magnetic field.

Cronstedtite is related to greenalite, but has some Fe^{3+} substituted in both octahedral and tetrahedral sites. These produce values of 46.7 T and 40.6 T, respectively, for B_{hf}, whereas the value for Fe^{2+} is 14.1 T (Coey, 1984). At higher temperatures the spectra of cronstedtite are complicated by electron hopping processes (Coey, Moukarika and McDonagh, 1982). There is rapid charge relaxation in the temperature range 300–500 K, involving Fe^{2+} and Fe^{3+} in edge-sharing octahedral sites and possibly also Fe^{3+} in tetrahedral sites.

3.5.3 2 : 1 layer silicates

3.5.3.1 Dioctahedral minerals

Layer charges covering the range 0 to 2 exist for the dioctahedral 2 : 1 layer silicates, although the expanding-layer clay minerals have values at the low end of this range. All minerals have roughly one-third of their octahedral sites vacant and there has been a great deal of effort directed at attempting to use Mössbauer spectroscopy to determine the distribution of iron over these sites and the location and extent of ordering of the vacancies. Unfortunately a great deal of the literature is based on unjustified assumptions concerning the assignment of Mössbauer spectral components to specific crystallographic sites (Dainyak, Drits and Heifitz, 1992). Even the interpretation of iron-rich specimens may not be straightforward (Goodman, 1987).

The spectrum of a typical nontronite consists of an asymmetrically broadened doublet (Figure 3.21), which requires three doublets for a statistically acceptable fit (Goodman, *et al.*, 1976). The isomer shifts for these components are all consistent with Fe^{3+}, one having tetrahedral and the other two having octahedral coordination. The two octahedrally coordinated components do not correspond to the two structurally distinct octahedral sites, because it can be shown by selected area electron diffraction that the M1 site (with a *trans* arrangement of hydroxyl groups) is essentially unoccupied (Mering and Oberlin, 1967). Thus the two octahedral components in Figure 3.21 could correspond to the same type of structural site, but with differences in the nature of neighbouring cations, i.e. whether or not there is a trivalent ion in a neighbouring tetrahedral site (Goodman, 1978a). This approach has been further developed by Besson *et al.* (1983), who modelled the distribution of trivalent tetrahedral ions and concluded that they were arranged in pyroxene-like chains, and by Dainyak and Drits (1987), who also

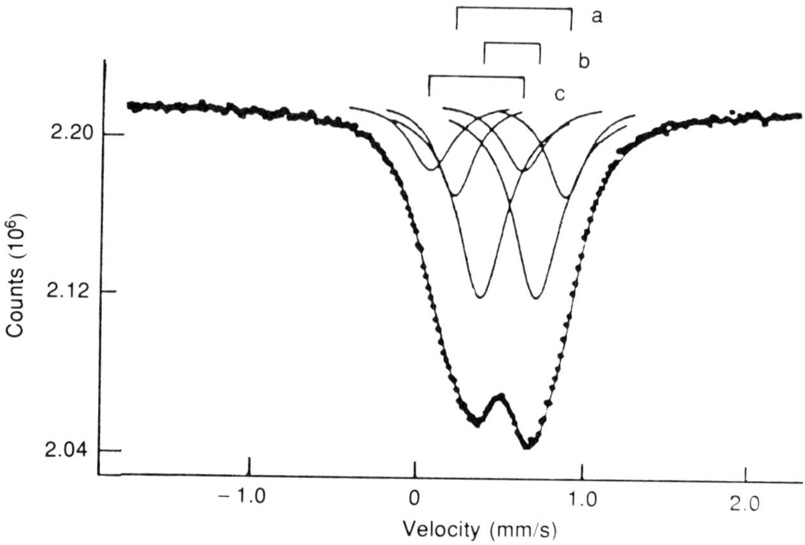

Figure 3.21 Mössbauer spectrum and computer fit to a typical nontronite specimen. Doublets (a) and (b) are assigned to octahedral Fe^{3+} and doublet (c) to tetrahedral Fe^{3+}. (Reproduced from Goodman *et al.*, 1976.)

performed electric field gradient calculations to support their spectral assignments.

In the older literature there was considerable debate as to whether there was any Fe in tetrahedral sites in nontronites, such as the Garfield nontronite, which has sufficient Al to accommodate any deficiency in Si. This was partly because spectra were acquired with inadequate signal-to-noise ratios for the resolution of minor tetrahedral Fe^{3+} components and partly because of a lack of consideration of the possible influence of the interlayer composition. Johnston and Cardile (1985) and Luca (1991a) have shown that the Mössbauer parameters of Fe in structural sites are influenced by the nature of the interlayer cation and the degree of hydration of the interlayer spaces. This is particularly noticeable for the tetrahedral ions, and Luca and Cardile (1989) and Luca (1991b) have presented a novel method for improved resolution of the spectral component from tetrahedral Fe^{3+} by measuring Ca^{2+}-exchanged specimens after dehydration at 200°C.

Some nontronites show magnetic ordering in their spectra at low temperatures and this can provide an alternative approach to the resolution of components from the different structural sites. As an example Cardile, Johnston and Dickson (1986) have derived B_{hf} values of 51 and 45 T for the two octahedral components and 41 T for the tetrahedral component in the spectrum of a specimen from Spokane County, Washington. Other nontronites require the application of large external magnetic fields before magnetic hyperfine structure can be resolved. Under these conditions it was possible to determine a Néel temperature of about 20 K for the Garfield nontronite (Murad, Cashion and Brown, 1990), even though there was no evidence of any magnetic structure in the absence of magnetic fields at temperatures as low as 2 K. These results are consistent with the model of Hartmann-Boutron, Aït-Bahammou and Meyer (1987) for magnetic clusters in disordered magnetic materials and indicate the presence of a distribution of cluster sizes each with its own ordering temperature. The frustration of magnetic behaviour in the absence of an applied magnetic field is further evidence for the presence of some tetrahedral Fe^{3+} ions, because this would not occur if all of the iron were in the M2 octahedral sites.

Ferripyrophyllite represents a much more straightforward case; at low temperature its spectrum consists of a single sextet with a B_{hf} of 51.8 T (Coey, Chukhrov and Zvyagin, 1984), whereas at room temperature a doublet with narrow linewidth and small quadrupole splitting (0.28 mm/s, Coey, Chukhrov and Zvyagin, 1984; 0.22 mm/s, Badaut, Decarreau and Besson, 1992) accounts for *c.* 85–90% of the total spectral intensity. Also, the isomer shift (0.35–0.36 mm/s relative to iron metal at room temperature) is indicative of octahedral coordination. This iron must, therefore, occur in a single structural site and this must be the M2 site because of the high Fe content of the mineral. The magnetic ordering temperature of ferripyrophyllite is considerably higher than that of nontronite, indicating a greater degree of order and absence of tetrahedral Fe^{3+} in the structure of the former.

The interpretation of the Mössbauer spectra of smectites with low iron contents has been a source of considerable debate. This is not surprising because electron diffraction work (Tsipursky and Drits, 1984) has shown that there are considerable sample-to-sample variations in the levels of occupancy of the M1 site. However, because Al is the predominant cation occupying the octahedral sites in these minerals, these results cannot be used to authenticate assignments of the Mössbauer spectra. Room-temperature spectra typically consist of a broad doublet, which is generally resolved into two components on computer fitting. The isomer shifts for both components (*c.* 0.37 mm/s relative to Fe metal at room temperature) are indicative of octahedral coordination, but quadrupole splittings are considerably larger than those of the iron-rich minerals discussed above. Assignments exist in the literature, where the two spectral components are assigned separately to the M1 and M2 sites, all to the M2 site, or all to the M1 site.

Few authors have questioned whether their specimens might consist of more than one phase. This is an important consideration in low-iron specimens because a small amount of an iron-rich phase (such as a poorly ordered oxide) could make a major contribution to the Mössbauer

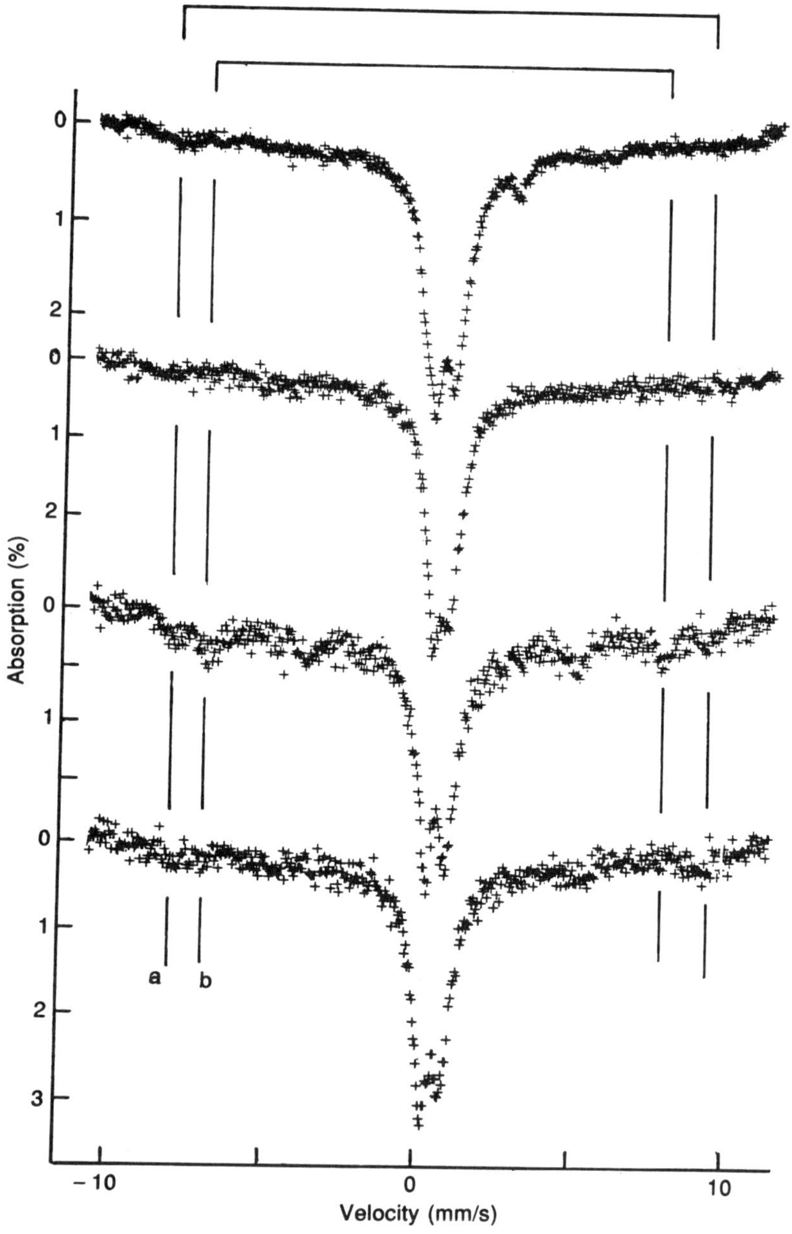

Figure 3.22 Mössbauer spectrum at 77 K of four bentonite clay specimens. The positions of the outermost peaks of the two sextet components that can be seen in some of these samples are shown; component (a) arises from slowly relaxing Fe^{3+} ions in a magnetically dilute environment, whereas component (b) probably corresponds to goethite. (Reproduced from Goodman, Nadeau and Chadwick, 1988.)

spectra (Goodman, 1978b). Indeed, EPR measurements often indicate the presence of iron-rich phases in montmorillonites (Chapter 5), in addition to the presence of more than one type of structural environment for the low-iron phase(s). Low-temperature Mössbauer spectra on several bentonite clays (Goodman, Nadeau and Chadwick, 1988) have shown that the iron is distributed between phases with distinctly different iron contents, the iron content of one phase being small enough for it to exhibit slowly relaxing paramagnetic hyperfine structure at 77 K (Figure 3.22). Thus assignment of Mössbauer spectra of montmorillonites and montmorillonitic clays is seldom unambiguous.

Illitic clays present a similar problem and Goodman and Nadeau (1988) have shown that many specimens have small amounts of crystalline iron oxide phases associated with them. Glauconites, which contain a mixture of expanding (montmorillonitic) and non-expanding (illitic) layers, presumably have the same impurity problem.

Within the series of dioctahedral 2 : 1 layer silicates there is a progressive increase in magnitude of the splitting of the inner quadrupole doublet in the order ferripyrophyllite < nontronite < glauconite < montmorillonite < illite < muscovite (Coey, 1980), which appears to be related to the size of the *b*-axis parameter (Rozenson and Heller-Kallai, 1977).

3.5.3.2 *Trioctahedral minerals*

Like their dioctahedral counterparts the trioctahedral 2 : 1 layer silicates have a wide range of charges per formula unit and a variety of different cations in their octahedral sites. In the majority of cases a high percentage of the iron is in the ferrous form and there are few if any octahedral vacancies. The problem of interpreting the Mössbauer spectra is at least as difficult as with the dioctahedral minerals. Doublets arising from Fe^{2+} have temperature-dependent quadrupole splittings, because of the small separation between the electronic ground state and the nearest excited states. In addition the q_{latt} term is considerably smaller than the single-electron q_{val} term. Thus it is necessary to obtain spectra at more than one temperature in order to obtain values for lattice contributions to electric field gradients (i.e. structural site distortions). In addition there is usually considerable overlap between the various Fe^{2+} components, with the low-velocity peaks also overlapping those of any Fe^{3+} in the specimen, as can be seen in the example of a biotite in Figure 3.23. Furthermore, because of the platy structure of these minerals there is a tendency for non-random orientation and the existence of unequal areas for the component peaks of the spectral doublets. Not surprisingly, therefore, there are examples in the literature of Mössbauer spectroscopy being used to support almost every conceivable arrangement of Fe^{2+} and Fe^{3+} ions in the structure.

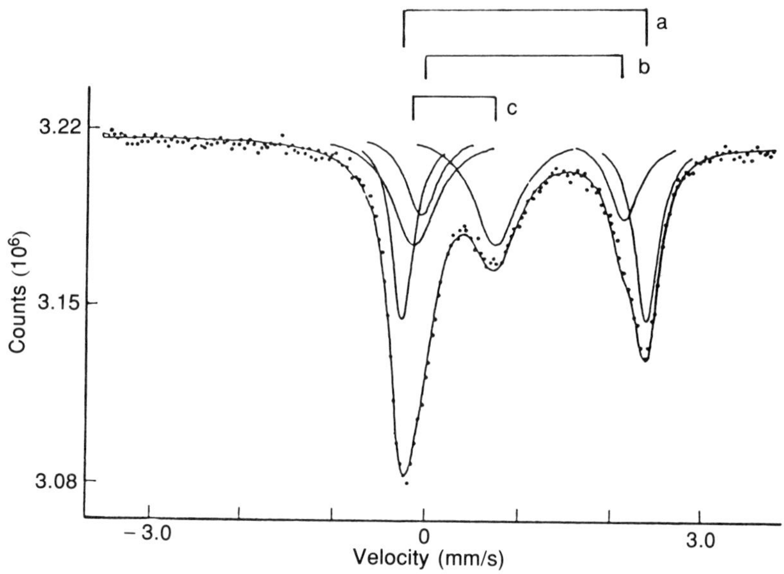

Figure 3.23 Mössbauer spectrum of a biotite showing the component peaks from a computer fit to three doublets; doublets (a) and (b) arise from Fe^{2+} and doublet (c) from Fe^{3+}, all with octahedral coordination.

There are situations where Mössbauer spectroscopy is able to give useful structural information on these minerals. The most obvious of these is in the determination of the extent of any Fe^{3+} substitution in the tetrahedral sheets (Annersten and Olesch, 1978; Sanz *et al.*, 1978), although it is often difficult to distinguish between small amounts of tetrahedral Fe^{3+} and texture effects, unless the specimen is oriented at the 'magic angle'.

Magnetic ordering in iron-rich specimens at low temperatures results in spectra in which magnetic hyperfine structure from both the Fe^{2+} and Fe^{3+} components is resolved (Ballet and Coey, 1982). The 2 : 1 layer silicates show a strong two-dimensional magnetic character in which there is ferromagnetic coupling between Fe^{2+}–O–Fe^{2+} and Fe^{2+}–O–Fe^{3+} groupings within the sheets and antiferromagnetic coupling between sheets.

3.5.4 Pillared clays

The recognition of pillared clays as potential catalysts for the chemical industry has led to considerable interest in their detailed characterization. They are prepared by the intercalation in smectites of large cations followed by calcination to effectively cross-link the silicate sheets. Tichit *et al.* (1988) have used several techniques to investigate the sintering of montmorillonites pillared by hydroxy-aluminium species.

They found that the Mössbauer spectrum from the mineral exchanged with the $[Al_{13}O_4(OH)_{24}(H_2O)_{12}]^{7+}$ ion was identical to that of the original (mainly K^+ exchanged) montmorillonite. Calcining the mineral above 680°C resulted in the progressive formation of a more distorted and more covalent environment for the Fe^{3+} ions.

The synthesis of clays with iron polycations as pillars is more difficult, but Doff *et al.* (1988) have shown that intercalation with the trinuclear Fe(III) acetato ion produces a suitable starting phase. At low temperatures the Mössbauer spectra of the calcined mineral shows magnetic hyperfine structure, which has been interpreted as suggesting that the pillars have a squat aspect with diameters in the region 30–40 Å.

3.5.5 Mixed-layer silicates

The Mössbauer spectra of mixed-layer silicates are potentially even more complex than the cases mentioned above because iron can occur in more than one type of layer. In practice there seems to be a considerable amount of peak overlap and it is difficult to assign components with any degree of confidence to the different types of structural site. The spectra of chlorites, which may be considered to have alternate layers of talc-like and brucite-like (magnesium hydroxide) structures, will be used to illustrate the difficulties in properly interpreting the results from this type of mineral. At room temperature, peaks are narrow and even with iron-rich specimens there is little difference in the parameters for the various Fe^{2+} components (Figure 3.24). Thus assignment to structural sites cannot be made with any confidence. In contrast to the uniformity of the Fe^{2+} parameters, the Fe^{3+} values show considerable variations from one specimen to another. Evidence for tetrahedral Fe^{3+} is sometimes apparent (Figure 3.24), even in specimens in which there is sufficient Al to compensate for any deficiency in Si.

3.5.6 Other silicate minerals

The other groups of silicate and aluminosilicate minerals will be mentioned briefly for completeness, although they are not generally considered to be clay minerals. The main families are the group silicates, chain silicates and framework silicates. In many of these, isomorphous substitution is common and, where this involves iron, Mössbauer spectra can be obtained and can contribute to the sample characterization.

3.5.6.1 Group silicates

Orthosilicates have structures based on isolated SiO_4 groups, the most common examples being the olivines and garnets. In olivines, which have the idealized composition R_2SiO_4, the octahedral cations, R, are

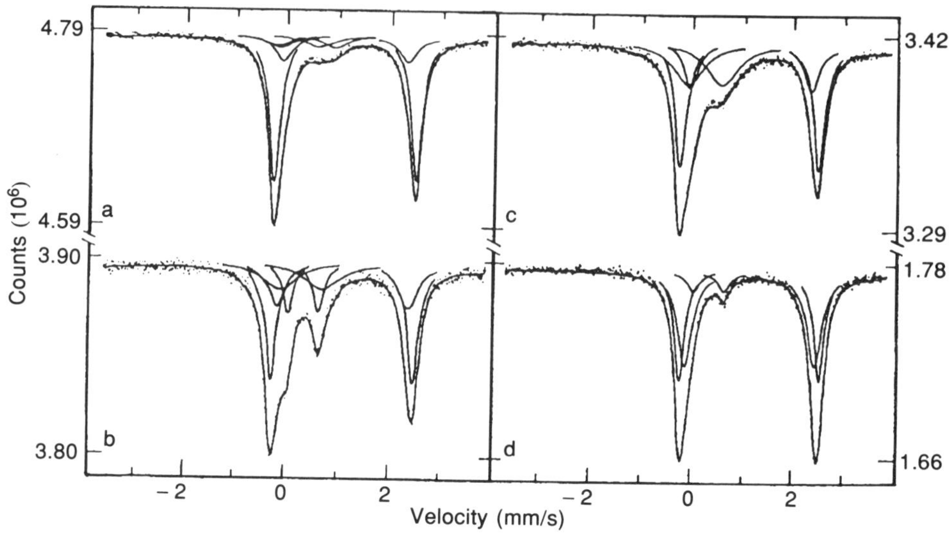

Figure 3.24 Mössbauer spectra at room temperature of four chlorite specimens. Note the small range of quadrupole splittings for Fe^{2+}. (Reproduced from Goodman and Bain, 1978.)

arranged in two distinct types of structural site. In the iron end-member, fayalite, these can be distinguished readily at low temperatures by the differences in their magnetic structures (Kündig *et al.*, 1967), but their quadrupole split components are not well separated at higher temperatures.

Although the structures of garnets, which have the general formula $R(II)_3R(III)_2(SiO_4)_3$, are more complex than those of olivines, Amthauer, Annersten and Hafner (1976) have shown that well-resolved quadrupole doublets can be obtained. These can be assigned unambiguously to Fe^{2+} at dodecahedral sites and Fe^{3+} at octahedral and tetrahedral sites, even though the parameters for the octahedral ions exhibit considerable variation with mineral composition. Amthauer, Annersten and Hafner (1976) have also determined the signs of the principal components of the electric field gradients for the various components in almandine and andradite by recording spectra with large external magnetic fields. At low temperatures andradite orders magnetically and spectra show two sextets with similar hyperfine fields and isomer shifts, but having different angles between their electric field gradients and directions of magnetization (Murad, 1984). The spectra of Ti-rich garnets, such as schlorlomites, have some major differences from those of other garnets, primarily because of significant substitutions of Ti, Al and Fe in tetrahedral sites. Although complex, these spectra are well-resolved and it has been possible to characterize Fe^{2+} dodecahedral, octahedral and tetrahedral sites and Fe^{3+} octahedral and tetrahedral sites (Amthauer, Annersten and Hafner, 1977; Huggins, Virgo and Huckenholz, 1977). Of particular note is the systematic increase in isomer shift from tetrahedral (*c.* 0.60 mm/s), through octahedral (c. 1.10 mm/s) to dodecahedral coordination (*c.* 1.25 mm/s) for Fe^{2+}.

Ilvaite has a structure based on isolated pyrosilicate (Si_2O_7) groups linked by double chains of octahedra, into which both Fe^{2+} and Fe^{3+} can substitute. At temperatures below *c.* 115 K iron end-members yield a complex spectral pattern with components from both Fe^{2+} and Fe^{3+} (Grandjean and Gerard, 1975). At higher temperatures spectra contain three doublets, one of which exhibits strong temperature dependence of its quadrupole splitting. Such spectra have been analysed by Litterst and Amthauer (1984) in terms of doublets from Fe^{2+} and Fe^{3+} at M2 sites, Fe^{3+} at M1 sites and a relaxation model for electron exchange between Fe^{2+} and Fe^{3+} in M2 sites.

There are many other examples of group silicates with more complex structures, e.g. beryl [$Be_3Al_2Si_6O_{18}$] and epidote [$Ca_2(Al, Fe)_3(SiO_4)$ $(Si_2O_7)O(OH)$], in which Fe can substitute for both Al and Si, and tourmaline [$NaMg_3Al_6Si_6B_3O_{27}(OH)_4$], in which a range of substitutions of both Fe^{2+} and Fe^{3+} occur.

3.5.6.2 Chain silicates

The basic structure of pyroxenes is based on chains of silicate groups of composition $(SiO_3)_n$, which are held together by cations bound to the non-bridging oxygen atoms. This structure gives rise to two crystallographically distinct sites, which are generally labelled M1 and M2. The M1 site is the least distorted, and Mössbauer spectra of orthopyroxenes [$(Mg, Fe)_2Si_2O_6$] suggest that there is a preference of Fe^{2+} for the M2 site (Virgo and Hafner, 1969). There does, however, appear to be a significant next-nearest-neighbour contribution to the quadrupole splittings (Dowty and Lindsley, 1973), which complicates the spectral analysis. Tetrahedral Fe^{3+} may be found occasionally (e.g. Hafner and Huckenholz, 1971), but it is not usually possible to assign octahedral Fe^{3+} to specific structural sites. In some other pyroxenes the structures are such that the number of inequivalent sites is increased (e.g. spodumene and omphacites) and analysis of the Mössbauer spectra becomes rather tentative.

In amphiboles the structures are based on double silicate chains, which again are held together by octahedral cations. In the basic structure there are four distinct types of octahedral site, which can contain Fe^{2+} or Fe^{3+}. Furthermore, there is the possibility of some tetrahedral Fe^{3+}. Thus spectra can be very complex and, with the possible exception of end-member compositions, very difficult to assign unambiguously.

3.5.6.3 *Framework silicates*

There are generally only very low levels of substitution of iron in the framework silicates. In feldspars, which have the basic formula (Na, Ca)(Al, Si)$_4$O$_8$, Fe^{2+} has been observed to substitute in both types of site (Hafner, 1975), whereas Fe^{3+} has been found in the tetrahedral sites in some potassium feldspars (Annersten, 1976).

3.6 PHOSPHATE MINERALS

The environmental chemistry of phosphorus is very complex and a large number of different iron-containing phosphates have been observed in soils and sediments. Many of these minerals are comparatively rare and this section will be limited to a brief consideration of the Mössbauer spectra of vivianite, a mineral that occurs in anaerobic conditions in soils and sediments, and some of its oxidation products.

Vivianite, with eight water molecules per $Fe_3(PO_4)_2$ formula unit is the best known ferrous phosphate. It contains two structurally distinct cation sites with octahedral coordination; Fe_A is coordinated to oxygens from two phosphate ions and four water molecules, whereas the Fe_B ions, which occur in pairs, are coordinated to the oxygens from four phosphate ions and two water molecules. The two types of site are distinguished by their Mössbauer spectra, with the A site giving a quadrupole splitting of *c.* 2.5 mm/s at room temperature compared with *c.* 3.0 mm/s for the B site (Stevens *et al.*, 1983).

Vivianite is oxidized readily on exposure to air to form mixed valence hydroxy phosphates, a reaction that involves the conversion of H_2O ligands to OH^- ions

$$2[Fe^{2+} - H_2O] + \frac{1}{2}O_2 \rightarrow 2[Fe^{3+} - OH] + H_2O.$$

Oxidized vivianites have a deep blue colour as a result of charge transfer transitions, but any electron hopping between Fe^{2+} and Fe^{3+} ions must be slower than *c.* 10^{-8} s, because no intermediate valence component is seen in the Mössbauer spectrum at room temperature (McCammon and Burns, 1980). The final product of vivianite oxidation is lipscombite, $Fe_x^{2+}Fe_{3-x}^{3+}(PO_4)_2(OH)_{3-x}$, whose Mössbauer spectrum yields two Fe^{2+} and two Fe^{3+} components (Vochten and de Grave, 1981), even though it has only one type of structural site available for iron. Thus this is yet another example of the major influence that neighbouring cations can have on the magnitudes of Mössbauer spectroscopic parameters.

3.7 APPLICATIONS OF MÖSSBAUER SPECTROSCOPY IN ARCHAEOLOGY AND ART

Over the past 20 years Mössbauer spectroscopy has played a small, but nevertheless valuable, role in the characterization of archaeological artefacts (see, e.g. Kostikas, Simopoulos and Gangas, 1976; Longworth, 1984b). Many of these studies have been concerned with the characterization of pottery. The key to the use of Mössbauer spectroscopy lies

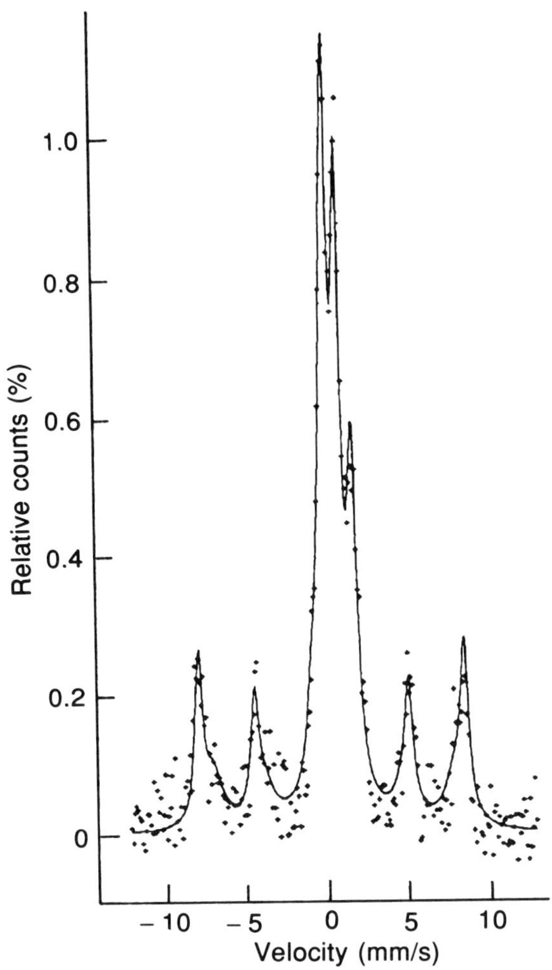

Figure 3.25 Conversion X-ray Mössbauer spectrum of a sample of Greek 'Etruscan' pottery at room temperature. (Reproduced from Longworth and Warren, 1975.)

in the effects on the forms of iron in the clays of firing temperature and whether oxidizing or reducing conditions pertained. For example Bouchez *et al.* (1974) have used Mössbauer spectroscopy to characterize the origin and formation of two types of pottery from Turkestan dating from the third millenium BC. The two types of pottery, one red and one grey, were found to coexist over archaeological levels covering several centuries. The authors were able to demonstrate that both pottery types were produced from the same clay deposits and that their differences related solely to different firing conditions.

Similar types of measurement have been made with surface glazes. For example, Hedges (1975) was able to relate the colours of Chinese glazed ceramics to the chemical states of their iron obtained from Mössbauer spectroscopy in the transmission mode. Other workers have used back-scattered radiation, either electrons or X-rays, to examine surface compositions down to depths of either 0.3 or 10 μm, respectively. An example of the conversion X-ray scattering spectrum of Greek 'Etruscan' black pottery from Cerveteri (Longworth and Warren, 1975) is shown in Figure 3.25. This consists of a combination of hematite (in the red body of the sherd) and magnetite (in the black surface 'gloss'), both thought to have been produced in a single firing by control of the kiln atmosphere.

Many of the pigments used in paints contain significant quantities of iron, usually in the form of oxides. Thus in an analogous way to the use of Mössbauer spectroscopy to characterize iron oxide phases in soils, which was discussed in section 3.4.9, the technique can be used to distinguish between different artists' pigments. This may be of particular importance in the resolution of authenticity problems. The spectra of many pigments have been determined, both in transmission mode (Keisch, 1974) and through back-scatterred X-rays (Cranshaw *et al.*, 1985). Examples of the latter are presented in Figure 3.26, where it can be seen that Indian red is primarily hematite, yellow ochre is mainly goethite, burnt umber is a microcrystalline oxide that produces a doublet spectrum at room temperature, Davy's grey contains Fe^{2+} in a silicate structure and Prussian blue is a ferrocyanide complex.

3.8 CONCLUSIONS

In the characterization of clay mineral species Mössbauer spectroscopy is a very important technique for the oxides, especially those of poor crystallinity. It is able to provide considerable detail on the magnetic properties and coordination environments of the iron in such phases, even when pure mineral phases cannot be obtained. Mössbauer spectroscopy is also able to identify and characterize iron oxide components that are present as very minor components in mineral samples, if the other phases are of low iron content. Similarly, Mössbauer spectroscopy is a

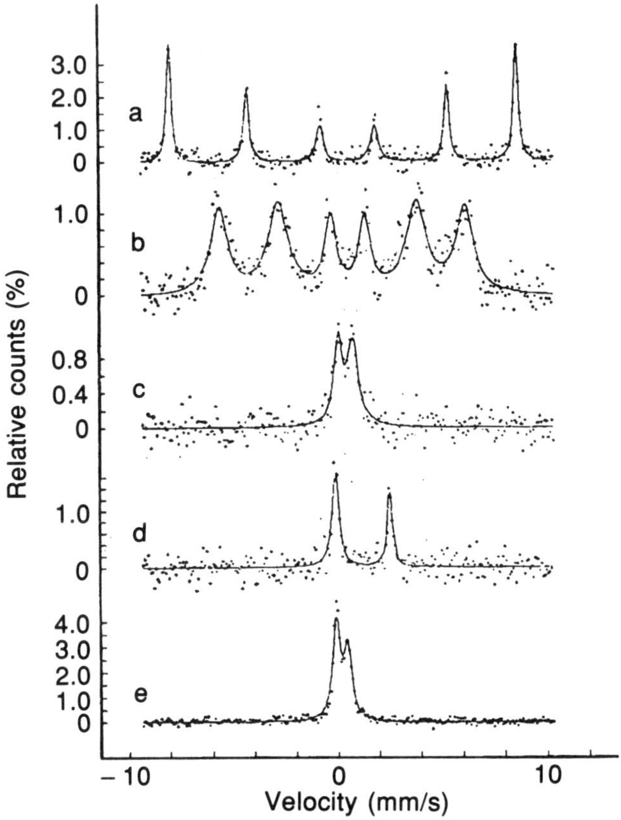

Figure 3.26 Mössbauer spectra from back-scatterred X-rays of a selection of artists' pigments at room temperature; (a) Indian red, (b) yellow ochre, (c) burnt umber, (d) Davy's grey and (e) Prussian blue. (Reproduced from Cranshaw *et al.*, 1985.)

useful technique in the characterization of air-sensitive minerals. Such samples can be studied in the frozen state without any chemical pretreatments. Oxidation states and coordination numbers can be determined readily and in some cases specific mineral phases can be identified. The technique is at its most powerful, however, in the characterization of the iron in mineral phases whose structures have been identified by other techniques, such as XRD.

The technique has been somewhat disappointing in its contribution to the characterization of aluminosilicate minerals. Although oxidation states and coordination numbers can usually be determined, along with magnetic properties of iron-rich phases, it is often not justifiable to equate spectral components to structural sites. Unfortunately, there are many papers in the primary literature where just such an exercise has been attempted and the majority of conclusions concerning the distribution of either Fe^{2+} or Fe^{3+} over structural sites with the same coordination number are invalid.

Quantitative measurements can, in principle, be made with Mössbauer spectroscopy, but they do require that the relative magnitudes of the *f* factors be known for the various iron-containing sites. Consequently, spectra should be acquired at at least three different temperatures for accurate quantitation. Unfortunately, the low sensitivity of Mössbauer spectroscopy means that such determinations are very time consuming. If only a single measurement can be made this should be at as low a temperature as possible, because the curves for the variations of *f* factors with temperature are closest at very low temperatures. Another factor that should be borne in mind when considering quantitative measurements is the fact that at best only the amounts of iron in each type of chemical environment can be determined. Thus with specimens that consist of a number of phases, the relative amounts of these phases can be estimated only if their iron contents are known.

ACKNOWLEDGEMENTS

Acknowledgement is given to the Scottish Office Agriculture and Fisheries Department for providing funding for this work. Figures 3.6, 3.9, 3.12, 3.15, 3.16, 3.17, 3.18, 3.20, 3.25 and 3.26 are reproduced with permission from the copyright holders.

3.6 Wickman (1966) Plenum.
3.9 Murad (1988), D. Reidel.
3.12 Murad and Taylor (1984) *Clay Minerals*, **19**, 77–83.
3.15–3.18 Longworth *et al.* (1979) *Journal of Social Science*, **30**, 93–110.
3.20 Ballet and Coey (1978) *Journal de Physique* (Paris) *Suppl. c6*, **39**, 765–766.
3.25 Longworth and Warren (1975) *Nature*, **255**, 625–627.
3.26 Cranshaw *et al.* (1985) Cambride University Press.

REFERENCES

Amthauer, G., Annersten, H. and Hafner, S.S. (1976) The Mössbauer spectrum of ^{57}Fe in silicate garnets. *Zeitschrift fuer Kristallographie, Kristallgeometrie, Kristallphysik, Kristallchemie*, **143**, 14–55.

Amthauer, G., Annersten, H. and Hafner, S.S. (1977) The Mössbauer spectrum of ^{57}Fe in titanium-bearing andradites. *Physics and Chemistry of Minerals*, **1**, 399–413.

Annersten, H. (1976) New Mössbauer data on iron in potash feldspar. *Neues Jahrbuch fuer Mineralogie, Monatshefte*, **8**, 337–343.

Annersten, H. and Olesch, M. (1978) Distribution of ferrous and ferric iron in clintonite and the Mössbauer characteristics of ferric iron in tetrahedral coordination. *Canadian Mineralogist*, **16**, 199–203.

Badaut, D., Decarreau, A. and Besson, G. (1992) Ferripyrophyllite and related Fe^{3+}-rich 2:1 clays in recent deposits of Atlantis II Deep, Red Sea. *Clay Minerals*, **27**, 227–244.

Ballet, O. and Coey, J. M. D. (1978) Greenalite – a clay showing two-dimensional magnetic order. *Journal de Physique (Paris) Colloque, Supplement C6*, **39**, 765–766.

Ballet, O. and Coey, J. M. D. (1982) Magnetic properties of sheet silicates; 2:1 layer minerals. *Physics and Chemistry of Minerals*, **8**, 218–229.

Bancroft, G. M. (1973) *Mössbauer Spectroscopy: an Introduction for Inorganic Chemists and Geochemists*, McGraw-Hill, London.

Bender Koch, C. and Morup, S. (1991) Identification of green rust in an ochre sludge. *Clay Minerals*, **26**, 577–582.

Besson, G., Bookin, A. S., Dainyak, L. G. *et al.* (1983) Use of diffraction and Mössbauer methods for the structural and crystallochemical characterization of nontronites. *Journal of Applied Crystallography*, **16**, 374–383.

Bouchez, R. *et al.* (1974) Mössbauer study of firing conditions used in the manufacture of the grey and red ware of Tureng-Tepe. *Journal de Physique (Paris) Colloque, Supplement C6*, **35**, 541–546.

Bowen, L. H. and Weed, S. B. (1984) Mössbauer spectroscopy of soils and sediments, in *Chemical Mössbauer Spectroscopy* (ed. R. H. Herber). Plenum, New York, pp. 217–242.

Cardile, C. M. (1988) Tetrahedral Fe^{3+} in ferrihydrite: ^{57}Fe Mössbauer spectroscopic evidence. *Clays and Clay Minerals*, **36**, 537–539.

Cardile, C. M., Johnston, J. H. and Dickson, D . P. E. (1986) Magnetic ordering at 4.2 and 1.3 K in nontronites of different iron contents: a ^{57}Fe Mössbauer spectroscopic study. *Clays and Clay Minerals*, **34**, 233–238.

Chadwick, J. *et al.* (1986) A Mössbauer study of ferrihydrite and aluminium substituted ferrihydrites. *Journal of Magnetism and Magnetic Materials*, **61**, 88–100.

Childs, C. W., Goodman, B. A. and Churchman, G. J. (1978) Application of Mössbauer spectroscopy to the study of iron oxides in some red and yellow/brown soil samples from New Zealand. *Developments in Sedimentology*, **27**, 555–565.

Childs, C. W., Dickson, D. P. E., Goodman, B. A. and Lewis, D. G. (1984) Mössbauer parameters for ferrihydrites at 4 K. *Australian Journal of Soil Research*, **22**, 149–154.

Childs, C. W., Goodman, B. A., Paterson, E. and Woodhams, F. W. D. (1980) A Mössbauer spectroscopic investigation of the nature of iron in akaganéite (β-FeOOH). *Australian Journal of Chemistry*, **33**, 15–26.

Coey, J. M. D. (1980) Clay minerals and their transformations studied with nuclear techniques: the contribution of Mössbauer spectroscopy. *Atomic Energy Review*, **18**, 73–124.

Coey, J. M. D. (1984) Mössbauer spectroscopy of silicate minerals, in *Mössbauer Spectroscopy Applied to Inorganic Chemistry*, Vol. 1 (ed. G. J. Long). Plenum, New York, pp. 443–509.

Coey, J. M. D. and Readman, P. W. (1973) Characterisation and magnetic properties of natural ferric gel. *Earth and Planetary Science Letters*, **21**, 45–51.

Coey, J. M. D., Chukhrov, F. V. and Zvyagin, B. B. (1984) Cation distribution, Mössbauer spectra, and magnetic properties of ferripyrophyllite. *Clays and Clay Minerals*, **32**, 198–204.

Coey, J. M. D., Morrish, A. H. and Sawatsky, G. A. (1971) A Mössbauer study of conduction in magnetite. *Journal de Physique (Paris) Colloque, Supplement C1*, **32**, 271–273.

Coey, J. M. D., Moukarika, A. and McDonagh, C. M. (1982) Electron hopping in cronstedtite. *Solid State Communications*, **41**, 797–800.

Cranshaw, T. E. (1974) Mössbauer spectroscopy. *Journal of Physics E: Scientific Instruments*, **7**, 497–505.

Cranshaw, T. E., Dale, B. W., Longworth, G. O. and Johnson, C. E. (1985) *Mössbauer Spectroscopy and Its Applications*. Cambridge University Press, Cambridge.

Cuttler, A. H., Man, V., Cranshaw, T. E. and Longworth, G. (1990) A Mössbauer study of green rust precipitates: I. Preparations from sulphate solutions. *Clay Minerals*, **25**, 289–301.

Dainyak, L. G. and Drits, V. A. (1987) Interpretation of Mössbauer spectra of nontronite, celadonite and glauconite. *Clays and Clay Minerals*, **35**, 363–372.

Dainyak, L. G., Drits, V. A. and Heifits, L. M. (1992) Computer simulation of cation distribution in dioctahedral 2 : 1 layer silicates using IR data: application to Mössbauer spectroscopy of glauconite. *Clays and Clay Minerals*, **40**, 470–477.

Dézsi, I., Keszthelyi, L., Kulgawczuk, D. *et al.* (1967) Mössbauer study of δ- and δ′-FeOOH and their disintegration products. *Physica Status Solidi*, **22**, 617–629.

Dibar de Ure, M. C. (1973) Study of diffusion by the Mössbauer effect. Technique for the removal of the blackness distortion of Mössbauer spectra. Effect of diffusion by the vacancy mechanism on the Mössbauer line broadening. Thesis, Carnegie-Mellon University, Pittsburgh, USA.

Doff, D. H., Gangas, N. H. J., Allan, J. E. M. and Coey, J. M. D. (1988) Preparation and characterization of iron oxide pillared montmorillonite. *Clay Minerals*, **23**, 367–377.

Dowty, E. and Lindsley, D. H. (1973) Mössbauer spectra of synthetic hedenbergite–ferrosilite pyroxenes. *American Mineralogist*, **58**, 850–868.

Eggleton, R. A. and Fitzpatrick, R. (1988) New data and a revised structural model for ferrihydrite. *Clays and Clay Minerals*, **36**, 111–124.

Fysh, S. A., Cashion, J. D. and Clark, P. E. (1983) Mössbauer effect studies of iron in kaolin. I. Structural iron. *Clays and Clay Minerals*, **31**, 285–292.

Goodman, B. A. (1976) The effect of lattice substitutions on the derivation of quantitative site populations from the Mössbauer spectra of 2 : 1 layer silicates. *Journal de Physique (Paris) Colloque, Supplement C6*, **37**, 819–823.

Goodman, B. A. (1978a) The Mössbauer spectra of nontronites: consideration of an alternative assignment. *Clays and Clay Minerals*, **26**, 176–177.

Goodman, B. A. (1978b) An investigation by Mössbauer and EPR spectroscopy of the possible presence of iron-rich impurity phases in some montmorillonites. *Clay Minerals*, **13**, 351–356.

Goodman, B. A. (1987) On the use of Mössbauer spectroscopy for determining the distribution of iron in aluminosilicate minerals. *Clay Minerals*, **22**, 363–366.

Goodman, B. A. and Bain, D. C. (1978) Mössbauer spectra of chlorites and their decomposition products. *Developments in Sedimentology*, **27**, 65–74.

Goodman, B. A. and Lewis, D. G. (1981) Mössbauer spectra of aluminous goethites (α-FeOOH). *Journal of Soil Science*, **32**, 351–363.

Goodman, B. A. and Nadeau, P. H. (1988) Identification of oxide impurity phases and distribution of structural iron in some diagenetic illitic clays as determined by Mössbauer spectroscopy. *Clay Minerals*, **23**, 301–308.

Goodman, B. A., Nadeau, P. H. and Chadwick, J. (1988) Evidence for the multiphase nature of bentonites from Mössbauer and EPR spectroscopy. *Clay Minerals*, **23**, 147–159.

Goodman, B. A., Russell, J. D., Fraser, A. R. and Woodhams, F. W. D. (1976) A Mössbauer and infrared spectroscopic study of the structure of nontronite. *Clays and Clay Minerals*, **24**, 53–59.

Grandjean, F. and Gerard, A. (1975) Analysis by Mössbauer spectroscopy of the electron hopping process in ilvaite. *Solid State Communications*, **16**, 553–556.

Hafner, S. S. (1975) Mössbauer spectroscopy in lunar geology and mineralogy, in *Topics in Applied Physics*, Vol. 5, *Mössbauer Spectroscopy* (ed. U. Gonser). Springer-Verlag, Berlin, pp. 167–199.

Hafner, S. S. and Huckenholz, H. G. (1971) Mössbauer spectra of synthetic ferridiopside. *Nature (London) Physical Science*, **233**, 9–11.

Häggström, L., Annersten, H., Ericsson, T., Wäppling, R., Karner, W. and Bjarman, S. (1978) Magnetic dipolar and electric quadrupolar effects on the Mössbauer spectra of magnetite above the Vervey transition. *Hyperfine Interactions*, **5**, 201–214.

Hargrove, R. S. and Kündig, W. (1970) Mössbauer measurements of magnetite below the Vervey transition. *Solid State Communications*, **8**, 303–308.

Hartmann-Boutron, F., Aït-Bahammou, A. and Meyer, C. (1987) Mössbauer detection of magnetic clusters in disordered magnetic materials. *Journal de Physique (Paris)*, **48**, 435–444.

Hedges, R. E. M. (1975) Mössbauer spectroscopy of Chinese glazed ceramics. *Nature*, **254**, 501–503.

Huggins, F. E., Virgo, D. and Huckenholz, H. G. (1977) Titanium-containing silicate garnets. I. The distribution of Al, Fe^{3+}, and Ti^{4+} between octahedral and tetrahedral sites. *American Mineralogist*, **62**, 475–490.

Johnston, J. H. and Cardile, C. M. (1985) Iron sites in nontronite and the effect of interlayer cations from Mössbauer spectra. *Clays and Clay Minerals*, **33**, 21–30.

Keisch, B. (1974) Mössbauer effect studies of fine arts. *Journal de Physique (Paris) Colloque, Supplement C1*, **35**, 151–164.

Kostikas, A., Simopoulos, A. and Gangas, N. H. (1976) Analysis of archaeological artifacts, in *Applications of Mössbauer Spectroscopy*, Vol. 1 (ed. R. L. Cohen). Academic Press, New York, pp. 241–261.

Kündig, W. and Hargrove, R. S. (1969) Electron hopping in magnetite. *Solid State Communications*, **7**, 223–227.

Kündig, W., Cape, J. A., Lindquist, R. H. and Constabaris, G. (1967) Some magnetic properties of Fe_2SiO_4 from 4 K to 300 K. *Journal of Applied Physics*, **38**, 947–948.

Litterst, F. J. and Amthauer, G. (1984) Electron delocalization in ilvaite, a reinterpretation of its ^{57}Fe Mössbauer spectrum. *Physics and Chemistry of Minerals*, **10**, 250–255.

Longworth, G. (1984a) Spectral data reduction and refinement, in *Mössbauer Spectroscopy Applied to Inorganic Chemistry*, Vol. 1 (ed. G. J. Long). Plenum Press, New York, pp. 43–56.

Longworth, G. (1984b) Studies of ceramics and archaeological materials, in *Mössbauer Spectroscopy Applied to Inorganic Chemistry*, Vol. 1 (ed. G. J. Long). Plenum Press, New York, pp. 511–526.

Longworth, G. and Warren, S. E. (1975) Mössbauer spectroscopy of Greek Etruscan pottery. *Nature*, **255**, 625–627.

Longworth, G., Becker, L. W., Thompson, R. *et al.* (1979) Mössbauer effect and magnetic studies of secondary iron oxides in soils. *Journal of Soil Science*, **30**, 93–110.

Luca, V. (1991a) ^{57}Fe Mössbauer spectroscopic study of structural changes during dehydration of nontronite: effect of different exchangeable cations. *Clays and Clay Minerals*, **39**, 478–489.

Luca, V. (1991b) Detection of tetrahedral Fe^{3+} sites in nontronite and vermiculite by Mössbauer spectroscopy. *Clays and Clay Minerals*, **39**, 467–477.

Luca, V. and Cardile, C. M. (1989) Improved detection of tetrahedral Fe^{3+} in nontronite SWa-1 by Mössbauer spectroscopy. *Clay Minerals*, **24**, 555–559.

McCammon, C. A. and Burns, R. G. (1980) The oxidation mechanism of vivianite as studied by Mössbauer spectroscopy. *American Mineralogist*, **65**, 361–366.

Meads, R. E., Place, B. M., Woodhams, F. W. D. and Clark, R. C. (1972) Effect of absorber thickness in measurement of quadrupole splitting in a Mössbauer experiment. *Nuclear Instruments and Methods*, **98**, 29–35.

Mering, J. and Oberlin, A. (1967) Electron optical study of smectites. *Clays and Clay Minerals*, **15**, 3–25.
Murad, E. (1982a) The characterization of goethite by Mössbauer spectroscopy. *American Mineralogist*, **67**, 1007–1011.
Murad, E. (1982b) Ferrihydrite deposits in an artesian fountain in lower Bavaria. *Neues Jahrbuch fuer Mineralogie, Monatshefte*, 45–56.
Murad, E. (1984) Magnetic ordering in andradite. *American Mineralogist*, **69**, 722–724.
Murad, E. (1988) Properties and behaviour of iron oxides as determined by Mössbauer spectroscopy, in *Iron in Soils and Clay Minerals* (eds J. W. Stucki, B. A. Goodman and U. Schwertmann). D. Reidel, Dordrecht, pp. 309–350.
Murad, E. and Johnston, J. H. (1987) Iron oxides and oxyhydroxides, in *Mössbauer Spectroscopy Applied to Inorganic Chemistry*, Vol. 2 (ed. G. J. Long). Plenum, New York, pp. 507–582.
Murad, E. and Schwertmann, U. (1983) The influence of aluminium substitution and crystallinity on the Mössbauer spectra of goethite. *Clay Minerals*, **18**, 301–312.
Murad, E. and Schwertmann, U. (1986) The influence of aluminium substitution and crystallinity on the room-temperature Mössbauer spectrum of hematite. *Clays and Clay Minerals*, **34**, 1–6.
Murad, E. and Schwertmann, U. (1988) The characterization of poorly crystalline Si-containing natural iron oxides by Mössbauer spectroscopy. *Hyperfine Interactions*, **41**, 835–838.
Murad, E. and Taylor, R. M. (1984) The Mössbauer spectra of hydroxycarbonate green rusts. *Clay Minerals*, **19**, 77–83.
Murad, E., Cashion, J. D. and Brown, L. J. (1990) Magnetic ordering in Garfield nontronite under applied magnetic fields. *Clay Minerals*, **25**, 261–269.
Murad, E., Bowen, L. H., Long, G. J. and Quin, T. G. (1988) The influence of crystallinity on magnetic ordering in natural ferrihydrites. *Clay Minerals*, **23**, 161–173.
Persoons, R. M., Chambere, D. G. and De Grave, E. (1986) Mössbauer effect study of the magnetic structure in iron oxyhydroxide (δ-FeOOH). *Hyperfine Interactions*, **28**, 647–650.
Pollard, R. J., Cardile, C. M., Lewis, D. G. and Brown, L. J. (1992) Characterization of FeOOH polymorphs and ferrihydrite using low-temperature, applied-field, Mössbauer spectroscopy. *Clay Minerals*, **27**, 57–71.
Quin, T. G., Long, G. J., Benson, C. G. *et al.* (1988) Influence of silicon and phosphorus on structural and magnetic properties of synthetic goethite and related oxides. *Clays and Clay Minerals*, **36**, 165–175.
Rozenson, I. and Heller-Kallai, L. (1977) Mössbauer spectra of dioctahedral smectites. *Clays and Clay Minerals*, **25**, 94–101.
Sanz, J., Meyers, J., Vielvoye, L. and Stone, W. E. E. (1978) The location and content of iron in natural biotites and phlogopites: a comparison of several methods. *Clay Minerals*, **13**, 45–52.
Sawatzky, G. A., Coey, J. M. D. and Morrish, A. H. (1969) Mössbauer study of electron hopping in the octahedral sites of Fe_3O_4. *Journal of Applied Physics*, **40**, 1402–1403.
Schwertmann, U. and Murad, E. (1990) The influence of aluminium on iron oxides: XIV. Al-substituted magnetite synthesized at ambient temperatures. *Clays and Clay Minerals*, **38**, 196–202.
Stevens, J. G., Pollak, H., Zhe, L. *et al.* (1983) *Mineral Data Handbook*. Mössbauer Effect Data Center, University of North Carolina.
Taylor, R. M. (1980) Formation and properties of Fe(II)Fe(III) hydroxy-carbonate and its possible significance in soil formation. *Clay Minerals*, **15**, 369–382.

Tichit, D., Fajula, F., Figueras, F. *et al.* (1988) Sintering of montmorillonites pillared by hydroxy-aluminium species. *Clays and Clay Minerals*, **36**, 369–375.

Tsipursky, S. I. and Drits, V. A. (1984) The distribution of octahedral cations in the 2:1 layers of dioctahedral smectites. *Clay Minerals*, **19**, 177–193.

Virgo, D. and Hafner, S. S. (1969) Fe^{2+}, Mg order-disorder in heated orthopyroxenes. *Mineralogical Society of America Special Paper*, **2**, 67–81.

Vochten, R. and De Grave, E. (1981) Petrographic, Mössbauer and electrokinetic study of synthetic lipscombite. *Physics and Chemistry of Minerals*, **7**, 197–203.

Wickman, H. H. (1966) Mössbauer paramagnetic hyperfine structure, in *Mössbauer Effect Methodology*, Vol. 2 (ed. I. J. Gruverman), Plenum, New York, pp. 39–66.

Window, B. (1971) Hyperfine field distributions from Mössbauer spectra. *Journal of Physics E: Scientific Instruments*, **4**, 401–402.

Wivel, C. and Morup, S. (1981) Improved computational procedure for evaluation of overlapping hyperfine parameter distributions in Mössbauer spectra. *Journal of Physics E: Scientific Instrument*, **14**, 605–610.

CHAPTER 4

Nuclear magnetic resonance spectroscopy

B. A. Goodman and J. A. Chudek

4.1 INTRODUCTION

Nuclear magnetic resonance (NMR) spectroscopy measures transitions between the spin states of nuclei in the presence of strong magnetic fields. Most of the major elements in minerals have isotopes with non-zero spin (Table 4.1). The spin, I, of the nucleus confers a magnetic moment on it and, in the presence of a magnetic field, the nature of the interaction between the field and the nuclear magnetic moment is determined by the magnetic properties of the nucleus and its environment. The nuclear spin may have any half-integer or integer value, including zero. The magnitude of I is determined by the numbers of protons and neutrons in the nucleus; for example, ^{12}C with six protons and six neutrons

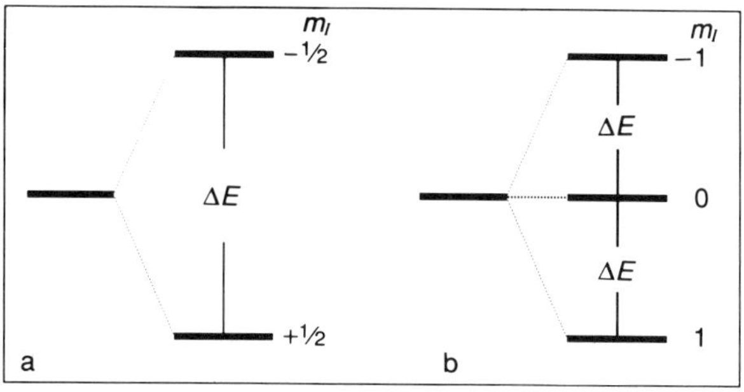

Figure 4.1 Nuclear spin energy level diagrams for nuclei with (a) $I = \frac{1}{2}$ and (b) $I = 1$.

Clay Mineralogy: Spectroscopic and Chemical Determinative Methods.
Edited by M. J. Wilson. Published in 1994 by Chapman & Hall, London.
ISBN 0 412 53380 4.

Table 4.1 Natural abundances and nuclear properties of magnetic nuclei of potential use in minerals

Isotope	*Natural Abundance (%)*	*Spin*	*Quadrupole moment (10^{-24} cm^2)*	*Frequency (MHz at 7.2 T)*
^{1}H	99.984	$\frac{1}{2}$	–	300
^{2}H	1.56×10^{-2}	1	0.0028	46.1
^{7}Li	92.58	$\frac{3}{2}$	– 0.03	116.6
^{10}B	19.58	3	0.074	32.2
^{11}B	80.42	$\frac{3}{2}$	0.0355	96.2
^{13}C	1.108	$\frac{1}{2}$	–	75.4
^{14}N	99.64	1	0.016	21.7
^{15}N	0.36	$\frac{1}{2}$	–	30.4
^{17}O	3.7×10^{-2}	$\frac{5}{2}$	– 0.026	40.7
^{19}F	100	$\frac{1}{2}$	–	282.2
^{23}Na	100	$\frac{3}{2}$	0.14	79.4
^{25}Mg	10.05	$\frac{5}{2}$	Not known	18.4
^{27}Al	100	$\frac{5}{2}$	0.149	78.2
^{29}Si	4.7	$\frac{1}{2}$	–	59.6
^{31}P	100	$\frac{1}{2}$	–	121.4
^{33}S	0.74	$\frac{3}{2}$	– 0.064	23.0
^{35}Cl	75.4	$\frac{3}{2}$	– 0.0789	29.4
^{39}K	93.1	$\frac{3}{2}$	0.11	14.0
^{79}Br	50.5	$\frac{3}{2}$	0.33	75.2
^{85}Rb	71.25	$\frac{5}{2}$	0.27	29.0
^{109}Ag	48.2	$\frac{1}{2}$	–	14.0
^{113}Cd	12.26	$\frac{1}{2}$	–	66.5
^{133}Cs	100	$\frac{7}{2}$	– 0.003	39.4
^{199}Hg	16.8	$\frac{1}{2}$	–	53.5
^{207}Pb	22.6	$\frac{1}{2}$	–	62.6

has $I = 0$, ^{13}C with six protons and seven neutrons has $I = \frac{1}{2}$, and ^{17}O with eight protons and nine neutrons has $I = \frac{5}{2}$.

Nuclear magnetic resonance is achieved by matching the energy of electromagnetic radiation (in the radiofrequency region of the spectrum) to the energy separation of the nuclear spin states (Figure 4.1). This can be carried out by varying the magnetic field in a spectrometer operating with a constant radiofrequency (RF) and exciting the various nuclei separately (this is comparable to the continuous wave EPR technique described in Chapter 5). These days though, most applications of NMR spectroscopy are performed with Fourier transform (FT) spectrometers, where all of the nuclei of a particular isotope are excited simultaneously by using short pulses of high power RF waves.

For nuclei with $I = \frac{1}{2}$ the difference in energy (ΔE) between the two states in an applied magnetic field (B_0) is given by

$$\Delta E = h\gamma B_0 / 2\pi,$$

where γ is the magnetogyric ratio. The difference in population of the two states is given by the Boltzmann distribution law as

$$n_\beta / n_\alpha = \exp\left(\frac{-\Delta E}{kT}\right),$$

where n_α and n_β are the numbers of nuclei in the lower and upper energy states, respectively. Promotion of nuclei from their ground to excited spin states in the NMR experiment must be done in a manner which ensures that a difference in population always remains. Essentially, this means that only a few nuclei in every million can be sampled at any one time, making NMR a very insensitive technique. The energy difference between the ground and excited spin states is proportional to the applied magnetic field, so that the larger the field the greater the energy difference and the fewer the number of nuclei in the higher energy state at equilibrium. This then results in greater spectral sensitivity and is one of the major reasons for the desire for ever greater magnetic fields in NMR spectroscopy. Higher fields also provide a greater dispersion of resonances (as their positions are field dependent), which can lead to a simplification of complex spectra. There may, however, be some drawbacks with higher magnetic fields, as will be seen later.

4.2 PRINCIPLES OF NMR SPECTROSCOPY

4.2.1 The Zeeman effect

In the presence of a magnetic field, nuclei with spin, I, have $2I + 1$ distinct energy states, the separation being proportional to the magnetic field. Figure 4.1 shows examples for $I = \frac{1}{2}$, where there are two states with quantum numbers, m_I, equal to $\frac{1}{2}$ and $-\frac{1}{2}$, and for $I = 1$, where the m_I values are 1, 0 and -1.

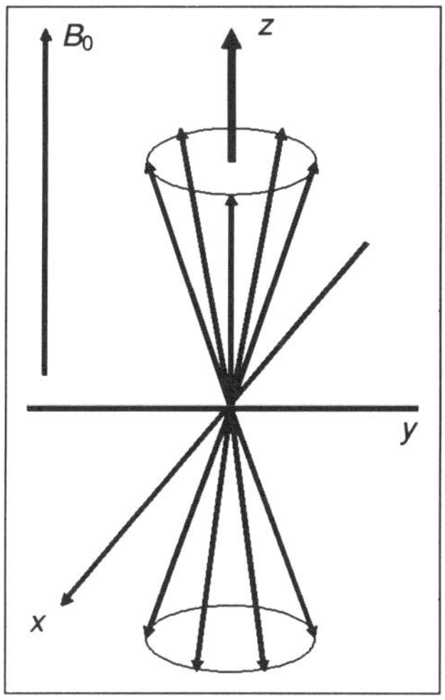

Figure 4.2 Aligment of magnetic vectors for a collection of $I=\frac{1}{2}$ nuclei in a magnetic field. The nuclear spins precess about the z-axis at the Larmor frequency, w_0, to produce a resultant magnetization vector parallel to B_0.

The magnetic vector of the nucleus precesses about the magnetic field at a frequency that is dependent upon the magnetogyric ratio of the nucleus, its orientation being dependent upon its m_I value. Figure 4.2 shows an ensemble of vectors for an $I=\frac{1}{2}$ nucleus; a small majority are aligned in the direction of the magnetic field B_0 (the ground state), whereas a minority are aligned in the opposite direction to B_0 (the excited state). Usually there are only very small differences in the populations of the ground and excited states. Conventionally the direction of B_0 is assigned to the z-axis, with the x-axis pointing into the page and the y-axis in the plane of the page.

In an NMR experiment a rotating magnetic field, equal to the precession frequency, is applied in the x–y plane and this causes the nuclei in the ground state to absorb energy and move into the excited state. This rotating magnetic field is in the form of electromagnetic radiation, in the RF region of the spectrum; its energy must match precisely the separations between the nuclear spin energy states.

4.2.2 The chemical shift

The separations between the energy states (i.e. the frequencies at which the nuclei absorb energy) are, as stated above, proportional to the magnetogyric ratio of the nucleus. They are also influenced by the electronic environment of the nucleus. Electrons shield nuclei from the magnetic field of the magnet, thus changing the precession frequency. These changes in frequency are known as **chemical shifts** and, in order to facilitate comparison of results from spectrometers operating at different frequencies, are usually expressed relative to a reference standard in parts per million (ppm) of the precession frequency. The formula

$$\delta = \frac{(v_{\text{sample}} - v_{\text{reference}}) \times 10^6}{\text{oscillator frequency}}$$

is used to calculate the value of the chemical shift. For example, a peak 600 Hz away from the reference in a proton spectrum measured using a spectrometer with a proton resonating frequency of 300 MHz would be 2 ppm away from the reference, usually designated $\delta_H = 2.0$ ppm. On a spectrometer with a 200 MHz resonating frequency, this peak would be 400 Hz from the reference.

The NMR chemical shift is determined by the local chemical environment of the nucleus and by bulk magnetic susceptibility effects. In clay minerals the latter are determined principally by paramagnetic ions, such as Fe^{2+} and Fe^{3+}, which result in major peak broadening processes in the spectra. These effects will be considered separately in section 4.2.12. The chemical shifts may be considered in terms of diamagnetic shielding and paramagnetic deshielding effects by electrons associated with the NMR nucleus and by electrons on other nuclei. According to Tossell (1984) the shielding constant for a nucleus A can be written as

$$\sigma_A = \sigma_A^{\text{dia}} + \sigma_A^{\text{para}} + \Sigma\sigma_{AB}$$

where σ_A^{dia} is the diamagnetic shielding and σ_A^{para} is the paramagnetic deshielding due to the electrons on atom A and σ_{AB} represents the combined effects due to electrons on atom B.

4.2.3 Relaxation processes

As the nature of the relaxation mechanisms give considerable information about the chemical and physical environment of the nucleus, it is worthwhile studying them in more detail. To do this the nuclei may be observed in a different reference frame. The discussion so far has been in the laboratory frame of reference. If, however, the x–y axis is rotated at the precession frequency and the ensemble is observed from the direction of the moving x-axis, the vectors of the individual nuclei appear stationary with respect to the moving x-axis (redefined x'-axis). This is called the **rotating frame of reference** (Figure 4.3). Further, the resultant magnetization (**M**) lies along the (now) z'-axis, pointing in the direction of the magnetic field, B_0. In the rotating frame of reference, application of energy at the correct frequency causes **M** to topple into the y'-axis, the angle through which **M** topples being proportional to the pulse length. A pulse of sufficient length to topple **M** through 90° will in a single pulse experiment give the maximum emitted signal. (Peak intensity is proportional to the displacement from z'; consequently a 180° pulse will give no signal.)

Many different mechanisms can affect the rate at which nuclei relax back to their ground states. Some of these are cumulative and can be

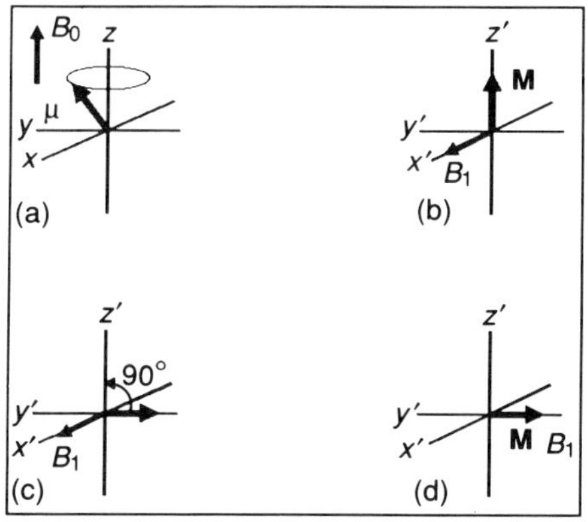

Figure 4.3 Conversion of coordinates from the laboratory frame of reference to the rotating frame: (a) precession of magnetic moment, μ, about laboratory coordinates; (b) change of coordinates to the rotating frame and application of a radiofrequency induced magnetic field, B_1, along x'; (c) after application of B_1 along x' the vector **M** moves through the z'–y' plane by 90° to become colinear with y'; (d) after changing the phase of the irradiation so that B_1 is colinear with **M** and y'.

considered collectively under the type of relaxation they produce. Two such relaxation times, the effects of which are described below, are the spin–lattice or longitudinal relaxation time, T_1, and the spin–spin or transverse relaxation time, T_2. In solution it is usual for $T_1 = T_2$, but in the solid state T_1 is often much greater than T_2 (T_2 is never greater than T_1).

T_1 is a measure of the time taken for the resultant magnetism to return to the z' axis after the cessation of the pulse, the mechanism of relaxation being via interaction of the nuclear spin with those of neighbouring nuclei. To visualize T_2 it is necessary to remember that the nuclei are precessing about an axis and that, although each nuclear species resonates at the same frequency, minor inhomogeneities in their magnetic environments lead to a small range of precession frequencies. Thus the precessing nuclei will be randomly distributed about the z-axis. With the application of the RF pulse all nuclei are rotated towards the y'-axis before beginning to relax back to the random distribution about z. Relaxation time T_2 is a measure of the time taken for this to take place. For T_2 the relaxation effect of neighbouring spins may be augmented with the effect of other electronic and nuclear interactions. This ensures that T_2 is always $\leqslant T_1$.

4.2.4 Linewidths

Spectral linewidths (i.e. the range of frequencies covered by a single nuclear species) are inversely proportional to the magnitudes of the T_2

relaxation times. In clay minerals short relaxation times (and hence large linewidths) are produced by two common types of ion. The first type includes paramagnetic species, such as transition metal ions or free radicals, because unpaired electrons have much larger magnetic dipoles than nuclei; as dipoles rotate they create magnetic fields which enhance relaxation rates. The second group is represented by nuclei with a quadrupole moment; here the distribution of the nuclear charge is in the form of an ellipsoid, which again leads to a stronger interaction.

4.2.5 Cross-polarization

In situations where no natural relaxing agents, such as some paramagnetic species, are present the spin–lattice relaxation time, T_1, can be very long. Thus long recovery times between pulses are required and, consequently, this leads to excessively long accumulation times for spectra. A technique called cross-polarization, which may under certain circumstances reduce this relaxation time, has been developed. Where a magnetically abundant isotope with a relatively short relaxation time, usually a proton, is bonded to a magnetically dilute species, e.g. ^{29}Si (natural abundance 4.7% – Table 4.1), with a long relaxation time, it is possible by 'matching the spins of the two species' to allow the protons to 'share' their shorter relaxation times with the ^{29}Si nuclei. This will take place where the sample is solid and where the structure has some form of order, if only short range. It is also necessary that the proton relaxation time is appreciably shorter than that of the second element and that there is a population difference between the energy states. Experimentally the protons are excited using a 90° pulse and kept excited by the application of a second longer pulse. Simultaneous to the protons being excited, a pulse of equal length to that of the proton pulse is applied at the resonating frequency of the dilute nucleus. In the set-up procedure the return signal from this pulse is observed as the power of the pulse is varied, until there is a sudden marked increase in signal size. This is the Hartmann-Hahn condition, where the proton power times the proton gyromagnetic ratio equals the power of the second nuclear frequency times its magnetogyric ratio.

$$\gamma_H B_{1H} = \gamma_X B_{1X}$$

At this point the proton energy is shared with that of the other nucleus, effectively reducing its T_1; this happens during the period when the second proton power is applied, usually from 1–5 ms. This procedure is shown diagrammatically in Figure 4.4. Also, although in this description the proton has been cited as the abundant nuclear spin, in practice other nuclei may be used if the necessary criteria are met.

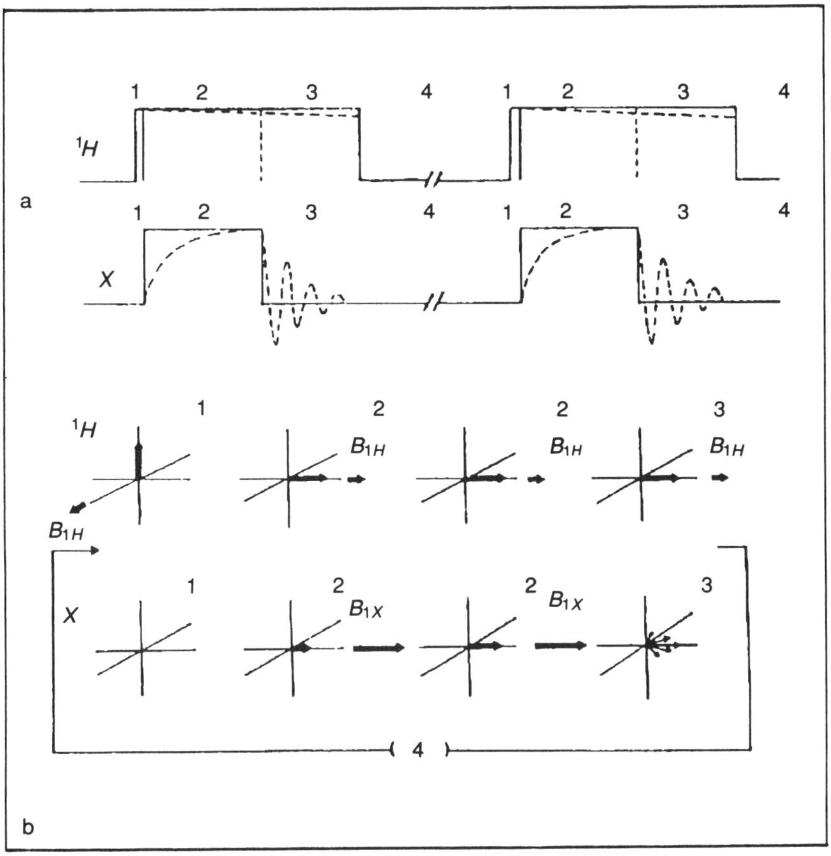

Figure 4.4 (a) Pulse sequence used for cross-polarization of a dilute nucleus X (e.g. ^{29}Si) from an abundant nucleus (^1H). (b) The behaviour of the spin magnetizations of the ^1H and X nuclei during the cross-polarization sequence.

4.2.6 Intramolecular spin–spin interactions

Neighbouring nuclei with non-zero spin also affect the precession frequency of a nucleus. For simplicity we shall limit the description to spin $\frac{1}{2}$ nuclei. For any population of these nuclei there is essentially equal probability for occupancy of the $+\frac{1}{2}$ and $-\frac{1}{2}$ states. Thus one half will augment the magnetic field in their vicinity and one half will decrement it. A proton, therefore, with a second non-equivalent nucleus in close proximity will have a 50% probability of the field affecting it being slightly higher than the applied field (with a consequent increase in its resonant frequency) and a 50% probability that the field will be less (and consequently a lower resonating frequency). The spectrum will thus have two lines for this proton (a doublet) and there will be a reciprocal effect on the second proton.

In the situation where a proton has two neighbours (equivalent to each other) it will have a three-line spectrum with intensities in the ratio 1 : 2 : 1 as:

- two spins can be augmenting the applied field;
- one spin augmenting and one decrementing the field;
- as the previous situation but with the first spin decrementing and the second augmenting the field;
- two spins decrementing the field.

This effect is termed spin–spin coupling and, in general, it produces one more peak than the number of equivalent neighbours, with the peak intensity ratios having a binomial distribution. Magnitudes of spin–spin interactions are measured in hertz and are field independent; for many nuclei they are smaller than the natural linewidth of the absorption, and hence are not seen.

4.2.7 Decoupling

Although the interaction between dissimilar magnetic nuclei (coupling) can disclose much useful information, there are occasions, for example when protons couple with silicon nuclei, when the effects of the coupling become detrimental to spectral analysis. Various techniques have been developed to remove the effects of this coupling, and these are generally referred to as decoupling techniques. These procedures involve irradiating the sample at a second frequency while the NMR experiment is in progress. In addition to its use in simplifying spectra, decoupling can also be used to provide additional information on interactions taking place within the sample. Examples are given in later sections of this chapter.

4.2.8 Intermolecular dipolar coupling

In NMR spectroscopy of solids the peaks are normally broad because of nuclear dipolar interactions and anisotropy in the chemical shift, which is determined by distortions from cubic symmetry in the chemical environment of the NMR nucleus. For each individual nucleus in a sample these produce a range of resonance frequencies, depending on the orientation of the main symmetry axis relative to the magnetic field (Figure 4.5). Thus in solids it is not usual for individual spectral transitions to be resolved and relatively uninformative spectra are generally obtained. In liquids these interactions are averaged to zero because the frequencies of molecular motion are usually greater than the resonance frequency. Consequently narrow lines are obtained and solution state NMR has become a major spectroscopic technique for chemical characterization.

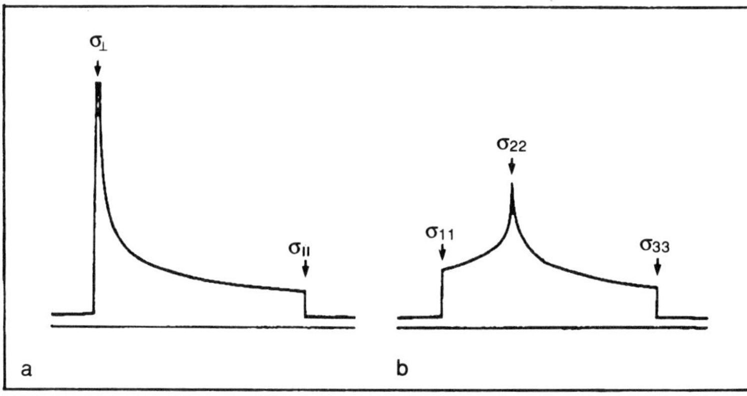

Figure 4.5 Typical spectral patterns for stationary solids exhibiting chemical shift anisotropy (σ); (a) with axial symmetry, and (b) without axial symmetry.

4.2.9 Magic-angle spinning (MAS)

The equations describing the anisotropic interactions in solids contain a factor of $(3\cos^2\theta - 1)$, where θ is the angle between the magnetic field and the principal axis of the nucleus. Rapid rotation of the specimen about an angle of $\cos^{-1}(1/(3)^{\frac{1}{2}})$, i.e. 54.7°, to the magnetic field causes this term to disappear with the result that linewidths approaching those for solutions can be obtained. This technique of 'magic angle' spinning (MAS), combined with the development of high field magnets has led to major advances in solid state NMR spectroscopy (Fyfe, 1983).

4.2.10 Suppression of dipolar interactions using multiple pulse sequences

An alternative to MAS is the use of multiple pulse sequences to suppress dipolar interactions (Gerstein, 1981). These sequences make use of a number of pulses of different angle or phase interspersed with delays of various lengths in order to attempt to rotate the nuclear spins by amounts of such magnitude that over the length of the sequence (at least four and possibly tens of pulses) the average dipolar interaction vanishes. One such sequence, consisting of four pulses, is called WAHUHA (Waugh, Huber and Haeberlen, 1968), a second, using eight, is MREV-8 (Mansfield *et al.*, 1973).

In proton work the difficulty encountered in attempting to resolve and measure resonances in the short chemical shift range (10–15 ppm) has led to the integration of the magic-angle spinning technique with that of multiple pulse sequences. This technique, which is known as CRAMPS (combined rotational and multiple pulse spectroscopy) (Haeberlen and

Waugh, 1968; Gerstein, 1981), can reduce proton linewidths to a few tens of hertz, thus allowing individual proton lines to be resolved.

4.2.11 Nuclei with spin > $\frac{1}{2}$

Nuclear magnetic resonance for nuclei with $I > \frac{1}{2}$ is complicated by quadrupolar interactions between the nucleus and any electric field gradient that it experiences. This produces a further splitting of the energy levels (Figure 4.6) and results in spectra that are potentially very complex when there is anisotropy in the transitions, although it is common for only some of these to be observable.

For nuclei with non-integer spin, often only the $\frac{1}{2} \rightarrow -\frac{1}{2}$ transition is observed and this produces spectra that resemble those obtained with spin

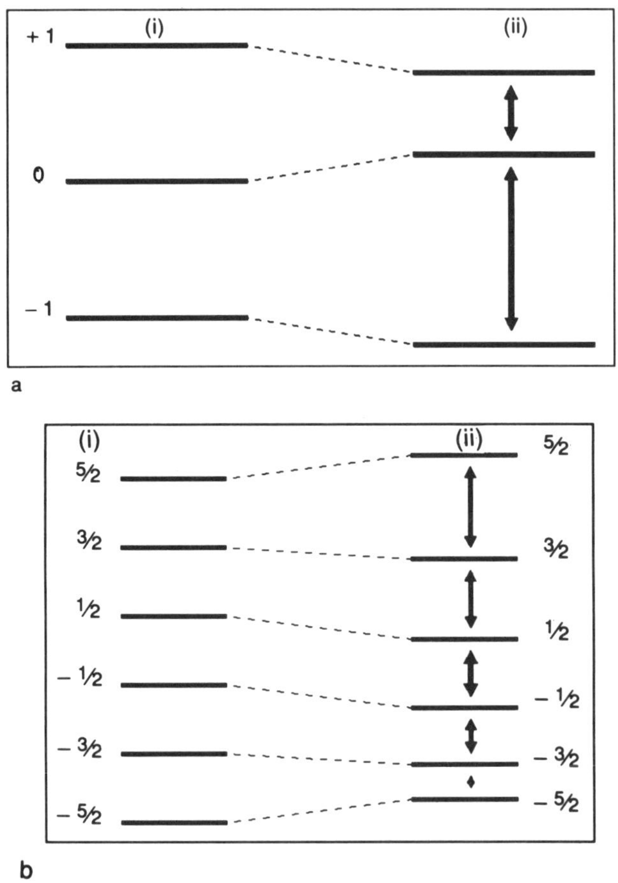

Figure 4.6 Schematic representations of the energy levels and transitions for (a) $I = 1$ and (b) $I = \frac{5}{2}$ nuclei in the absence (i) and presence (ii) of electric field gradients.

$\frac{1}{2}$ nuclei. When $I > 1$ there is a second-order interaction which is not averaged to zero by 'magic angle' spinning, and for small electric field gradients this produces a small shift in the centre of the resonance relative to its true chemical shift. This shift is magnetic field dependent on and decreases with increasing field. For spectra from ions in environments with relatively small distortions from cubic symmetry it is possible to distinguish between features caused by chemical shifts and those due to quadrupolar interactions by making use of variations in the excitation RF pulse length (Samoson and Lippmaa, 1983). This allows different components with overlapping spectra to be distinguished and is especially valuable at low to medium magnetic field strengths, where the second-order quadrupolar interaction makes a significant contribution to spectral patterns.

With highly distorted environments, such as with penta-coordinated aluminium, the electric field gradient may be so large that the second-order quadrupolar interaction dominates the spectra at even the largest magnetic fields that are currently available. In these circumstances very broad peaks are obtained and the spectra are essentially unobservable.

This problem of broad peaks in samples with large quadrupolar terms can be overcome if the sample is spun rapidly and simultaneously about two axes that are related to one another and the external magnetic field by the 'magic angle'. This may be done in one of two ways. The first method mechanically moves the spinning sample from one angle to the other, whereas the second involves a double rotor, one within the other, each spinning at different angles and at different speeds.

Nuclei with integer spins present a different problem. Deuterium, for example, with a spin of 1 produces two peaks that are shifted equal amounts on either side of the Larmor frequency. The positions of these peaks in solids are orientation dependent, but in a randomly oriented powder, features are seen at the singularities in the single crystal spectra (Figure 4.7). The separations of the peaks in these powder spectra give information on the local environments of the quadrupole nuclei, a technique that has been used extensively in polymer chemistry. 'Magic angle' spinning can produce significant line narrowing to yield chemical-shift information, and with $I = 1$ nuclei these spectra are not subjected to complications arising from second-order quadrupolar effects, which occur with $I > 1$.

4.2.12 The paramagnetic ion problem

Most minerals contain some paramagnetic ions, either in their structures or associated with their surfaces. This is well illustrated in Chapters 3 and 5, on Mössbauer spectroscopy and electron paramagnetic resonance (EPR) respectively, which are concerned with their characterization. These ions are often a major problem in work with NMR spectroscopy,

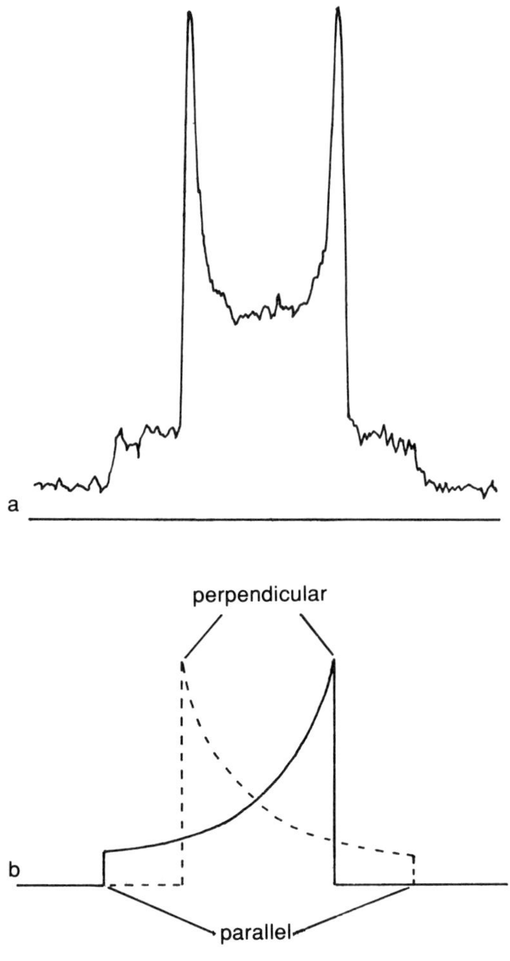

Figure 4.7 (a) A typical ^2H NMR spectral pattern for a randomly oriented solid sample, and (b) diagrammatic representation of the variation in position of the ^2H spectral peaks with alignment relative to the magnetic field.

because of large interactions between the unpaired electrons and magnetic nuclei. Their effects are twofold.

First, the very large dipolar interactions between unpaired electrons and near nuclear neighbours result in large changes in the magnetic fields experienced by them. These interactions can be observed with broad-band NMR, which can be used to determine the spatial distribution of paramagnetic ions around the protons within a crystal structure. Paramagnetic ions, however, seriously interfere with high-resolution structural investigations because the dipolar interactions may render near neighbour nuclei unobservable in the NMR spectra.

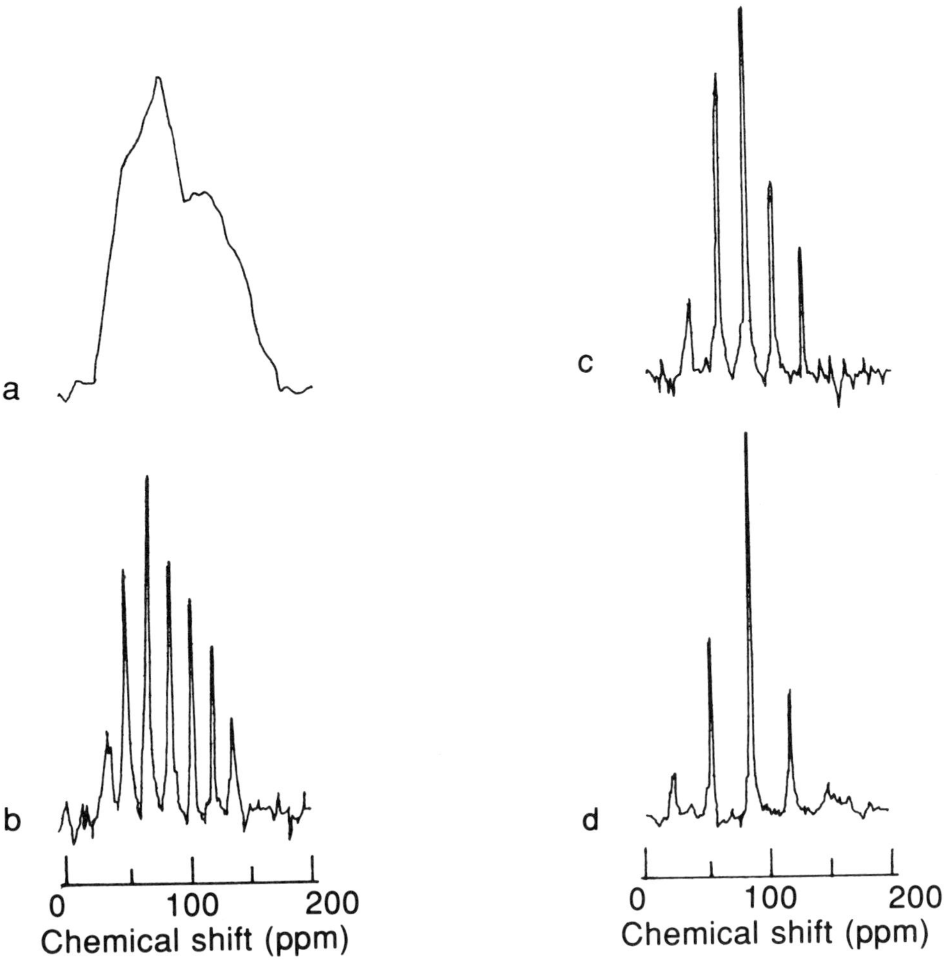

Figure 4.8 Effects of spinning speed on the ^{29}Si MAS NMR spectrum of a synthetic diopside; static sample (a), and sample rotated at rates of (b) 1.2 kHz, (c) 1.6 kHz and (d) 2.2 kHz. Note that the spinning sidebands move further apart as the spinning speed is increased, whereas the positions of true peaks remain unaltered. (Reproduced from Kirkpatrick, 1988.)

Second, the interactions of unpaired electrons with more remote nuclei produce magnetic field inhomogeneities across the sample and result in a broadening of high-resolution spectra that cannot be completely narrowed by 'magic angle' spinning. In such cases the spectra have extensive sideband structure, where the peak separations are equal to the spinning frequency.

Distinction between spinning sidebands and spectral peaks can be made by acquiring spectra at more than one spinning frequency because the positions of the sidebands change, whereas those of genuine peaks do not (Figure 4.8). There are also pulse sequences which invert spinning sideband peaks so that suppression of sidebands can be achieved by summing spectra with normal and inverted sidebands.

Spectra can be obtained from minerals with iron contents up to a few percent before line broadening renders the spectra essentially undetectable. However, except when the iron is present as a separate external phase, there is likely to be a selective loss of spectral information, with serious consequences for detailed spectral analysis. On the other hand, small amounts of paramagnetic ions can produce major decreases in spin–lattice relaxation times, which means that shorter delay times between pulses can be used. This is very valuable with insensitive nuclei such as ^{29}Si, when times for spectral acquisition can often amount to many hours.

4.3 EXPERIMENTAL TECHNIQUES

4.3.1 Basic instrumentation for high-resolution NMR

Nuclear magnetic resonance spectrometers may be either continuous wave or Fourier transform. The first type, which is now mainly used for low-field proton spectroscopy is a scanning instrument. Only one frequency is observed at a time. After the sample has been placed in the magnetic field the transmitter frequency is slowly scanned through the frequency range of the nucleus under examination. In a manner analogous to that used by many spectrophotometers the strength of the RF frequency is measured before and after transmission through the sample. As the scan passes through the frequency at which a nucleus resonates, the sample absorbs the required energy from the RF radiation, moving from the ground state to the excited state and producing a peak on a recording device. Continuous-wave NMR spectroscopy is both slow and insensitive. A typical scan through a proton spectrum may take 500 s. Furthermore, not many nuclei are sufficiently sensitive to be observed above the background noise. Devices have been developed to store many accumulations and average them in order to enhance the signal, but with the slow scan rate, even relatively sensitive nuclei may need many hours of accumulation to give an acceptable signal-to-noise ratio.

Electronically, the continuous-wave spectrometer is very similar to that used for Fourier transform (FT) spectroscopy, which uses very short pulses of RF radiation to simultaneously excite all the nuclei and which has now largely superseded the continuous-wave spectrometer. With FT instruments the nuclei are excited by very short (in the order of microseconds) pulses of high power RF radiation, which emit an envelope of frequencies and simultaneously excite all nuclei resonating within their range. As they relax back to their equilibrium states, each nucleus emits RF radiation at the frequency at which it was absorbed. The emitted frequency is in the form of a dampened sine wave and each nucleus transmits simultaneously its resonant signal; the total signal thus takes the form of a complicated interferogram. This is picked up by a receiver, amplified, passed through an analogue-to-digital converter, and finally stored in a computer. These accumulations are often on a time-scale of <1 s (but see section 4.2.3) as compared with the tens of seconds for continuous wave spectroscopy. Experimentally several (often thousands) of transients (i.e. pulses followed by accumulations) are summed to give an acceptable signal-to-noise ratio for the final spectrum. The interferogram or free induction decay (FID) is finally converted to give a spectrum in the frequency domain by means of a Fourier transform (hence the name) carried out by the computer.

The basic components of an FT spectrometer are shown in Figure 4.9. The sample is placed in a probe, which is then inserted into the magnet.

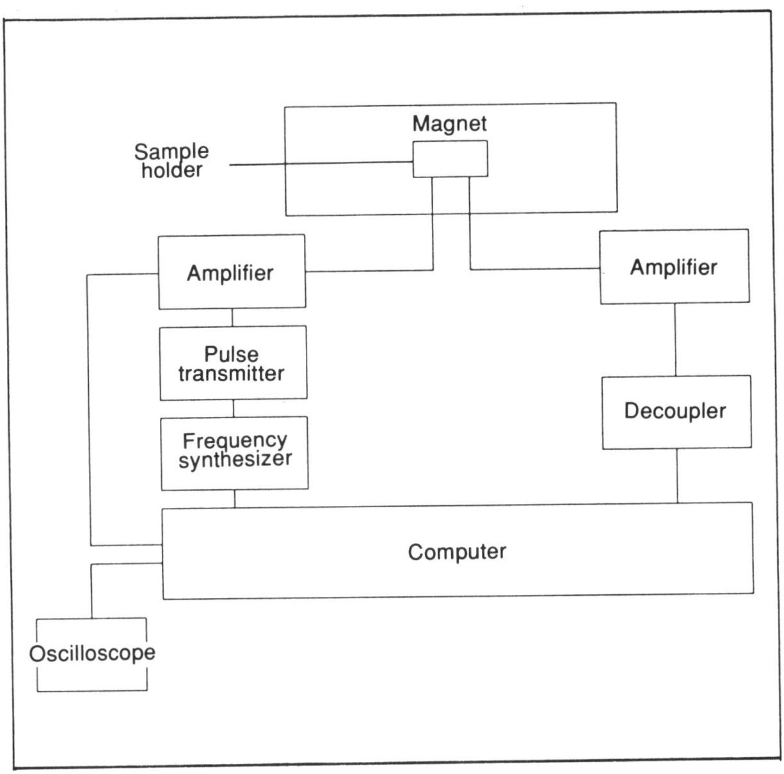

Figure 4.9 Basic components of an FT NMR spectrometer.

The probe is connected via amplifiers to an RF transmitter and receiver, which are usually housed in a separate console. With most modern spectrometers the instrument is controlled by an in-built computer, which is used to set instrumental parameters and to manipulate output.

4.3.1.1 Magnets

Magnet type is dependent upon the projected field strength required. Permanent magnets are used for low fields (below 1.5 T), electromagnets for intermediate fields (1.5–2.5 T) and superconducting magnets for high fields (2.5–14 T). In permanent and electromagnet systems the poles are designed to sit parallel, and each pole is shaped in such a manner that the magnetic field between them is uniform. Superconducting magnets are solenoidal, with a central bore into which the probes are placed.

All magnet types have an additional device to shape the magnetic field, the shims. These consist of a series of adjustable electromagnets, either round the pole pieces of permanent or electromagnets, or in the bore of superconducting magnets. By careful adjustment of the shims, the homogeneity of the magnetic field experienced by the sample can be better than one part in 10^9.

4.3.1.2 Probes

The NMR probes normally consist of an assembly containing the transmitter–receiver coil or coils, the ancillary tuning circuits, a means of heating or cooling the sample and a mechanism (often as simple as two screws) for reproducibly mounting the probe in the magnet.

In MAS NMR spectroscopy the design of the transmitter–receiver coils and their housing within the probe is extremely critical and it is worthwhile specifying the four requisites of such a probe before describing the transmitter–receiver in detail. These are:

- It must be capable of spinning the solid sample, usually contained in a hollow rotor, although occasionally machined into a rotor shape, at speeds in excess of 2 kHz while keeping it within the confines of the transmitter–receiver coil.
- It must be possible to reproducibly set the angle at which the axis of the rotor is spinning to the 'magic angle' (and occasionally to other angles).
- The transmitter–receiver coils must be able to withstand the high-power pulses of RF energy used in the experiments.
- It is convenient if sample changing is relatively easy.

4.3.1.3 MAS probes

There is a wide range of MAS probes on the market, all of which fulfil the first three criteria and, to a greater or lesser extent, the fourth. In the

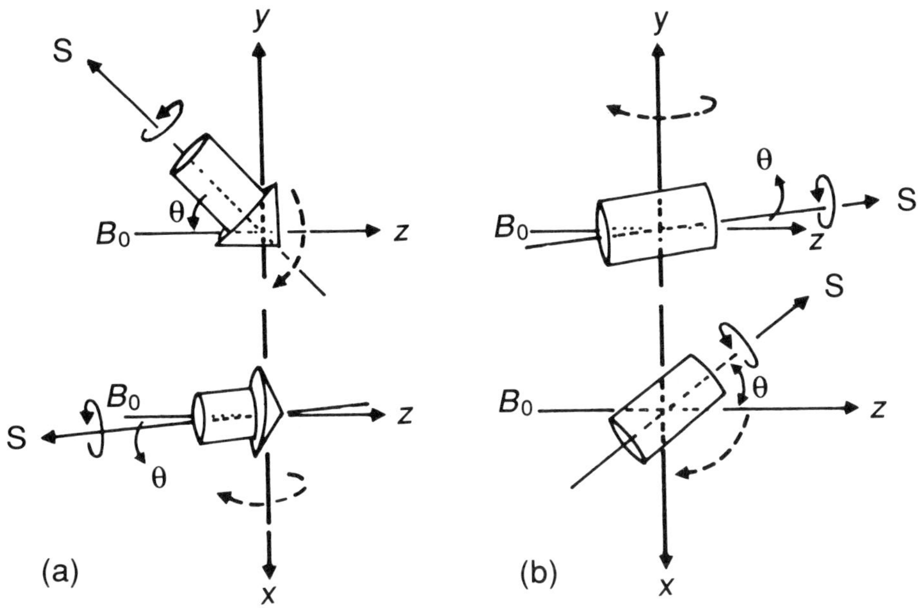

Figure 4.10 Diagrammatic representation of the geometric arrangement and design of two of the most commonly used rotors for MAS NMR; (a) the Andrew-Beams model and (b) the Lowe model.

majority of modern systems use is made of a cylindrical rotor (Figure 4.10) into which the sample can be packed. The rotor, or its cap, is machined into a turbine and when in the probe the rotor is balanced on either a single or double air-bearing and is spun by striking the turbine with a jet of air. With careful packing these rotors are capable of high rotation speeds, the maximum being dependent on the rotor diameter. Some older systems make use of the same air-jets to both suspend and spin the rotor.

A major problem for spinning rotors is uniform packing. It is preferable for the sample to be in powder form or in small uniform granules, and it is essential for this to be carefully packed into the rotor, either by 'tapping' or by using a 'ram-rod'. When the sample is spun faster than 2 kHz the force exerted on the rotor wall is in excess of 100 000 Gs. Recently, a 'pencil rotor' has been developed with what amounts to flywheels at each end so that it is able to spin holding less uniform samples. Again an air-bearing supports the rotor while an air-jet strikes a turbine at one end.

It is usual for the 'magic angle' to be set by means of a Vernier screw adjustment. Actual settings are carried out by optimizing a signal from a sample being spun in the probe. Various methods for setting the angle are used by different research groups. The method used in our laboratory involves observing the ^{79}Br signal from KBr and adjusting the angle to maximize the number of sidebands, i.e. satellite peaks that appear at multiples of the spinning frequency on both sides of the main resonance peak. Some probes need to be recalibrated every few sample changes, others less often. In early probes a trace of KBr was added to each sample in order to allow the 'magic angle' to be set for each individual experiment.

Although manufacturers' experience ensures that the coils are now capable of handling the required currents, it is still necessary to have a suitable 'dead time' in order to avoid overheating. Sample changing is very much probe dependent and many systems require that the probe be removed from the spectrometer in order to do this; there are a few designs, however, that have a feed mechanism. Finally, a probe is only as good as the rotor. Manufacturing materials vary depending on the nuclei to be observed. Popular materials include Kel-F (a plastic) or a ceramic such as zirconia.

For nuclei with half-integer spin greater than 1, the recently developed double-bearing probes offer considerable advantages, especially for situations in which the electric field gradient is large enough to generate a sizeable second-order contribution to the quadrupole interaction. Figure 4.11 illustrates the dramatic improvement in spectral quality that can be achieved with such a probe. With conventional MAS at 4.7 T the spectrum of $NaNO_2$ (Figure 4.11(b)) is completely dominated by the quadrupole interaction and it is impossible to determine the number of different environments for the sodium nucleus. With double-orientation rotation, however, these quadrupolar terms are completely removed and the spectra (Figure 4.11(c), (d)) reveal that only a single type of environment exists for the sodium.

4.3.2 Broad-band NMR

Broad-band NMR also requires specialized equipment. Many of the results cited using this technique have been gathered with continuous-wave instruments, although pulse instruments capable of performing such measurements have now been developed. In broad-band (or wide-line) NMR spectroscopy, signals with bandwidths of hundreds or even thousands of hertz are common and require detection apparatus that is capable of measuring and quantifying them accurately.

Continuous-wave spectrometers almost universally make use of superheterodyne receivers, which are capable of varying the 'window' of frequencies measured through a wide range. In some continuous-wave instruments it is actually the magnetic field that is varied (a constant frequency spectrometer), which has given rise to the expressions 'upfield' and 'downfield' that are commonly used to describe resonances to the right and left side of spectra.

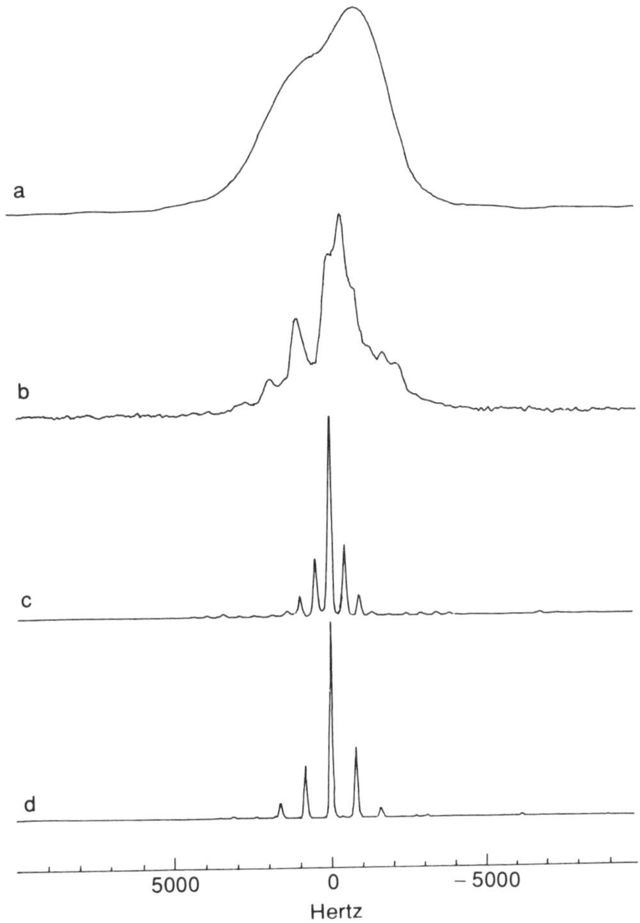

Figure 4.11 Effects of double orientation rotation on the spinning sidebands in MAS NMR spectra; ^{23}Na resonance at 52.9 MHz for $NaNO_2$ (a) static sample, (b) slow spinning MAS (note the overlap between the pattern from the second-order quadrupolar interaction and the spinning sidebands), (c) standard double-orientation rotation (note the efficient removal of the second-order quadrupolar structure), and (d) double-orientation rotation with synchronization of the outer rotor (note the loss of the odd-numbered sidebands). (Reproduced with permission of Bruker Analytische Messtechnik Gmbh.)

Pulse spectrometers designed to measure solution spectra have neither the transmitter power to give pulses capable of exciting a sufficient frequency range nor a receiver capable of collecting the wide-line signals (or analogue-to-digital converters fast enough to sample it); a specialized spectrometer or an extensive modification is needed.

To summarize, with all instrument types both the transmitter and the receiver must be capable of handling the large frequency ranges over which resonances are measured.

4.3.3 Instrumental parameters

Nuclear magnetic resonance spectroscopy has a plethora of instrumental parameters that must be considered before producing a spectrum. As it is likely that all accessible instruments capable of performing multinuclear experiments on clays will be pulse spectrometers, the details given here will be limited to this instrument type. Continuous-wave (CW) spectrometers for this type of work are conceptually either very simple (such as, for example, for broad-band work) or highly specialized.

In pulse spectrometry it is necessary to set parameters before acquisition of the FID and before transformation. The objective of this section is to explain some of the parameters that we consider to be most important; many parameters are interconnected, which makes NMR a very versatile technique (but often very confusing for the non-specialist).

4.3.3.1 Acquisition

(i) Spectrometer frequency (resonating frequency)
This is the frequency at which the nucleus of interest resonates in the magnetic field of the spectrometer being used. It is usually found in a table of frequencies supplied with the spectrometer. Where an element has more than one isotope with a spin it is worth carefully considering which will be the best for the experiment in question.

(ii) Pulse frequency
This is selected to ensure that the pulse is situated near to the centre of the frequency range to be examined. (For certain older instruments, to one side of the range to be examined).

(iii) Sweep width
This is the range in hertz or parts per million with a small margin of error that represents the frequencies of interest.

If either the pulse frequency or sweep width are miss-set, totally misleading spectra, e.g. back-to-front, may be produced.

(iv) Time domain
This represents the number of memory words into which the FID is collected. It is best if the FID is allowed to decay to noise. If it is truncated it will lead to distortion of the resultant spectrum.

(v) Acquisition time

This is the time taken for the above collection to take place. It is often calculated by the computer because it is based on a combination of the sweep width and the size of the time domain.

(vi) Receiver delay

This is a delay between the end of the acquisition and the next pulse. It allows the spins to regain equilibrium before re-excitation. It also ensures that the system does not overheat and melt when high-powered pulsing/decoupling is being used. This delay should not be confused with the dead time between the end of the pulse and activation of the receiver, which ensures that there is no 'pulse breakthrough', i.e. digitization of the tail of the pulse, which can cause grossly distorted baselines. It is usually set by the computer and is seldom if ever quoted in experimental sections. However, where T_2 is short and this delay is long (e.g. 100 μs), the FID can be passed before the receiver has activated.

(vii) Pulse width

In a one-acquisition experiment a 90° pulse gives a maximum signal. It then takes $5 \times T_1$ for the system to return to equilibrium before another pulse can be applied. If T_1 is short this does not cause any problems, but if it is of any appreciable length the total acquisition times may become impossibly long. Although some pulse sequences need pulses of a definite length, for straightforward acquisition it can be advantageous to use a pulse of $< 30°$. Pulse widths are usually quoted as the angle followed by the time in microseconds taken to turn the magnetization through that angle.

(viii) Number of scans

This represents the number of accumulations in the spectrum. The total acquisition time is the product of the number of scans and the combined acquisition time and receiver delay.

(ix) Contact time

This is the period during which energy is 'shared' between nuclei during cross-polarization. It is material dependent and should be ascertained for all samples individually.

(x) Spin rate

In MAS NMR this is the rotor speed (usually in kHz).

4.3.3.2 Transformation

(i) Size

This is the size in words, usually in K (thousand), into which the data in the time domain is to be transformed. There are several implications; the size is set before transformation and on transformation half contains the spectrum and the other half the 'imaginary' part of the answer. Thus the digital resolution is the sweep width/$0.5 \times$ size, meaning that the size must be large enough to give the required digital resolution. The size could be the same as the time domain or, if the time domain is small, empty data space may be added. This is called 'zero filling'. An advantage of zero filling is that the Fourier transform of zero is zero, whereas the transform of the noise of the FID is noise. Consequently, a zero-filled FID will give a spectrum with less noise than one from a pure time domain.

(ii) Exponential multiplication

Before transformation, the data in the FID may be manipulated in a manner which will affect the final spectrum. The most common manipulation is exponential multiplication. The FID is multiplied using an exponential function designed to slightly broaden the lines of the spectrum while drastically reducing the noise. Other functions may be used, e.g. Gaussian, trapezeoidal. Thus resolution may be sacrificed in order to remove noise or, if ultra-high resolution is required, this may be gained at the cost of increased noise.

4.4 GENERAL FEATURES OF MAS NMR SPECTRA OF MINERALS

This section introduces the main features of the spectra of the most important NMR nuclei in clay minerals. These are presented by nucleus, which is also the way that an NMR experiment is conducted, and contains some examples from the mineral literature in general.

4.4.1 ^{29}Si chemical shifts

The ^{29}Si isotope is the best characterized nucleus of importance to clay minerals. With spin $I = \frac{1}{2}$ it produces relatively straightforward spectra with high resolution. The low natural abundance of the isotope (4.7%), however, means that several hours are usually required to record a spectrum of a clay mineral specimen. Magic-angle spinning techniques are able to reduce linewidths to < 200 Hz, except when minerals contain significant amounts of iron. In such cases there is loss of resolution and spectral sensitivity because of the large dipolar interactions and overall magnetic susceptibility of the sample (see section 4.2.12).

The ^{29}Si isomer shift is strongly influenced by the chemical environment of the nucleus. For tetrahedral silicon there are good correlations between isomer shift and degree of polymerization, and with the number of next-nearest-neighbour Al atoms (Figure 4.12). There are a number of good review articles to which the reader is referred for further informa-

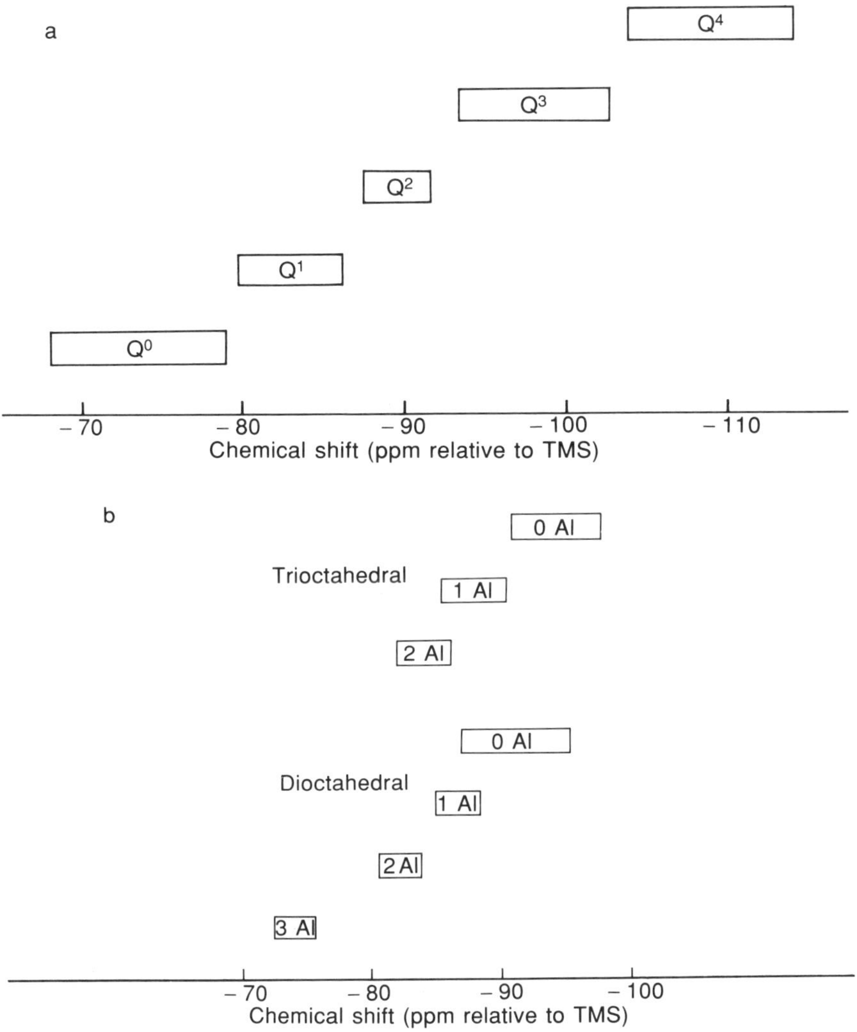

Figure 4.12 Isomer shift ranges for ^{29}Si as a function of (a) degree of polymerization, and (b) the number of next-nearest-neighbour Al in layer silicates.

tion (Magi *et al.*, 1984; Kirkpatrick *et al.*, 1985; Kirkpatrick, 1988). The shielding of the silicon nucleus is progressively decreased (and hence the isomer shift becomes less negative relative to TMS) with decreasing polymerization (Lippmaa *et al.*, 1980) and with increasing numbers of next-nearest-neighbour Al atoms for framework (Lippmaa *et al.*, 1981) and layer silicates (Kinsey *et al.*, 1985). There is a correlation with group electro-negativity, which is calculated by considering the total electro-negativity of the oxygen atoms of a tetrahedron and the other cations to which they are coordinated (James and Oldfield, 1985) and, for framework silicates, the isomer shift is also correlated with the mean Si–O–*T* (*T* = Si or Al) bond angle (e.g. Ramdas and Klinowski, 1984).

4.4.2 Al/Si order/disorder

When an element occupies more than one type of site in a crystal structure, separate resonances are usually observed for each distinct co-ordination environment. Resonance position can also be influenced by the nature of next-nearest-neighbour ions. For example, in an aluminosilicate where the tetrahedral site occupancy is shared between Si^{4+} and Al^{3+} ions, the ^{29}Si NMR spectrum will consist of a series of peaks corresponding to 0, 1, 2, 3, or 4 Al^{3+} ions in next-nearest-neighbour sites (Figure 4.12(b)). A great deal of work has been carried out with the objective of determining Si/Al distributions and also to test the aluminium avoidance rule (Lowenstein, 1954), which states that Al–O–Al linkages will not occur. This can be done by comparing ^{29}Si NMR spectra and chemical analyses, because the rule predicts that the Si/Al ratio of a site will be equal to $\Sigma I(\mathrm{Si}_n)/0.25\Sigma nI(\mathrm{Si}_n)$, where n is the number of next-nearest-neighbour Al (0–4), and $I(\mathrm{Si}_n)$ is the relative intensity of the ^{29}Si NMR peak with n next-nearest-neighbour Al.

A large number of measurements have been made with synthetic zeolites and the aluminium avoidance rule seems to hold. Indeed Klinowski *et al.* (1982) have concluded that there is also a tendency to minimize the number of Al–O–Si–O–Al linkages. Although there have been a smaller number of measurements with layer silicates, there has been a general consensus that the rule also holds here (see e.g. Barron, Slade and Frost, 1985; Herrero, Sanz and Seratosa, 1985a, b), although intensity distributions seem to indicate a rather more ordered distribution than that which would come from simple aluminium avoidance. However, the presence in such minerals of any paramagnetic ions that are not randomly distributed would seriously affect the validity of any conclusions that could be drawn from the detailed analyses of the ^{29}Si peak intensities.

4.4.3 ^{27}Al NMR

The principal value of ^{27}Al NMR spectroscopy to the clay mineralogist lies in its ability to distinguish readily and unambiguously between ions with octahedral and ions with tetrahedral coordination. As shown in Table 4.1, ^{27}Al is a quadrupolar nucleus with spin $I = \frac{5}{2}$, which can interact with any external electric field gradients. The central $\frac{1}{2} \rightarrow -\frac{1}{2}$ transition is observed, but it is shifted relative to the true chemical shift by an amount equal to $(e^2qQ/h)^2/\nu \times F$, where the components of the quadrupolar coupling constant, e^2qQ/h, are as defined in Chapter 3, ν is the frequency of observation, and F is a function of crystallite orientation with respect to the magnetic field and is ≈ 1. The frequency shifts of all of the other transitions are proportional to the quadrupole interaction, which means that in rotating powders they are smeared into a large number of spinning sidebands and are unobservable except when the crystal has pure cubic symmetry.

Dipolar interactions and chemical-shift anisotropy are removed by MAS as with spin $\frac{1}{2}$ nuclei, but the quadrupolar contribution to the linewidth is only reduced by about 80%. The remaining line-shape is determined by the magnitude and nature of crystal distortions, its variation with the asymmetry parameter, η, being given in Figure 4.13.

The positions of the resonance peaks are strongly influenced by coordination number and there is no difficulty in distinguishing octahedral and tetrahedral coordination, as can be seen in Figure 4.14. The

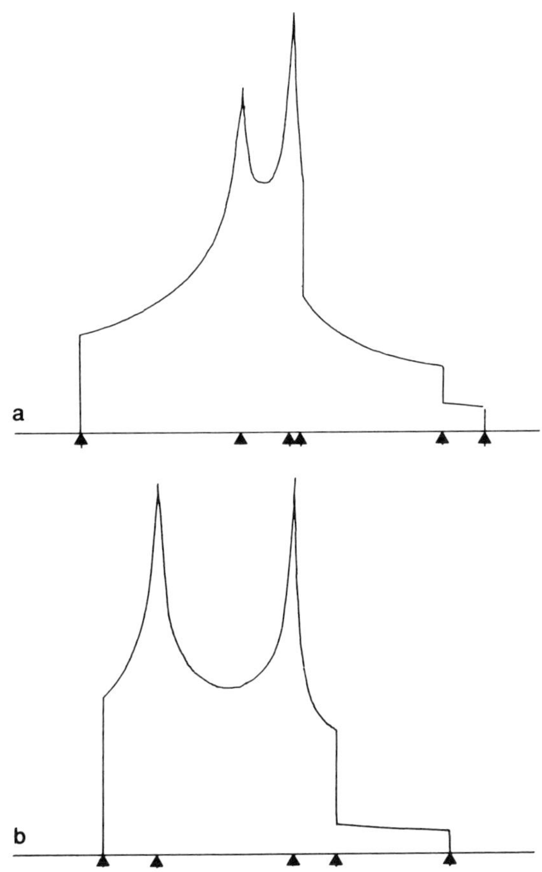

Figure 4.13 Idealized shapes for the $\frac{1}{2} \rightarrow -\frac{1}{2}$ transition in ^{27}Al; (a) asymmetry parameter, $> \frac{3}{7}$ (b) $< \frac{3}{7}$. The positions of the spectral turning points are indicated by arrows.

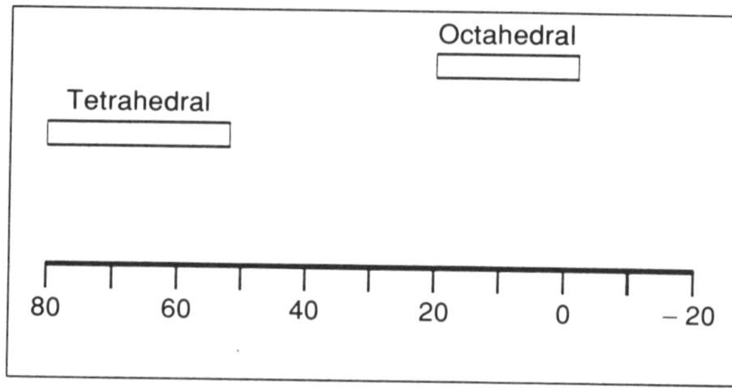

Figure 4.14 Relationship between chemical shift and coordination number for ^{27}Al. All shifts are expressed relative to α-Al_2O_3.

next-nearest-neighbour environment for tetrahedral Al can also be identified. Typical values of *c.* 80, 74, 69 and 64 ppm, respectively, have been reported for 0, 1, 2 and 3 nearest-neighbour Si atoms. No chemical shift range for fivefold Al coordination is included in Figure 4.14, because the large electric field gradient usually makes observation of spectra impossible with MAS. As mentioned in section 4.2.10, double rotation techniques are effective in removing large quadrupolar contributions to the spectral linewidths, but to date few measurements have been made by using this approach.

In many cases of MAS NMR, the quadrupolar line-shape is not seen clearly because of a non-uniform distribution of electric field gradients and a field-dependent line broadening is observed. Resolution is improved by operating at the highest practical magnetic fields, because the observed chemical shift (in hertz, see section 4.2.2) is proportional to the magnitude of the magnetic field. This is illustrated in Figure 4.15 for a clay, where a dramatic improvement in spectral quality was obtained when the field of observation was increased from 4.7 T (200 MHz ^1H) to 11.7 T (500 MHz ^1H). The tetrahedral resonance also split into two distinct components at the higher field.

Quantitative estimation of spectral contributions can be very difficult with quadrupolar nuclei, especially when there are mixtures of nuclei that are with and without quadrupole interactions. In the latter case all transitions are excited and contribute to the spectral area, whereas in the former the spectrum arises solely from the $\frac{1}{2} \rightarrow -\frac{1}{2}$ transition. This error can be up to a factor of four in the limit of short excitation pulses. However, the magnitudes of the 90° pulse may differ significantly for these two situations, making it difficult to estimate the relative errors for longer pulse widths.

4.4.4 ^{17}O chemical shifts

The natural abundance of ^{17}O is very low (0.037%) and its NMR spectra cannot be observed in natural minerals. Consequently the major applica-

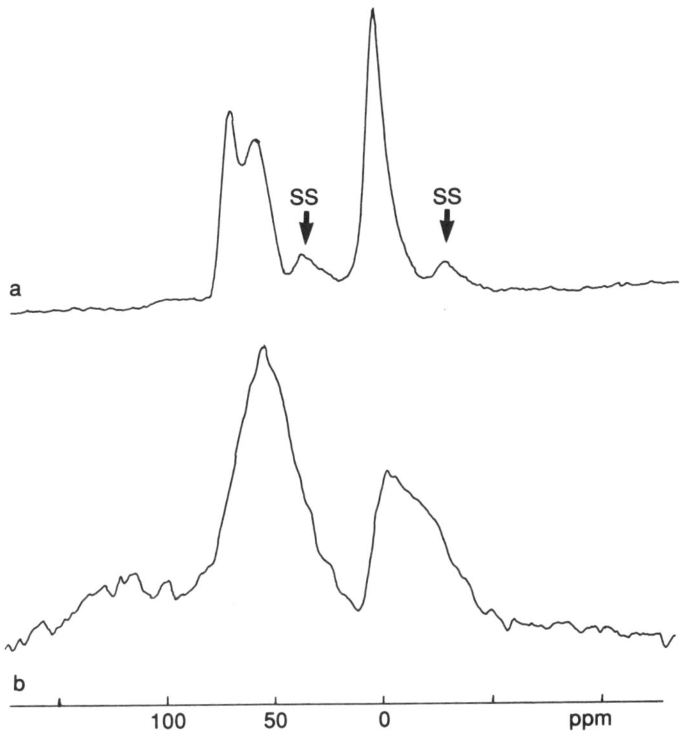

Figure 4.15 ^{27}Al MAS NMR spectra of a synthetic layer silicate clay mineral in magnetic fields of (a) 11.7 T and (b) 4.7 T. Note the resolution of two components with tetrahedral coordination in the spectrum obtained at the higher field. Spinning sidebands (SS) are indicated in spectrum (a). (Spectra reproduced with permission of Bruker Analytische Messtechnik Gmbh.)

tions of ^{17}O NMR are in characterizing the bonding in oxide phases, examining the products of hydrothermal reactions and identification of the structure of amorphous phases, where the various products can be synthesized with significant ^{17}O enrichment. Thus there has in recent years been an increasing interest in developing relationships between the chemical environments of oxygen and its NMR parameters.

For bridging oxygens there are only small variations in the magnitudes of their isotropic chemical shifts (Kirkpatrick, 1988). In fully polymerized framework structures the shifts for oxygens in Si–O–Si combinations (cristobalite) are in the range 44–52 ppm, those for Si–O–Al (zeolites) are in the range 31–40 ppm, and those for P–O–Al (aluminium phosphates) are in the range 61–63 ppm. In alkaline earth metasilicates the shifts are in the range 62–87 ppm, with the shielding decreasing with increasing cation size and decreasing cation electro-negativity.

The ^{17}O chemical shifts for non-bridging oxygens cover a much larger range than for bridging oxygen atoms, but are less well understood. For example, with alkaline earth metasilicates the shifts are in the range 42–169 ppm (Timken *et al.*, 1987), with the shielding decreasing from Mg to Ba.

The quadrupole coupling constants seem to vary systematically with chemical environment; they decrease with increasing ionic character of the bonds in which the oxygen is involved. Values of 3.7–5.3 MHz are reported for Si–O–Si sites, 3.1–3.4 MHz for Si–O–Al sites and 1.0–3.2 MHz for non-bridging oxygens (Kirkpatrick, 1988).

4.4.5 ^{31}P

Like ^{29}Si, ^{31}P has a nuclear spin of $\frac{1}{2}$ (Table 4.1), but because of its 100% natural abundance its NMR spectra are relatively quick to record. The typical ranges for the chemical shifts for phosphate minerals are shown in Figure 4.16. High-resolution spectra of a variety of orthophosphates have been obtained by Turner *et al.*, (1986), who found that there was an empirical relationship between the isomer shift and $Z[(r)^{\frac{1}{2}}]$, where Z is the charge of the associated cation and r is its ionic radius. With hydroxy-apatite and its non-stoichiometric forms Rothwell, Waugh and Yesinowski (1980) found that there was an upfield shift on protonation of the phosphate group, but the differences were small and resonances of

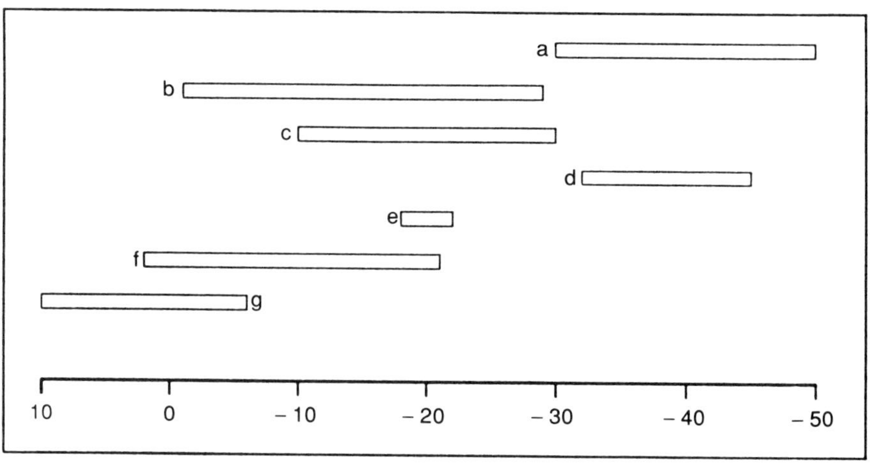

Figure 4.16 Ranges of ^{31}P chemical shifts for different types of phosphate mineral; (a) middle and ring groups in aluminophosphates (Q^2), (b) alkali and alkaline earth alumino-orthophosphates, (c) $AlPO_4 \cdot n(H_2O) \cdot m(OH)$ orthophosphates, (d) branching groups (Q^3), middle and ring groups (Q^2), (e) end groups (Q^1) and (f) alkali and alkaline earth orthophosphates (Q^0). (Adapted from Kirkpatrick, 1988.)

the amorphous phases were relatively broad. The chemical-shift anisotropy parameters, as revealed by slow MAS, varied appreciably and could be used to distinguish components with overlapping isomer shifts but different shift anisotropies.

In the study of bulk soils, spectral resolution tends to be poor, partly because of line broadening as a result of the presence of paramagnetic components and the heterogeneity of the samples and partly because of the relatively small range of chemical shifts for the main inorganic phosphorus phases.

4.5 HIGH-RESOLUTION NMR SPECTROSCOPY OF CLAY MINERALS

4.5.1 Introduction

The major limitation to the characterization of natural clay mineral samples by NMR spectroscopy is the amount of paramagnetic species (especially Fe^{2+} and Fe^{3+} ions) in their structures. As pointed out in section 4.2.12 their effects are twofold: firstly, spectral broadening is produced as a result of an overall increase in magnetic susceptibility within the specimen (the limit for the observation of an NMR spectrum would appear to be a few percent iron, although there are variations depending on whether or not there is any coupling between paramagnetic ions); and secondly, dipolar interactions between unpaired electrons and neighbouring nuclei may effectively eliminate any contribution of the latter to NMR spectra. This is particularly serious in quantitative work because paramagnetic ions will generally not be randomly distributed through structures. They will, therefore, selectively suppress the contributions of spectral components and, as a general rule of thumb, it will not be possible to obtain reliable quantitative information from any specimens that give significant EPR signals. Nevertheless, high-field NMR spectroscopy has made important contributions to the characterization of clay mineral structures in recent years. A selection of results will be discussed below.

4.5.2 Layer silicates

One of the most successful applications of the technique has been in the distinction between octahedral and tetrahedral coordination for aluminium (e.g. Müller *et al.*, 1981; Kirkpatrick *et al.*, 1985; Kinsey *et al.*, 1985). This is illustrated in Figure 4.17, where the spectra of kaolinite and saponite show large shifts in the positions of the ^{27}Al resonance between the minerals with octahedrally and tetrahedrally coordinated aluminium (Goodman and Stucki, 1984). In minerals where both types of

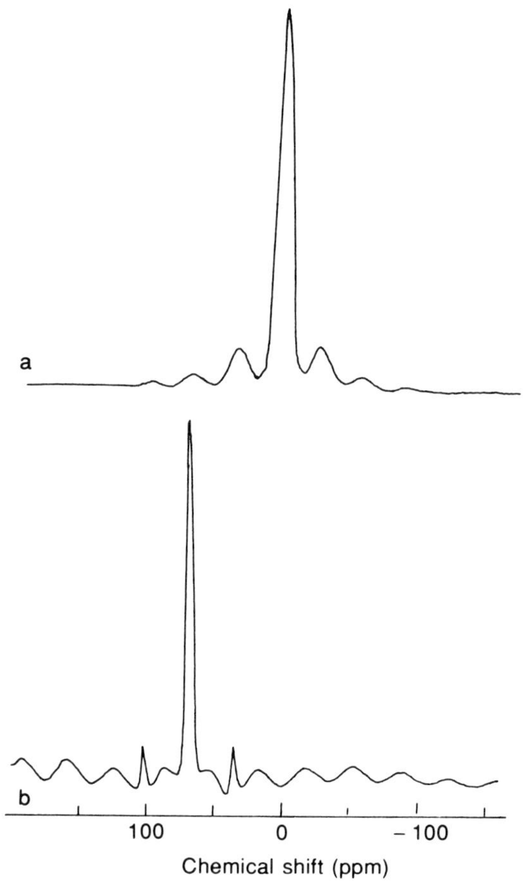

Figure 4.17 ^{27}Al Fourier transform NMR spectra at 130.3 MHz and MAS rate of *c.* 4.0 kHz of (a) kaolinite and (b) saponite. The chemical shift is expressed relative to α-Al_2O_3. (Reproduced from Goodman and Stucki, 1984.)

coordination exist it is necessary to carefully select the spinning frequency in order to avoid obliteration of a minor peak by spinning sidebands from a major peak. The ^{27}Al spectra of three montmorillonites (Figure 4.18) show definitively that all contain some tetrahedral aluminium (Goodman and Stucki, 1984). In that paper Goodman and Stucki found good agreement between the tetrahedral aluminium content and that calculated from chemical analytical data. However, Sanz and Serratosa (1984), in a similar determination of octahedral and tetrahedral aluminium in aluminosilicate minerals, reported poor agreement between NMR and chemical analyses.

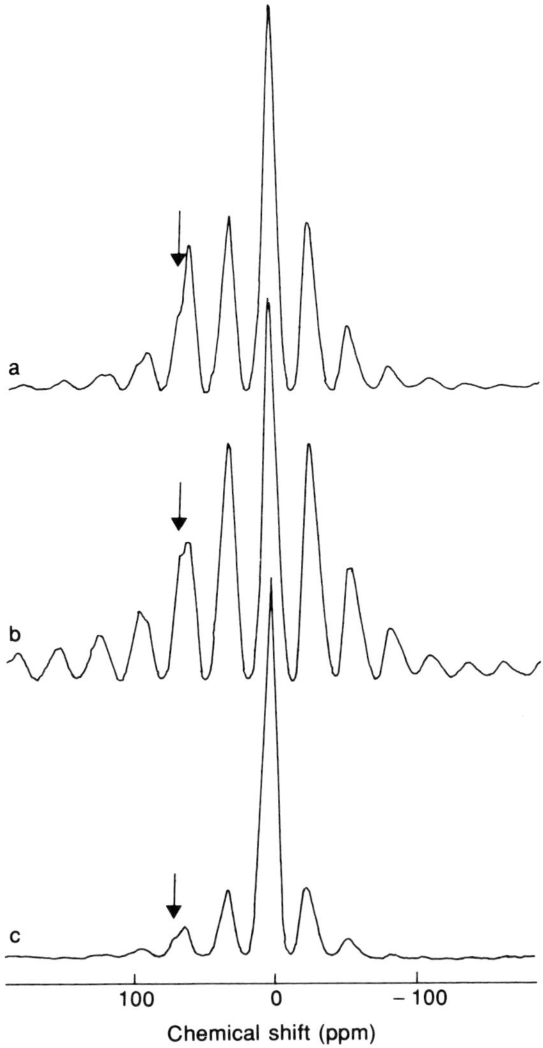

Figure 4.18 ^{27}Al Fourier transform NMR spectra at 130.3 MHz and MAS rate of *c.* 4.0 kHz of three montmorillonites having the following cation compositions per $O_{20}(OH)_4$ structural unit: (a) $(Si_{7.34}Al_{0.66})$ $(Al_{3.24}Fe^{3+}_{0.19}Mg_{0.77}Mn_{0.01})$; (b) $(Si_{7.70}Al_{0.30})$ $(Al_{3.21}Fe^{3+}_{0.37}Fe^{2+}_{0.06}Mg_{0.39})$; (c) $(Si_{7.95}Al_{0.05})$ $(Al_{2.69}Fe^{3+}_{0.11}Mg_{1.20})$. The positions of the centres of the resonances from tetrahedrally coordinated Al are indicated by arrows. (Reproduced from Goodman and Stucki, 1984.)

Structural iron will always present major problems in any attempts at quantitative work with clay minerals, and as mentioned above, if it is non-randomly distributed through the structure, it will selectively affect resonances from those nuclei that are closest to it. Thus in muscovite, where Sanz and Serratosa (1984) report an overestimation of tetrahedral Al by NMR, this could reflect the influence of Fe^{2+} and Fe^{3+}, which have been shown by Mössbauer spectroscopy to occur exclusively in octahedral sites.

The effects of structural iron on the derivation of quantitative site populations from NMR spectral data is progressively more serious with increasing iron contents of the minerals, and, because of its potential for selectivity in addition to the overall broadening that results from bulk magnetic susceptibility effects, any such measurements should be treated with caution.

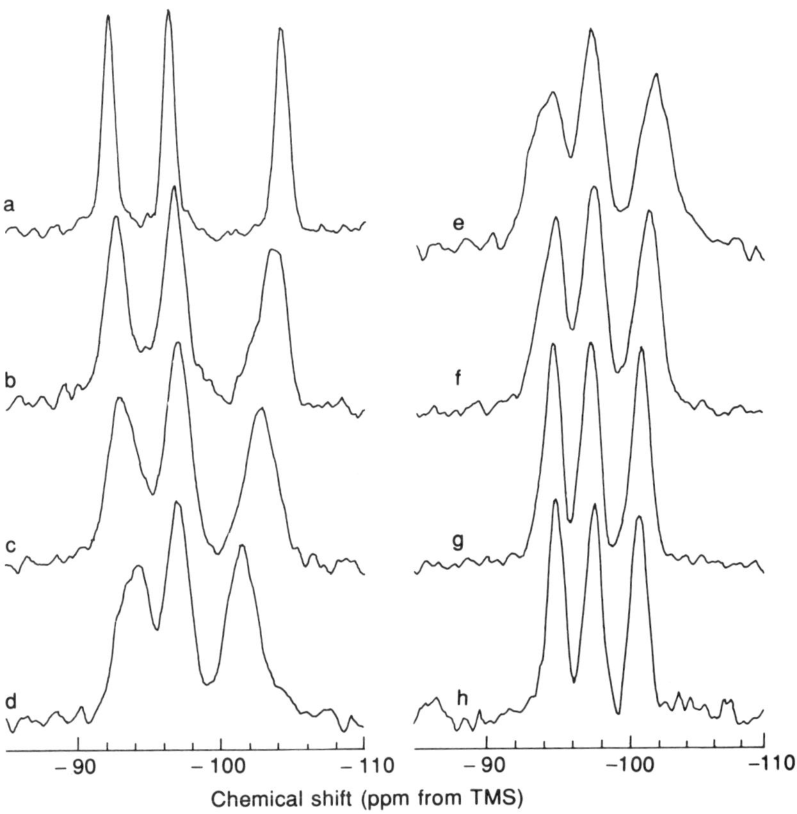

Figure 4.19 ^{29}Si MAS NMR spectra of Si, Al ordered feldspars along the albite–microcline join. Spectra correspond to (a) 0, (b) 17, (c) 33, (d) and (e) 49, (f) 66, (g) 83, and (h) 100% microcline. The larger peak widths for intermediate compositions are consistent with local variations in Na/K distributions. (Reproduced with permission from Phillips, Kirkpatrick and Hovis, 1988.)

4.5.3 Framework silicates

In anorthite-rich plagioclase feldspar, Kirkpatrick *et al.* (1987) have shown that the distributions of aluminium and silicon are different from those predicted simply by the aluminium avoidance rule. From analyses of spectral peak intensities it appears that any Si excess over the 1 : 1 Al : Si ratio of anorthite enters specific sites that are occupied by Al in pure anorthite, with the anorthite Si sites remaining unaltered.

For alkali feldspars it is possible to use NMR of ^{29}Si, ^{27}Al and ^{23}Na nuclei to obtain a comprehensive understanding of the structural relationships that exist within family members (Phillips, Kirkpatrick and Hovis, 1988). In the albite–microcline series (in which Na is replaced by K), the ^{29}Si NMR spectra consist of three peaks (Figure 4.19), which correspond to three crystallographically distinct Si sites. The fourth tetrahedral site is completely occupied by Al and only one component is seen in the ^{27}Al spectra (Figure 4.20). The interpretation of the central $(+\frac{1}{2}, -\frac{1}{2})$ transition in the ^{27}Al spectra of samples with end-member compositions is complicated by the presence of well-resolved second-order quadrupole patterns, and although poorer resolution was seen with the samples of intermediate composition, their spectra also appeared to contain more than one peak for this transition. Phillips, Kirkpatrick and Hovis (1988) were, however, able to demonstrate that there was only one type of Al site, because the spinning sidebands of the satellite $\pm(\frac{1}{2}, \frac{3}{2})$ transitions contained only one peak. In a similar way the ^{23}Na spectra (Figure 4.21) could also be fitted to one site, with the peaks for intermediate compositions again being less well resolved as a result of inhomogeneous broadening.

Nuclear magnetic resonance spectroscopy can provide valuable information on the nature of structural changes that are induced by chemical or physical processes in aluminosilicate minerals with low iron content. As an example, Kodama, Kotlyar and Ripmeester (1989) have shown that extensive grinding of kaolinites results in a progressive increase in the amount of aluminium with tetrahedral coordination. This was accompanied by an increase in linewidths of the ^{29}Si resonance, which indicates that distortions of the SiO_4 tetrahedra also occurred. The NMR spectra obtained from kaolinite after extensive grinding were similar to those of metakaolin (MacKenzie *et al.*, 1985; Watanabe *et al.*, 1987), which suggests that dehydroxylation of $Al^{VI}(OH)_4O_2$ occurred with the progressive formation of $AlO_4\square_2$ groups, where $\square$ is a vacancy produced by loss of H_2O.

4.5.4 Amorphous and microcrystalline minerals

The determination of short-range structure in poorly crystalline or amorphous minerals represents a major area of success for applications of the NMR technique, and a vast amount of work has been published on

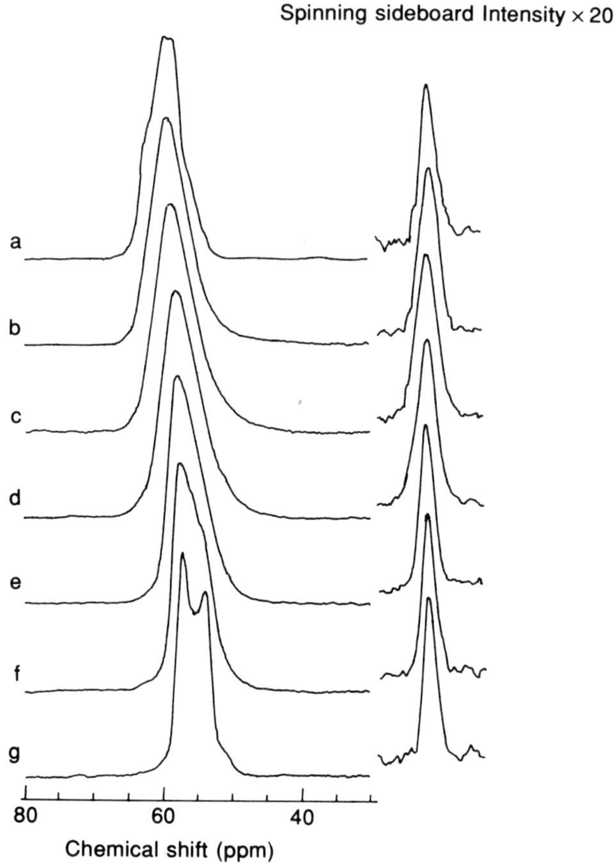

Figure 4.20 ^{27}Al MAS NMR spectra at 11.7 T of Si, Al ordered feldspars along the albite–microcline join. Spectra correspond to (a) 1, (b) 17, (c) 33, (d) 49, (e) 66, (f) 83 and (g) 100% microcline. (Reproduced from Phillips, Kirkpatrick and Hovis, 1988.)

the characterization of natural and synthetic glasses. Here we shall consider the specific application to the characterization of natural and synthetic allophanes. These minerals are essentially solid solutions of silica, alumina and water and are considered to be related to the mineral imogolite, which has a tubular structure (Figure 4.22) containing an orthosilicate outer surface and a gibbsite-like inner surface (Cradwick *et al.*, 1972). ^{29}Si NMR spectra of imogolite, proto-imogolite, which is considered to be a non-tubular aluminosilicate with an incompletely formed imogolite structure, and four allophane specimens (Goodman, *et al.*, 1985) are shown in Figure 4.23 along with their chemically determined Al/Si ratios. To a first approximation the allophane spectra are

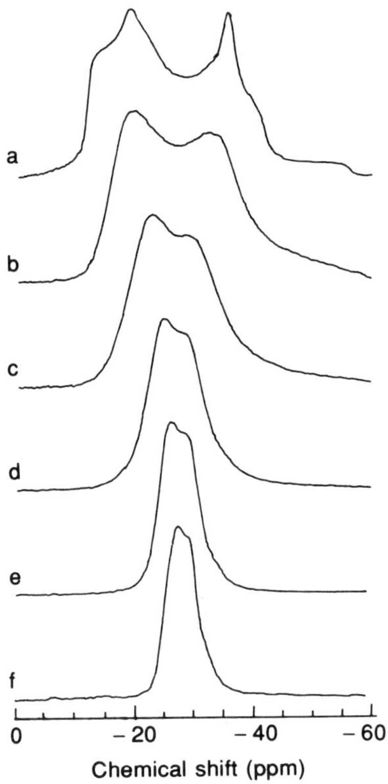

Figure 4.21 ^{23}Na MAS NMR spectra at 8.45 T of Si, Al ordered feldspars along the albite–microcline join. Spectra correspond to (a) 1, (b) 17, (c) 33, (d) 49, (e) 66, and (f) 83% microcline. (Reproduced from Phillips, Kirkpatrick and Hovis, 1988.)

composed of two components; a sharp peak (*c.* 3 ppm width) at about −79 ppm, which corresponds to the orthosilicate in imogolite-like structures, and a broad resonance (*c.* 20 ppm width) centred at about −90 ppm, which corresponds to more highly polymerized silicate species. The chemical shifts (−80 to −115 ppm) in this broad peak cover virtually the entire range of shifts reported for ^{29}Si in framework aluminosilicates (−75 to −120 ppm) (Smith *et al.*, 1983). Thus the full range of from 0 to 4 next-nearest-neighbour Al atoms is represented in these structures. This broad resonance peak appears to correlate with a component in the IR spectra that has been assigned to feldspathoid-like structures and its shifts encompass those that have been reported (−92.5 to −104.2 ppm) for feldspars (Lippmaa *et al.*, 1980).

The ^{27}Al NMR results from the same specimens are shown in Figure 4.24 and are consistent with those from ^{29}Si. Both imogolite and proto-imogolite produced a single peak at *c.* 5 ppm flanked by a pattern of

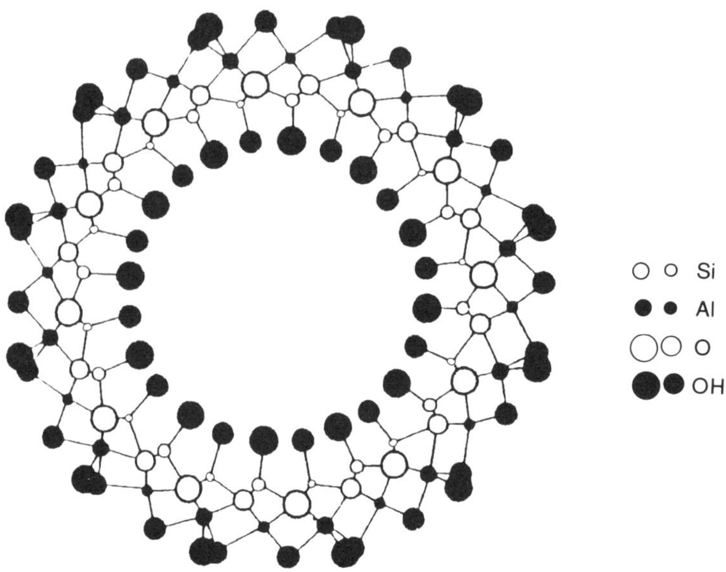

Figure 4.22 A cross-section of the structure of an imogolite tube. (Reproduced from Goodman *et al.*, 1985.)

spinning sidebands. This resonance is characteristic of aluminium bound octahedrally to oxygens. The allophane specimens, however, showed various levels of tetrahedral Al, as evidenced by the peak at *c.* 60 ppm, with the highest amounts occurring in those specimens with the most intensities in the broad ^{29}Si resonance at − 90 ppm. Thus the NMR results indicate that natural allophanes contain two distinct types of structure; one is imogolite-like, whereas the other contains more highly polymerized Si, probably as a framework aluminosilicate phase.

Thermal transformations of imogolite have been studied in detail by Wilson *et al.* (1988) and MacKenzie *et al.* (1989) using ^{29}Si and ^{27}Al NMR spectroscopy. It appears that on dehydroxylation many of the Al sites in imogolite become too distorted to produce an NMR spectrum (i.e. they probably have five-coordination), and others show a decrease in coordination number from six to four. Changes are also produced in the ^{29}Si resonance, in which with increasing temperature the original sharp peak is replaced progressively by a broader upfield resonance. Such a behaviour is consistent with the formation of a more condensed structure, probably through the cross-linking of Si–O–Si bonds.

The NMR spectra of heated synthetic allophanes showed that the thermal decomposition reactions were similar to those in pure imogolite, but the imogolite-like units in allophanes were less stable thermally than in imogolite (Wilson *et al.*, 1988; MacKenzie, Bowden and Meinhold,

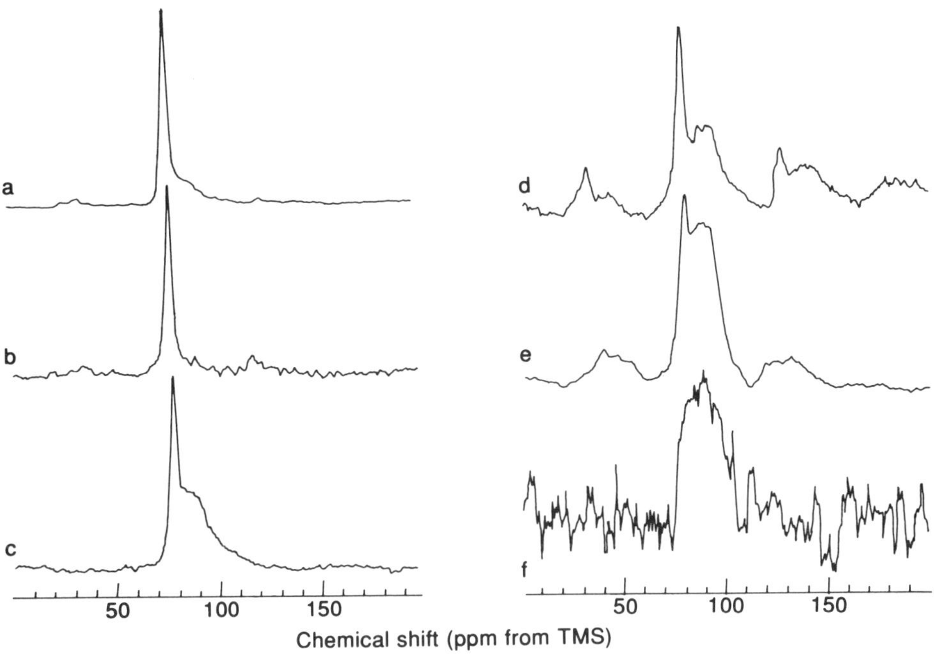

Figure 4.23 99.3 MHz ^{29}Si MAS NMR spectra of (a) natural imogolite, (b) natural proto-imogolite and (c)–(f) natural allophanes (from Goodman *et al.*, 1985). Al/Si ratios are (a) 1.7, (b) 2.3, (c) 2.7, (d) 1.1, (e) 1.2, and (f) 1.4. Note the spinning-sidebands in (d) probably arise from the presence of magnetic impurities in the specimen. (Reproduced from Goodman *et al.*, 1985.)

1991). The latter authors have shown that the allophane NMR spectra can be accounted for in terms of a two-sheet kaolinite-like structure, in which tetrahedral holes result in curvature to produce a sphere. The imogolite-like orthosilicate units anchor into the octahedral sheet and fit into defects in the tetrahedral sheet.

4.6 CLAY SURFACE STUDIES

4.6.1 Interactions of water with clay surfaces

A great deal of work has been performed on the characterization of water in contact with clay surfaces. The physical properties of water are affected by interaction with the surface and these can be probed effectively by NMR. The motion of protons in water molecules can be described in terms of three distinct processes; molecular rotation, translational diffusion and intermolecular proton exchange.

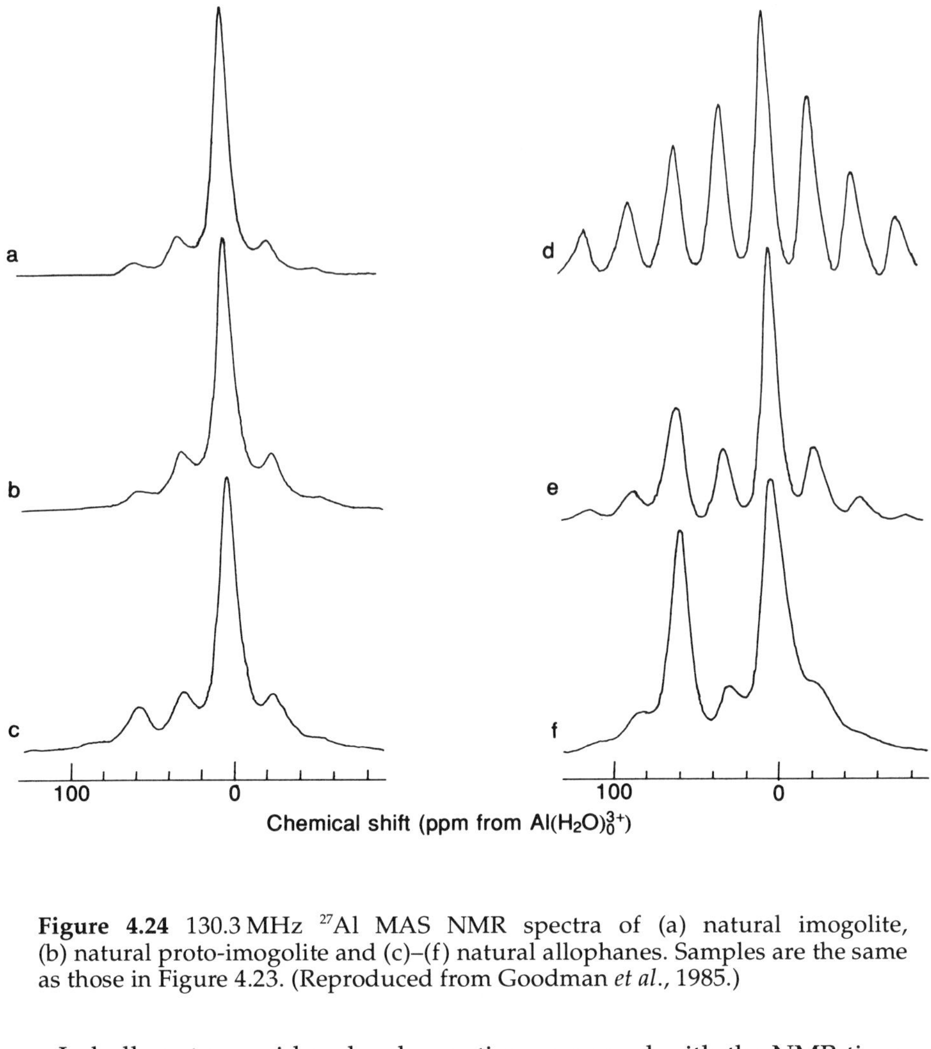

Figure 4.24 130.3 MHz ^{27}Al MAS NMR spectra of (a) natural imogolite, (b) natural proto-imogolite and (c)–(f) natural allophanes. Samples are the same as those in Figure 4.23. (Reproduced from Goodman *et al.*, 1985.)

In bulk water, rapid molecular motion compared with the NMR timescale results in an isotropic resonance. However, those molecules in contact with a surface experience a highly anisotropic interaction, the potential of which may remain constant over a period of many molecular reorientations. The rate of exchange of molecules between bulk water and surface-bound sites can, therefore, be investigated by NMR spectroscopy.

When a water molecule is oriented preferentially with respect to a surface the dipolar interaction between its protons is no longer averaged to zero as it is in the bulk liquid. The spectrum from these molecules is split into two peaks with a separation proportional to the magnitude of the dipolar interaction. However, because of motional averaging, this splitting is much less than that observed in the solid state. Proton exchange between neighbouring molecules also decreases this splitting, because of the virtually random distribution of their spin states, and the splitting vanishes completely when there is rapid intermolecular proton exchange.

This is not the case for deuterium, where the spectral splitting results from the interaction of the quadrupole moment of the nucleus with the electric field gradient around it. Hence ^{2}D NMR is more effective than ^{1}H NMR for the observation of preferential ordering of water molecules on surfaces. Also the spin–lattice relaxation time T_1 for deuterium is determined primarily by intramolecular interactions and is much less sensitive to intermolecular effects than it is for protons. Thus minor paramagnetic components in clay structures have much less influence on deuterium relaxation properties. In the situation of fast molecular exchange a weighted value for T_1 is observed and this may be written as

$$1/T_1 = a/T_{1f} + b/T_{1b},$$

where T_{1f} and T_{1b} represent the T_1 values for free and bound water, respectively, and a and b are the fractions of each form.

Water–clay systems have been investigated in detail by Woessner and co-workers (Woessner and Snowden, 1969a, b; Woessner, 1974, 1975, 1977). For Na-hectorite a linear relationship was observed between the deuterium splitting, B, and the clay concentration (Figure 4.25); proton splitting in this system was observed only for low water contents (< 0.2 cm^3/g dry clay) or low temperature. The splitting B is given by the equation

$$B = (3/8)\,B_0\,(e^2qQ/h)\,(3\cos^2\theta - 1),$$

where B_0 is the external magnetic field, e^2qQ/h is the quadrupole coupling constant and θ is the angle between B_0 and the symmetry axis of the electric field gradient.

The product of B and the clay/water content was found to be essentially constant (Woessner, 1975). This observation can be explained by rapid diffusion of water molecules, as a result of which all molecules experience the same average preferential orientation that exists only for the water layers closest to the surface. Similar relationships have been observed for other clays, both expanding and non-expanding, although in the latter the splittings must be normalized to unit surface area. As a consequence of these results, Woessner (1975) proposed that the preferential orientation of the water molecules is determined by the mere presence of a static interface and is independent of both the chemical and charge characteristics of the surface.

Measurements of deuterium T_1 as a function of water-layer thickness in NaFl hectorite also support the conclusion that only one or two layers of water molecules are influenced by the surface. In an experiment Woessner (1980) found that the rotational rate was about five times slower for molecules at the clay surface than in bulk water. In contrast, Ovcharenko

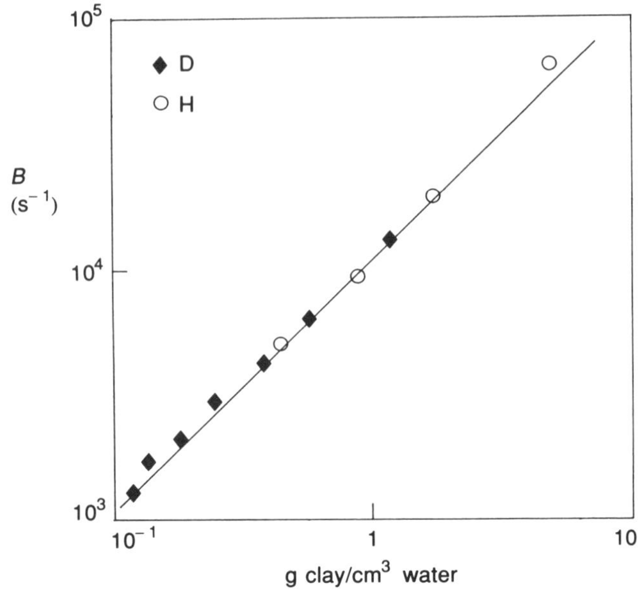

Figure 4.25 Relationship between deuterium and proton spectral splitting constants and clay concentration in clay–water mixtures for oriented Na-hectorite. (Reproduced from Woessner, 1975.)

et al. (1974) found that in kaolinite dispersions there was a break in the curve of proton spin–lattice relaxation versus clay concentration at about 65% kaolinite.

By examination of the NMR line-shapes and relaxation times of clay–water systems as a function of temperature, it appears that there is a proportion of the water that freezes at a lower temperature than in pure water (Annanyan, 1978). A model for ice nucleation has been proposed by Anderson (1967), in which a process is envisaged whereby the lifetimes and sizes of H-bonded domains increase progressively. By investigating the proton NMR spectra of Na montmorillonite with a range of water contents, Anderson and Morgenstern (1973) concluded that the water molecules at the clay–water interface had liquid-like properties, but were more structured than interlamellar water. An empirical relationship between temperature and thickness of the water layer has been derived by Banin and Anderson (1975) to describe the amount of unfrozen water associated with a mineral surface.

4.6.2 Determination of coordination of clay-bound water

In mineral systems with low water contents the states of water may be considered to be intermediate between those in ionic solutions and crystalline hydrates. Three major forms of water in expanding layer silicates can be envisaged: water in the hydration sphere of an interlayer cation; water not linked to a cation but still within the interlayer space; and water associated with external surfaces of the clay. The relative amounts of these forms of water are influenced by both the nature of the silicate layer and the interlayer cation. If rates of exchange between these different forms of water is slow and their relaxation times sufficiently different, it is possible to estimate their relative abundances from NMR spectra. In particular the nature and origin of any surface charge (i.e. as a result of tetrahedral or octahedral charge deficiences) may be expected to affect the properties of any interlayer water.

The application of NMR spectroscopy to the characterization of interlayer water in a layer silicate is illustrated in the work of Hougardy, Stone and Fripiat (1976, 1977) with the hydrates of vermiculites. Results obtained with oriented samples of a two-layer hydrate of a Na-vermiculite are illustrated in Figure 4.26. Because of its high cation-exchange capacity (200 meq./100 g) most of the water in this specimen was in the coordination sphere of the Na^+ cation. The NMR spectra showed a $(3\cos^2\theta - 1)$ dependence of the doublet splitting on the alignment of the specimen in the magnetic field, which is indicative of preferential orientation of water molecules around the sodium. An analysis of line shapes supported an octahedral arrangement in which there was rapid rotation about the C_2 and C_3 axes, as indicated in Figure 4.27. The situation was, however, more complex with the one-layer Na-hydrate and the two-layer Li-hydrate, where it was not possible to interpret the spectra in terms of dipolar doublets (Hougardy, Stone and Fripiat, 1977).

The one-layer state of Li-hectorite has been studied by Fripiat *et al.* (1980), who found that both ^1H and ^2D NMR spectra showed the usual $(3\cos^2\theta - 1)$ relationship for orientation of specimens in the magnetic field. Here, line-shape analysis suggested that there were several contributions, with possibly a dynamic pyramidal ordering of three water molecules around each Li and two different types of rotational motion that are roughly analogous to those of Na-vermiculite in Figure 4.27. For the one-layer Ba-hectorite a distinct central line was also observed (Kadi-Hanifi, 1980). The rapid diffusion of the monovalent cation is also an important factor in averaging the water structural framework, and a detailed study (Conrad, 1976a, b) of Li-hectorite with low water content (*c.* 5%) has concluded that the nearly flat trihydrated Li^+ ion is held by three hydrogen bonds to the six oxygens of the hexagonal cavity. In order to reconcile the known acidity of these systems with the proposed double rotation of the three water molecules, it was suggested that the protons were in fact tunnelling among the various possible sites.

In halloysite, where there is no charge balancing cation, the NMR spectra of one-layer hydrates showed no doublets for either H_2O or D_2O (Cruz, Letellier and Fripiat, 1978). Thus on the NMR time-scale

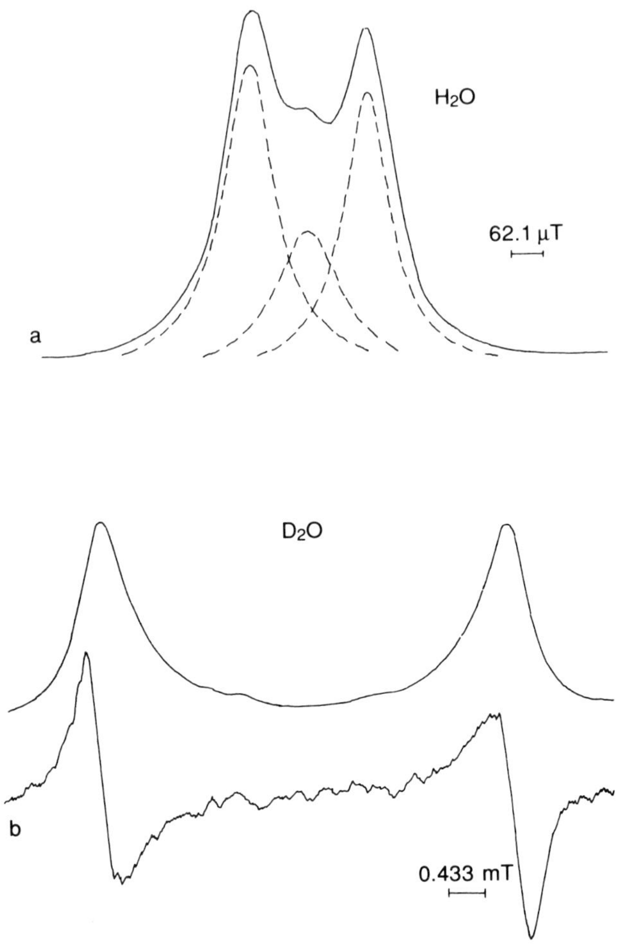

Figure 4.26 Nuclear magnetic resonance spectra for the two-layer hydrate of Na-Llano vermiculite at room temperature; (a) ^1H absorption spectrum, obtained with the sample oriented at 0° to the magnetic field, showing deconvolution into a doublet and singlet components, and (b) ^2D spectrum, obtained with the sample oriented at 90° to the magnetic field, displayed as absorption and first derivative. (Reproduced from Hougardy, Stone and Fripiat, 1976.)

there is no preferential orientation of the water molecules, which demonstrates that interactions with clay surfaces are less important than interactions with cations in determining water structure in the interlayer regions of clay minerals.

Several clay specimens exhibit a central peak between the dipolar doublets (e.g. Figure 4.26). In vermiculite this was seen in the ^1H spectrum but not in that from ^2D. This result supports the hypothesis that the singlet arises from a proton exchange process, which collapses the

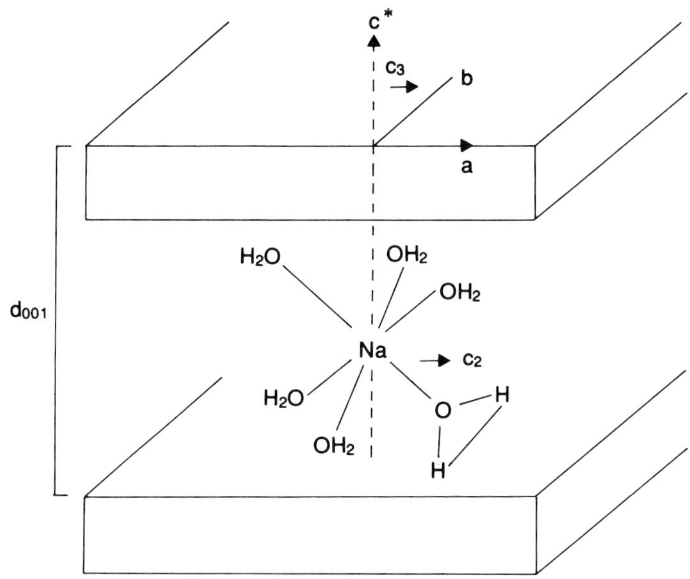

Figure 4.27 Arrangement of water molecules around the sodium ion in the two-layer hydrate of sodium vermiculite and the axes of molecular rotation. (Adapted from Hougardy, Stone and Fripiat, 1976.)

proton doublet in some of the water molecules, because exchange of deuterium nuclei does not affect the deuterium doublet. Although this explanation may be satisfactory for vermiculite, other factors may be important for other minerals, especially where there is disorganization of microcrystals or where micropores are present in the structure.

In summary, NMR is able to demonstrate that coordination to interlayer cations is of primary importance in the preferential organization of water molecules and that this order is dynamic in nature.

4.6.3 Determination of motional characteristics of absorbed water

Nuclear magnetic resonance spectroscopy can be used to determine the correlation times that characterize the various types of motion. This is done by measuring spin–lattice relaxation times, T_1, as a function of temperature. However, because of the complexities of clay–water systems, interpretation of these relaxation times is often not straightforward. Plots of T_1 as a function of temperature often show several minima (i.e. different motional mechanisms), a number of which are shallow and may simply be indicative of heterogeneities in the systems. In hectorite, Fripiat *et al.* (1980) were able to to distinguish between coordinated and non-coordinated water because of sizeable differences in their respective

T_1s. They found that the population of non-coordinated water increased with decreasing temperature (from 20% at room temperature to 50% at 200 K) and that the rate of exchange between the two types was slow on the NMR time-scale. In the two-layer hydrate of Na-vermiculite, Hougardy, Stone and Fripiat (1976) estimated that the rate of proton exchange between adjacent water molecules at room temperature was in the region of 10^{-4} to 10^{-5} s, which is some 10–100 times faster than in bulk water. Such fast exchange processes were also found by Woessner (1974) for a series of clay–water systems, where a dependence on the nature of the exchange cation was observed. Thus the enhanced proton exchange rates for water in clay minerals is associated with the polarizing effect of the interlayer cation and is probably connected with the higher degree of dissociation of interlayer water (Touillaux *et al.*, 1968).

4.6.4 Surface adsorbed species, intercalation complexes and their reactions

In a manner analogous to that discussed above for water it is possible to use NMR spectroscopy to determine the degree of mobility of other molecules associated with mineral surfaces. Thus ^{13}C NMR can be used to study the dynamics of a vast range of organic species. Furthermore, with oriented clays it is possible to deduce the surface orientations of adsorbed molecules from a combination of a knowledge of the chemical shifts of the types of carbon in the molecule with variation of the orientation of the clay surface with respect to the direction of the magnetic field.

Nuclear magnetic resonance spectroscopy is also a powerful technique for the study and characterization of intercalation complexes. For example, Thompson (1985) has reported that increased shielding by the organic molecule in organic intercalates of kaolinite causes the ^{29}Si resonance to be shifted to a higher field than in untreated kaolinite. A similar shift was seen by Sugahara *et al.* (1989) in intercalates of kaolinite with pyridine and monosubstituted pyridine derivatives. By combining CP/MAS NMR with IR and XRD, these authors were able to show that the pyridine molecules were aligned approximately perpendicular to the kaolinite sheet, with the nitrogen atom facing the gibbsitic layer.

Sugahara *et al.* (1990) have shown that the NMR technique is also very powerful for studying reactions involving intercalated molecules. In the preparation of kaolinite–polyacrylamide compounds they were able to follow the progress of the polymerization of acrylamide by monitoring the disappearance of the C=C bonds using ^{13}C CP/MAS NMR. The NMR and IR spectra also showed that hydrogen bonding and surface organic reactions changed after polymerization.

Nuclear magnetic resonance has important contributions to make towards understanding the synthesis of aluminium oxide cross-linked (pillared) clays and chemical reactions that can take place in them; such species are now becoming recognized as potentially important catalysts. Two examples of the use of NMR spectroscopy will be briefly mentioned here. Firstly, Endo *et al.* (1988) have reported that ^{27}Al NMR spectra of Al-pillared montmorillonites can be valuable in characterizing the chemical nature of the pillars that may be formed under different preparative conditions. They also showed that ^{1}H NMR could be used to study the polymerization of ethanol with such catalysts. Secondly, Carrado *et al.* (1990) have recently used solid-state ^{13}C NMR spectroscopy to investigate the reactivity of anisoles on pillared smectites, where O–methyl bond cleavage led to the formation of phenolic products. The spectra showed that extensive dealkylation and *trans*-alkylation reactions occured and that conversion was largely complete after two days. They also showed that the catalytic reaction was accompanied by a marked loss of molecular motion of the complexes in the clay micropores.

4.7 STRUCTURAL DISTRIBUTION OF PARAMAGNETIC IONS

By measuring the large dipolar couplings between unpaired electrons and magnetic nuclei it is possible to determine the relative arrangements of paramagnetic ions in a clay-mineral lattice. Interpretation of results is straightforward in homogeneous samples, but there are often complications in natural systems. Application of this approach is well illustrated in two papers on micas by Sanz and Stone (1977, 1979), from which the main findings are summarized below.

In phlogopites and biotites the ^{1}H broad-line NMR spectra consist of a main central peak that is flanked by smaller satellite peaks (Figure 4.28). These satellite peaks arise from OH groups that are coordinated to 1, 2, or 3 Fe^{2+} ions in octahedral structural sites and their positions vary with the orientation of the sample with respect to the magnetic field. There is, therefore, an increase in the number and intensity of these satellites with increasing Fe content of the structure and their positions can be analysed in terms of the macroscopic distribution of Fe within the specimens. Because in 1M micas the sheets are aligned in the same orientation as those above and below, Sanz and Stone were able to interpret the spectrum in Figure 4.28 as demonstrating that there were twice as many Fe^{2+} ions in M2 sites as there were in M1 sites, i.e. the Fe showed no preference for one type of site. However, such a conclusion may not be tenable because the existence of just a few misalignments of sheets would negate this approach. The biotite spectrum in Figure 4.28(b) shows peaks that arise from 0, 1, 2, or 3 Fe^{2+} ions in octahedral sites adjacent to the OH group. Satellite peaks associated with vacancies have also been observed and, in the case of biotites, it appears that there is a

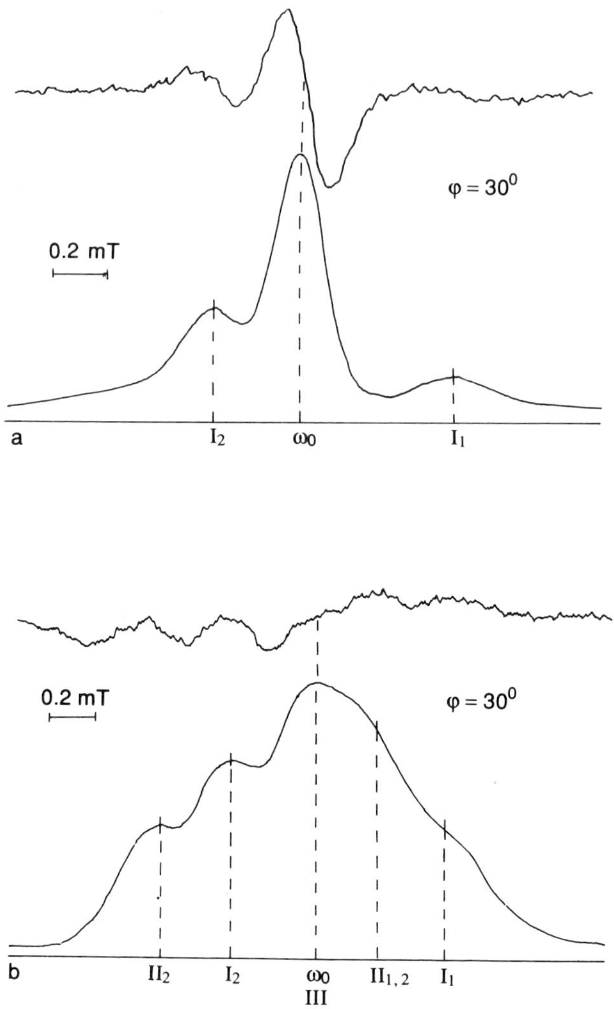

Figure 4.28 14 MHz ^{1}H NMR spectra, shown as absorptions and first derivatives, at room temperature of mica specimens at the same orientations about the b-axis; (a) phlogopite (2.6% Fe) and (b) biotite (13.8% Fe). The symbols I_1 and I_2 correspond to OH groups with one neighbouring Fe^{2+} ion in the M1 or M2 site, respectively; $II_{1,2}$ and II_2 correspond respectively to two neighbouring Fe^{2+} ions shared equally between M1 and M2 sites or located entirely on M2 sites; III corresponds to three neighbouring Fe^{2+} ions. (Reproduced from Sanz and Stone, 1977.)

tendency for Al cations to cluster around the OH groups. With samples having sufficient fluorine for OH substitutions to enable both ^{1}H and ^{19}F NMR spectra to be obtained, it appears that there is a tendency for Fe^{2+} to cluster around OH groups (Stone, 1981). Fe^{2+} avoids direct coordination to F, because only peaks corresponding to second and third neighbours are seen. There is also a tendency for the F^- ions to associate with one another (Sanz and Stone, 1979), a result that has been interpreted as indicating the presence of homogeneous F-rich domains from which Fe^{2+} ions are excluded.

In powdered glauconites, which are dioctahedral Fe^{3+}-rich micas, the ^1H NMR spectra consist of asymmetric signals, which can be decomposed into two peaks of approximately equal intensity (Kohler and Burkert, 1976). The separation between the two peaks increased with increasing Fe^{3+} content of the mineral. However, it was considered that the narrower peak corresponded to structural OH groups, whereas the broader one was assigned to either Fe-containing impurities or to interlayered hydroxy iron complexes.

4.8 CONCLUSIONS

The full potential of NMR spectroscopy for clay-mineral characterization has only been fully appreciated in recent years with the ready availability of high-field spectrometers. These have led to the development of mineralogical applications of the high-resolution and high-sensitivity techniques that have been responsible for the major advances in the characterization of chemical and biochemical molecules. The technical advances in NMR spectroscopy in recent years have been so profound that it is now an essential technique for any analytical laboratory that is seriously involved in the characterization of the chemical properties of any type of organic or inorganic material. The costs of such equipment remain appreciable (*c.* £400 000 for a spectrometer with an 11.7 T magnet, i.e. operating at 500 MHz for protons), despite the growing international market and competition between instrument manufacturers. Lower-field systems are much less expensive, because the costs of superconducting magnets are strongly field dependent and account for a sizeable proportion of the total cost of a high-field instrument.

Nuclear magnetic resonance spectroscopy is in essence three techniques rolled into one. Conventional solution procedures can be used to probe the identity and physical properties of any liquid phase molecules associated with clay minerals. Both high- and low-field spectrometers have been used effectively, the former for detailed chemical characterization of organic molecules, which are weakly adsorbed on to clay surfaces or intercalated into interlayer spaces. Low-field systems may, however, be adequate to probe the physical properties of water or other molecules that may be associated with clays in appreciable quantities. Under these conditions any solid or immobilized components do not contribute to the spectra.

The structural components of clay minerals can be investigated by using solid-state techniques, which involve either rapid spinning of the sample or complex pulse sequences, or both. For this type of work it is usually beneficial to use as large a magnetic field as possible. This serves two purposes; it increases the sensitivity and resolution, which is of particular value for nuclei of low natural abundance, such as ^{29}Si, and it decreases the influence of electric field gradients on the shapes of spectra from quadrupolar nuclei, which is of great value for understanding the spectral details of nuclei such as ^{27}Al, ^{23}Na and ^{17}O. Paramagnetic ions produce a broadening of the basic spectral lines as a result of large dipolar interactions between the nuclear and electronic magnetic moments. These are generally too great to collapse completely when MAS techniques are used and result in the presence of spinning sidebands in the spectra. In such circumstances the spectral contributions from those nuclei that are directly associated with the unpaired electron are often not seen, because of the large shifts involved. For this reason it is generally not possible to observe MAS NMR spectra from minerals that contain more than a small percentage of iron randomly distributed in their structures.

The distribution of paramagnetic ions in mineral structures can be investigated by using broadband techniques, which operate at low fields usually by using electromagnets to achieve a wide fieldspan. Such measurements are now performed only rarely, largely because the appropriate instrumentation is generally considered to be obsolete and is no longer readily available.

ACKNOWLEDGEMENTS

Funding for part of this work was provided by the Scottish Office Agriculture and Fisheries Department. Figures 4.8, 4.11, 4.15, 4.19, 4.20, 4.21, 4.25, 4.26, 4.27 and 4.28 are reproduced with permission from the copyright holders:

4.8 Kirkpatrick (1988) Mineralogical Soc. Am.

4.11 Bruker Analytische Messtechnik Gmbh.

4.15 Bruker Analytische Messtechnik Gmbh.

4.19–4.21 Phillips, Kirkpatrick and Hovis (1988) *Physics and Chemistry of Clay Minerals*, **16**, 262–275.

4.25 Woessner (1975) Plenum, New York.

4.26 and 4.27 Hougardy, Stone and Fripiat (1976) *Journal of Chemical Physics*, **64**, 3840–3851.

4.28 Sanz and Stone (1977) *Journal of Chemical Physics*, **67**, 3739–3743.

REFERENCES

Anderson, D. M. (1967) The interface between ice and silicate surfaces. *Journal of Colloid and Interface Science*, **25**, 174–191.

Anderson, D. M. and Morgenstern, N. R. (1973) Physics, chemistry and mechanics of frozen ground: a review, in *Permafrost. Second International Conference*. National Academy of Sciences, Washington, DC, pp. 257–288.

Annanyan, A. A. (1978) Study of non-freezing water in clays. *Colloid Journal of the USSR*, **40**, 1165–1168.

Banin, A. and Anderson, D. M. (1975) A similar law may govern water freezing in minerals and living organisms. *Nature*, **255**, 261–262.

Barron, P. F., Slade, P. and Frost, R. L. (1985) Ordering of aluminium in tetrahedral sites in mixed-layer 2:1 phyllosilicates by solid-state high-resolution NMR. *Journal of Physical Chemistry*, **89**, 3880–3885.

Carrado, K. A., Hayatsu, R., Botto, R. E. and Winans, R. E. (1990) Reactivity of anisoles on clay and pillared clay surfaces. *Clays and Clay Minerals*, **38**, 250–256.

Conrad, J. (1976a) Structure of H_2O and hydrogen bonding on clays studied by Li^7 and H^1 NMR, in *Magnetic Resonance in Colloid and Interface Science* (eds H. A. Resing and C. G. Wade). *American Chemical Society Symposium Series*, **34**, 85–93.

Conrad, J. (1976b) Etude structurale de l'eau adsorbée sur l'hectorite Li par RMN, in *Proceedings, International Clay Conference 1975* (ed. S. W. Bailey). Appl. Publ. Ltd., pp. 221–230.

Cradwick, P. D. G., Farmer, V. C., Russell, J. D. *et al.* (1972) Imogolite, a hydrated aluminium silicate of tubular structure. *Nature (London), Physical Science*, **240**, 187–189.

Cruz, M. I., Letellier, M. and Fripiat, J. J. (1978) NMR study of adsorbed water. II. Molecular motions in the monolayer hydrate of halloysite. *Journal of Chemical Physics*, **69**, 2018–2027.

Endo, T., Ikawa, N., Sato, T. and Shimada, M. (1988) The synthesis of aluminium oxide cross-linked montmorillonite and its catalytic properties. *Clay Science*, **7**, 59–72.

Fripiat, J. J., Kadi-Hanifi, M., Conrad, J. and Stone, W. E. E. (1980) NMR study of adsorbed water. III. Molecular orientation and protonic motions in the one-layer of a Li-hectorite, in *Magnetic Resonance in Colloid and Interface Science* (eds J. P. Fraissard and H. A. Resing). D. Reidel, Dordrecht, pp. 529–535.

Fyfe, C. A. (1983) *Solid State NMR for Chemists*. C.F.C. Press, Guelph, Ontario.

Gerstein, B. C. (1981) High resolution n.m.r. in solids with strong homonuclear dipolar broadening: combined multiple-pulse decoupling and magic angle spinning. *Philosophical Transactions of the Royal Society of London*, Series A, **299**, 521–546.

Goodman, B. A. and Stucki, J. W. (1984) The use of nuclear magnetic resonance (NMR) for the determination of tetrahedral aluminium in montmorillonite. *Clay Minerals*, **19**, 663–667.

Goodman, B. A. Russell, J. D., Montez, B. *et al.* (1985) Structural studies of imogolite and allophanes by aluminium-27 and silicon-29 nuclear magnetic resonance spectroscopy. *Physics and Chemistry of Clay Minerals*, **12**, 342–346.

Haeberlen, U. and Waugh, J. S. (1968) Coherent averaging effects in magnetic resonance. *Physical Review*, **175**, 453–467.

Herrero, C. P., Sanz, J. and Seratosa, J. M. (1985a) Si, Al distribution in micas: analysis by high resolution ^{29}Si NMR spectroscopy. *Journal of Physics C: Solid State Physics*, **18**, 13–22.

Herrero, C. P., Sanz, J. and Seratosa, J. M. (1985b) Tetrahedral cation ordering in layer silicates by ^{29}Si NMR spectroscopy. *Solid State Communications*, **53**, 151–154.

Hougardy, J., Stone, W. E. E. and Fripiat, J. J. (1976) NMR study of adsorbed water. I. Molecular orientation and protonic motions in the two-layer hydrate of a Na vermiculite. *Journal of Chemical Physics*, **64**, 3840–3851.

Hougardy, J., Stone, W. E. E. and Fripiat, J. J. (1977) Complex proton NMR spectra in some ordered hydrates of vermiculites. *Journal of Magnetic Resonance*, **25**, 563–567.

James, N. and Oldfield, E. (1985) Prediction of silicon-29 nuclear magnetic resonance chemical shifts using a group electro-negativity approach: applications to silicate and aluminosilicate structures. *Journal of the American Chemical Society*, **107**, 6769–6775.

Kadi-Hanifi, M. (1980) Proton NMR studies of one layer hydrates of oriented hectorite. *Clays and Clay Minerals*, **28**, 65–66.

Kinsey, R. A., Kirkpatrick, R. J., Hower, J. *et al.* (1985) High resolution aluminium-27 and silicon-29 nuclear magnetic resonance study of layer silicates, including clay minerals. *American Mineralogist*, **70**, 537–548.

Kirkpatrick, R. J. (1988) MAS NMR spectroscopy of minerals and glasses, in *Spectroscopic Methods in Mineralogy and Geology* (ed. F. C. Hawthorne). Mineralogical Society America, Washington, DC, pp. 341–403.

Kirkpatrick, R. J., Smith, K. A., Schramm, S. *et al.* (1985) Solid state nuclear magnetic resonance spectroscopy of minerals. *Annual Review of Earth and Planetary Sciences*, **13**, 29–47.

Kirkpatrick, R. J., Carpenter, M. A., Yang, W.-H. and Montez, B. (1987) ^{29}Si magic-angle NMR spectroscopy of low-temperature ordered plagioclase feldspar. *Nature*, **325**, 236–238.

Klinowski, J., Ramdas, S., Thomas, J. M. *et al.* (1982) A re-examination of Si, Al ordering in zeolites Na_X and Na_Y. *Journal of the Chemical Society, Faraday Transactions*, **278**, 1025–1050.

Kodama, H., Kotlyar, L. S. and Ripmeester, J. A. (1989) Quantification of crystalline and noncrystalline material in ground kaolinite by X-ray powder diffraction, infrared, solid-state nuclear magnetic resonance, and chemical-dissolution analyses. *Clays and Clay Minerals*, **37**, 364–370.

Kohler, E. E. and Burkert, P. K. (1976) Kernmagnetische resonanzmessungen an glaukoniten. *Clay Minerals*, **11**, 303–310.

Lippmaa, E., Magi, M., Samoson, A. *et al.* (1980) Structural studies of silicates by solid-state high-resolution ^{29}Si NMR. *Journal of the American Chemical Society*, **102**, 4889–4893.

Lippmaa, E., Magi, M., Samoson, A. *et al.* (1981) Investigation of the structure of zeolites by solid-state high-resolution ^{29}Si NMR spectroscopy. *Journal of the American Chemical Society*, **103**, 4992–4996.

Lowenstein, W. (1954) The distribution of aluminium in the tetrahedra of silicates and aluminates. *American Mineralogist*, **39**, 92–96.

MacKenzie, K. J. D., Bowden, M. E. and Meinhold, R. H. (1991) The structure and thermal transformations of allophanes studied by ^{29}Si and ^{27}Al high-resolution solid-state NMR. *Clays and Clay Minerals*, **39**, 337–346.

MacKenzie, K. J. D., Brown, I. W. N., Meinhold, R. H. and Bowden, M. E. (1985) Outstanding problems in the kaolin–mullite reaction sequence investigated by ^{29}Si and ^{27}Al solid-state nuclear magnetic resonance: I. Metakaolinite. *Journal of the American Ceramic Society*, **68**, 293–297.

MacKenzie, K. J. D., Bowden, M. E., Brown, I. W. M. and Meinhold, R. H. (1989) Structure and thermal transformations of imogolite studied by ^{29}Si and ^{27}Al high-resolution solid-state nuclear magnetic resonance. *Clays and Clay Minerals*, **37**, 317–324.

Magi, M., Lippmaa, E., Samoson, A., *et al.* (1984) Solid state high-resolution silicon-29 chemical shifts in silicates. *Journal of Physical Chemistry*, **88**, 1518–1522.

Mansfield, P., Orchard, M. J., Stalker, D. C. and Richards, K. H. B. (1973) Symmetrized multipulse nuclear magnetic resonance experiments in solids.

Measurement of the chemical shift shielding tensor in some compounds. *Physical Review*, **B7**, 90–105.

Müller, D., Gessner, W., Behrens, H. J. and Scheler, G. (1981) Determination of the aluminium coordination in aluminium–oxygen compounds by solid-state high-resolution ^{27}Al NMR. *Chemical Physics Letters*, **79**, 59–62.

Ovcharenko, F. D., Brekhunets, A. G., Mank, V. V. and Pereverzeva, I. N. (1974) Relaxation and filtration characteristics of aqueous dispersions of kaolinite. *Colloid Journal of the USSR*, **36**, 1177–1179.

Phillips, B. L., Kirkpatrick, R. J. and Hovis, G. L. (1988) ^{27}Al, ^{29}Si and ^{23}Na MAS NMR study of an aluminium silicon ordered alkali feldspar solid solution series. *Physics and Chemistry of Minerals*, **16**, 262–275.

Ramdas, S. and Klinowski, J. (1984) A simple correlation between isotropic ^{29}Si NMR chemical shifts and T–O–T angles in zeolite frameworks. *Nature*, **308**, 521–523.

Rothwell, W. P., Waugh, J. S. and Yesinowski, J. P. (1980) High resolution variable temperature ^{31}P NMR of solid calcium phosphates. *Journal of the American Chemical Society*, **102**, 2637–2643.

Samoson, A. and Lippmaa, E. (1983) Excitation phenomena and line intensities in high resolution NMR powder spectra of half-integer quadrupolar nuclei. *Physical Review*, **B28**, 6567–6570.

Sanz, J. and Serratosa, J. M. (1984) Distinction of tetrahedrally and octahedrally coordinated Al in phyllosilicates by NMR spectroscopy. *Clay Minerals*, **19**, 113–115.

Sanz, J. and Stone, W. E. E. (1977) NMR study of micas. I. Distribution of Fe^{2+} ions on the octahedral sites. *Journal of Chemical Physics*, **67**, 3739–3743.

Sanz, J. and Stone, W. E. E. (1979) NMR study of micas. II. Distribution of Fe^{2+}, F^- and OH^- in the octahedral sheet of phlogopites. *American Mineralogist*, **64**, 119–126.

Smith, K. A., Kirkpatrick, R. J., Oldfield, E. and Henderson, D. M. (1983) High resolution silicon-29 nuclear magnetic resonance spectroscopic study of rock-forming silicates. *American Mineralogist*, **68**, 1206–1215.

Stone, W. E. E. (1981) The use of NMR in the study of clay minerals, in *Advanced Techniques for Clay Mineral Analysis* (ed. J. J. Fripiat). Elsevier, Amsterdam, pp. 77–112.

Sugahara, Y., Satokawa, S., Yoshioka, K.-I., *et al.* (1989) Kaolinite–pyridine intercalation compound derived from hydrated kaolinite. *Clays and Clay Minerals*, **37**, 143–150.

Sugahara, Y., Satokawa, S., Kuroda, K. and Kato, C. (1990) Preparation of a kaolinite–polyacrylamide intercalation compound. *Clays and Clay Minerals*, **38**, 137–143.

Thompson, J. G. (1985) Interpretation of solid state ^{13}C and ^{29}Si nuclear magnetic resonance of kaolinite intercalates. *Clays and Clay Minerals*, **33**, 173–180.

Timken, H. K. C., Schramm, S. E., Kirkpatrick, R. J. and Oldfield, E. (1987) Solid state oxygen-17 nuclear magnetic resonance spectroscopic studies of alkaline earth metasilicates. *Journal of Physical Chemistry*, **91**, 1054–1058.

Tossell, J. A. (1984) Correlation of ^{29}Si nuclear magnetic resonance chemical shifts in silicates with orbital energy differences obtained from X-ray spectra. *Physics and Chemistry of Minerals*, **10**, 137–141.

Touillaux, R., Salvador, P., Van der Meersche, C. and Fripiat, J. J. (1968) Study of H_2O layers adsorbed on Na and Ca montmorillonite by NMR. *Israel Journal of Chemistry*, **6**, 337–348.

Turner, G. L., Smith, K. A., Kirkpatrick, R. J. and Oldfield, E. (1986) Structure and cation effects on phosphorus-31 NMR chemical shifts and chemical shift anisotropies of orthophosphates. *Journal of Magnetic Resonance*, **70**, 408–415.

Watanabe, T., Shimizu, H., Nagasawa, K., *et al.* (1987) ^{29}Si- and ^{27}Al-MAS/NMR study of the thermal transformations of kaolinite. *Clay Minerals*, **22**, 37–48.
Waugh, J. S., Huber, L. M. and Haeberlen U. (1968) Approach to high resolution NMR in solids. *Physics Review Letters*, **20**, 180–182.
Wilson, M. A., Wada, K., Wada, S. I. and Kakuto, Y. (1988) Thermal transformations of synthetic allophane and imogolite as revealed by nuclear magnetic resonance. *Clay Minerals*, **23**, 175–190.
Woessner, D. E. (1974) Proton exchange effects on pulsed NMR signals from preferentially oriented H_2O molecules. *Journal of Magnetic Resonance*, **16**, 483–501.
Woessner, D. E. (1975) NMR studies of preferentially oriented water at interfaces, in *Mass Spectrometry and NMR Spectroscopy in Pesticide Chemistry* (eds R. Hague and F. J. Biros). Plenum, New York, pp. 279–304.
Woessner, D. E. (1977) Nuclear magnetic relaxation and structure in aqueous heterogeneous systems. *Molecular Physics*, **34**, 899–920.
Woessner, D. E. (1980) An NMR investigation into the range of the surface effect on the rotation of water molecules. *Journal of Magnetic Resonance*, **39**, 297–308.
Woessner, D. E. and Snowden, B. S. (1969a) NMR doublet splitting in aqueous montmorillonite gels. *Journal of Chemical Physics*, **50**, 1516–1523.
Woessner, D. E. and Snowden, B. S. (1969b) A study of the orientation of adsorbed water molecules on montmorillonite clays by pulsed NMR. *Journal of Colloid and Interface Science*, **30**, 54–68.

CHAPTER 5

Electron paramagnetic resonance spectroscopy

B. A. Goodman and P. L. Hall

5.1 INTRODUCTION TO ELECTRON PARAMAGNETIC RESONANCE

5.1.1 Background

Electron paramagnetic resonance (EPR) spectroscopy is concerned with the detection of unpaired electrons and with characterizing their chemical environments. It is able to provide information on materials in both liquid and solid phases. The technique is extremely sensitive and under favourable circumstances the detection limit for paramagnetic centres can be as low as 10^{11}–10^{12} spins/g, i.e. well below 1 part per billion (Wertz and Bolton, 1986). Spectral resolution is lost, however, when adjacent paramagnetic centres are close enough for there to be a significant dipolar interaction between them, making this a technique primarily for the characterization of magnetically dilute specimens.

Electron paramagnetic resonance spectroscopy is not a technique that can be used for clay-mineral identification, but it does have the ability to characterize in clays the minor structural or surface components that contain unpaired electrons and which may be of importance in understanding specific properties of the minerals. It is a powerful technique for identifying the various types of association of paramagnetic metal ions with clay minerals, on the basal surfaces, within the interlamellar spaces, or as part of the aluminosilicate structures. Such information may be of vital importance in understanding clay chemistry from both academic research and industrial points of view.

In most experimental set-ups EPR spectroscopy involves measurement of the magnetic fields at which paramagnetic molecules come into

Clay Mineralogy: Spectroscopic and Chemical Determinative Methods.
Edited by M. J. Wilson. Published in 1994 by Chapman & Hall, London.
ISBN 0 412 53380 4.

resonance with applied monochromatic microwave radiation. Much rarer is the technique of FT EPR (analogous to FT NMR; Chapter 4) in which excitation of electrons is achieved with a pulse of microwave energy in the presence of a constant magnetic field.

Free electrons (in the absence of a crystal or magnetic field) are aligned at random, but in the presence of an external magnetic field they can take up $2S+1$ (where S is the net spin) different orientations, distinguished by the quantum number m_S. The relative energies of these two states are proportional to the applied magnetic field B and are given by

$$Em_s = g\mu_B B m_s,$$

where μ_B is the Bohr magneton (9.2731×10^{-24} J/T). The splitting of the energies of the electron spin states is known as the Zeeman splitting and the constant of proportionality, g, is known as the spectroscopic splitting factor, or just the g-value. It has a value of 2.0023 for a free electron, but this may vary when the electron is bound in a molecule or mineral structure (see later).

Electrons in the lower energy state, $m_S = -\frac{1}{2}$, can be excited to the upper state, $m_S = +\frac{1}{2}$ (Figure 5.1) by electromagnetic radiation (microwaves) of the appropriate frequency to satisfy the resonance condition,

$$h\nu = g\mu_B B,$$

where ν is the frequency of the applied radiation (Hz) and h is Planck's constant. When this condition is satisfied there is strong coupling be-

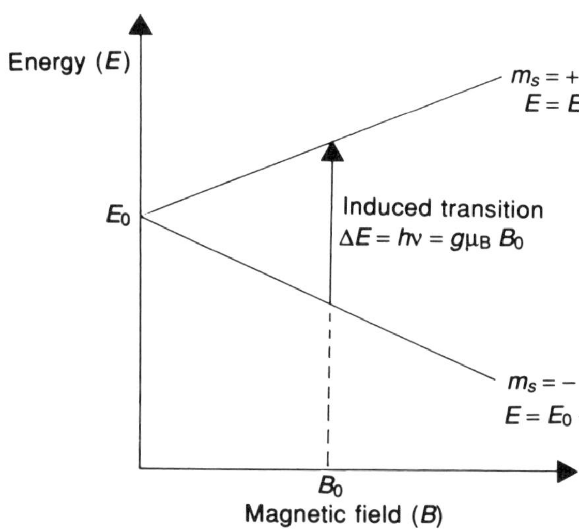

Figure 5.1 Diagrammatic representation of the variations in the energy levels of an electron in the presence of an external magnetic field and the EPR transition induced by electromagnetic radiation.

tween the electron spin and the radiation and strong absorption occurs as the electron spin makes the transition between the two states.

The above condition implies that the energy (and frequency) at which resonance occurs is proportional to the applied magnetic field B. At magnetic fields typical of those attainable by large electromagnets, the energy of the electron spin transition corresponds to a frequency of *c.* 10 GHz and a wavelength of a few centimetres, which means that it is located in the microwave region of the spectrum. Most EPR work is carried out at *c.* 9 GHz, which is known as the X-band, but there are occasions when experiments are performed at other microwave frequencies, such as L-band (*c.* 1 GHz), S-band (*c.* 4 GHz), K-band (*c.* 24 GHz) and Q-band (*c.* 35 GHz) frequencies. For comparison, the energies associated with the spin transitions of atomic nuclei are about 1000 times smaller for a given magnetic field value, and correspond to radiofrequencies in the MHz range (Chapter 4).

In general the g-value varies according to the orientation of the orbital containing the unpaired electron relative to the direction of the external magnetic field. Three characteristic values, g_x, g_y and g_z are obtained and these correspond to the values along the principal symmetry axes of the orbital. (Note that for axial symmetry $g_x = g_y$ and is designated $g_\perp$ and g_z is designated $g_{\parallel}$.)

5.1.2 Spectral acquisition and display

Most commercial spectrometers (Figure 5.2) operate with monochromatic radiation, and electron paramagnetic resonance (EPR) spectroscopy involves observation of the magnetic fields at which paramagnetic molecules come into resonance with this radiation. Experimentally, this is performed by placing the substance containing unpaired electrons in an absorption cell in the form of a resonance cavity, which itself is between the poles of the magnet. The sample is adjusted within the cavity in such a way that the microwave magnetic field is at its maximum and the electric field is at its minimum. The microwave energy is modulated and the absorption is measured by detector coils, amplified and ultimately displayed on a visual display unit (VDU) or chart recorder, usually as the first derivative of the absorption.

Care must be taken not to saturate the system, i.e. raise too many electrons to the upper level so that there are as many there as in the lower level. From the Boltzmann function, the relative populations of the upper, n^+, and lower, n^-, energy levels at equilibrium is given by

$$n^+/n^- = \exp(-h\nu/kT) = 1 - g\mu_B B/kT,$$

where k is the Boltzmann constant and T the absolute temperature. Thus the rate of excitation from the lower to the upper level must not exceed the rate of relaxation back to the ground state. As with NMR these relaxation

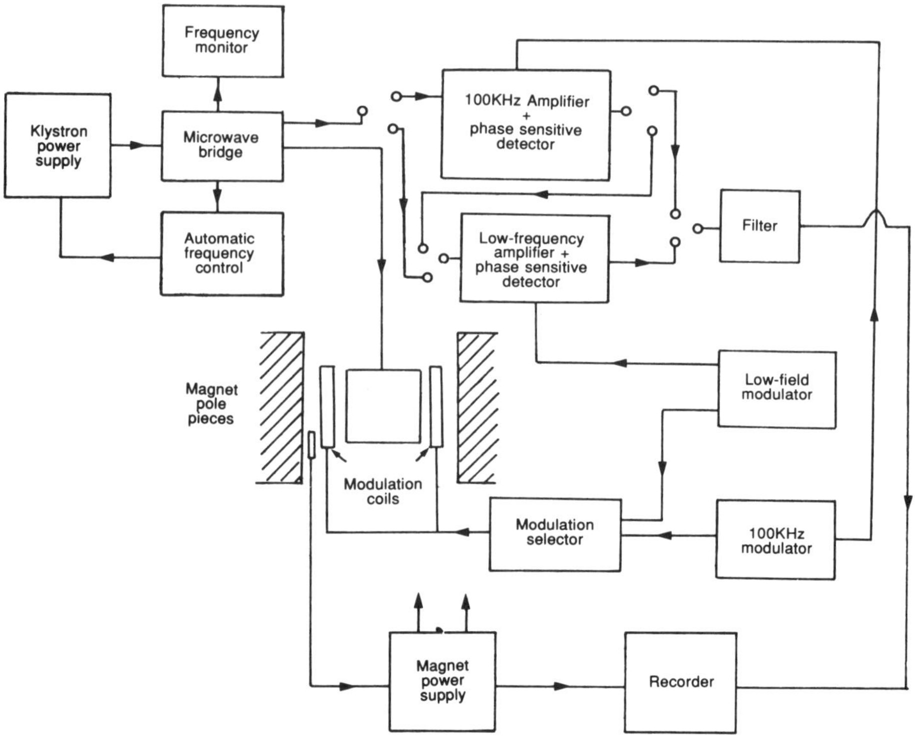

Figure 5.2 Block diagram illustrating the major components of an EPR spectrometer.

processes involve either the vibrational modes of the lattice (spin–lattice) or neighbouring spins (spin–spin). Spin–lattice relaxation is temperature dependent and is reduced by lowering the temperature. Consequently spectra saturate more readily at low temperatures. Spin–spin relaxation is temperature-independent and is controlled simply by the separations between the spins. It can be reduced if the sample can be made more magnetically 'dilute', i.e. by decreasing its paramagnetic ion content.

5.1.3 Spectral intensity

The intensity, or area beneath an EPR absorption curve, is proportional to the number of unpaired electron spins in the sample. Experimentally it is often difficult to measure these intensities in samples where there is a large spread of resonances from more than one type of paramagnetic species.

For instrumental purposes an EPR spectrum is usually recorded as the first derivative of the absorption spectrum instead of the absorption itself. This means that the point where the derivative crosses the baseline corresponds to the peak of the absorption. Improved resolution is obtained compared with the absorption, but in poorly resolved spectra it is often helpful to record second derivative spectra, the peaks of which are at the same positions as the original absorption peaks (Figure 5.3). Also, with higher derivative recording the spectral contributions from narrow peaks are accentuated relative to those from broader components, and this phenomenon increases as the order of the derivative increases. Thus it is possible to remove the spectral contributions from broad components by recording high-derivative spectra.

Numerical techniques for spectral deconvolution have been described in Chapter 1 of this book. One word of caution concerning their use for the improvement of spectral signal-to-noise ratios is that any residual noise develops a shape that resembles a genuine signal and it is possible to produce spectral artefacts if procedures are used uncritically.

As mentioned in the previous section, *g*-values may be anisotropic, i.e. they vary with the orientation of the molecular axes relative to the external magnetic field. In fluid solutions rapid molecular motions result in an averaged isotropic spectrum, but in the solid state each orientation

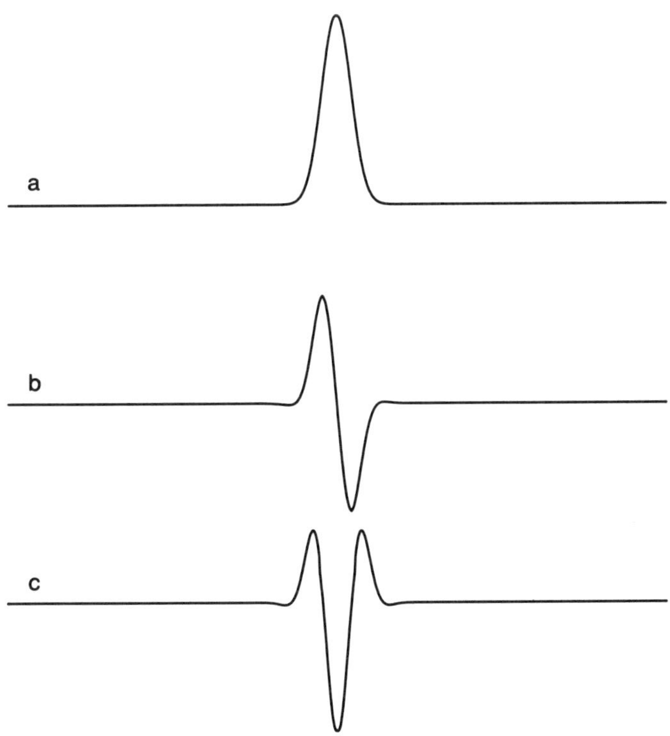

Figure 5.3 A Lorentzian line-shape displayed as (a) absorption, (b) first derivative and (c) second derivative.

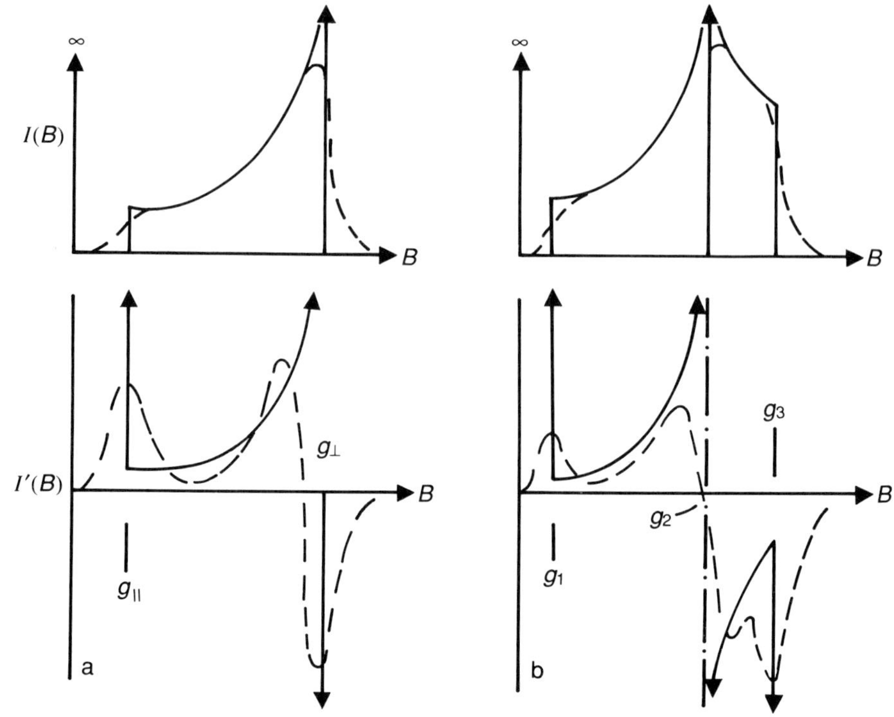

Figure 5.4 Spectral patterns for a paramagnetic ion in a polycrystalline specimen with (a) an axially symmetric and (b) a rhombic g-tensor, shown as absorption (upper curves) and first derivative (lower curves) line-shapes. Solid curves indicate the idealized line-shapes and the dashed curves are typical experimental spectral patterns.

produces its own spectrum. For a randomly oriented powder the weighted sum from these various orientations results in an envelope with features corresponding to the extreme resonance positions. Such powder envelopes are illustrated in Figure 5.4 for ions both with and without axial symmetry.

5.2 ELECTRONS IN TRANSITION METAL IONS

In clay-minerals research many applications of EPR spectroscopy involve transition metal ions, either substituted in the structures or adsorbed on surfaces. These ions can have up to five unpaired electrons, which can interact with one another and any associated nuclei that have non-zero spin, with the result that there may be several different magnetic interactions. The following sections will, therefore, introduce some basic properties of transition metal ions that are essential for understanding their EPR spectra.

5.2.1 Electronic configurations

The major paramagnetic ions of the first-row transition metals, together with their numbers of d electrons, are listed in Table 5.1. Partly filled electronic shells have a net orbital angular momentum which results from the combined orbital angular momenta of the individual electrons. In a filled shell the net orbital momentum is zero. The nomenclature used to describe the net orbital angular momentum of an electronic configuration is similar to that used for orbitals, except that a capital letter is used. Thus, net orbital angular momenta, L, of 0, 1, 2, 3, etc. are labelled S, P, D, F, etc. in an analogous way to the labelling of the azimuthal quantum number for electronic orbitals (s, p, d, f). Also the net spin, S, of a group of electrons can take any integral or half-integral value, as the spin of a single electron is $\frac{1}{2}$. The spin multiplicity, which is equal to $2S+1$, is often used to describe the spin state of an ion. Thus for a single electron $S=\frac{1}{2}$ and the spin multiplicity is equal to 2, and we call this a doublet state. The combined spin multiplicity and orbital angular momentum is written as ^{2S+1}L, e.g. the ^6S state of Mn^{2+} and Fe^{3+} represents a net spin angular momentum $S=\frac{5}{2}$ and orbital angular momentum $L=0$.

5.2.2 Spin–orbit coupling

As stated in the previous section the angular momentum of an electron in an atom is made up of two components, the spin and the orbital

Table 5.1 Electronic configurations for the main paramagnetic ions of 1st row transition metals

Ion	*Number of d electrons*	*Ground state quantum number* S	L	*Spectroscopic symbol*
V^{4+} (VO^{2+})	1	$\frac{1}{2}$	2	^2D
V^{3+}	2	1	3	^3F
Cr^{3+}	3	$\frac{3}{2}$	3	^4F
Mn^{2+}, Fe^{3+}	5	$\frac{5}{2}$	0	^6S
Fe^{2+}	6	2	2	^5D
Co^{2+}	7	$\frac{3}{2}$	3	^4F
Ni^{2+}	8	1	3	^3F
Cu^{2+}	9	$\frac{1}{2}$	2	^2D

angular momenta. If in an atom there are two electrons, 1 and 2, with spin angular momenta, s_1 and s_2, then there is the possibility of the two spin components combining separately (s_1s_2), as well as the spin of one electron combining with the orbital component of the same electron (s_1l_1) or of the other (s_1l_2). For transition metal ions, the magnitudes of the interactions are in the order

$$s_1s_2 >> l_1l_2 >> s_1l_1 = s_2l_2 >> s_1l_2 = s_2l_1.$$

The energy due to spin–orbit coupling in first-row transition metal ions is relatively small compared with the energy of electron interactions. In second- and third-row transition ions, spin–orbit coupling is more important, but does not become greater in magnitude than spin–spin or orbital–orbital interactions.

Spin–orbit coupling provides a mechanism by which a contribution from the orbital angular momentum can be added to (or mixes with) the spin angular momentum, thus changing any magnetic parameter arising from pure spin angular momentum (spin-only case). It is also the mechanism whereby energy levels of different symmetries may mix.

5.2.3 Basic crystal field theory

In a free ion the energies of all of the 3d orbitals are the same, but in a crystal field this degeneracy is reduced or lost completely. In a cubic crystal field the d_{xy}, d_{xz} and d_{yz} (e, e_g)orbitals have different energies to the $d_{x^2-y^2}$

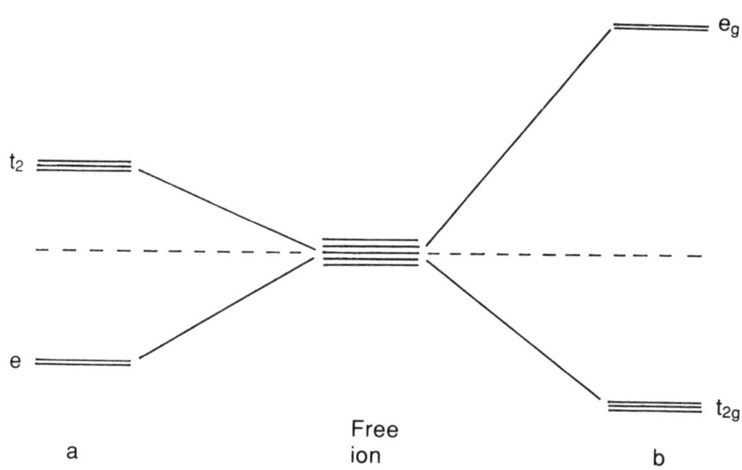

Figure 5.5 Splitting of the energy levels of d-orbitals in (a) tetrahedral and (b) octahedral crystal fields.

and d_{z^2} (t_2, t_{2g}) orbitals, the former being of lower energy for octahedral and higher energy for tetrahedral symmetries (Figure 5.5). Very few ions, however, have chemical environments with perfect octahedral or tetrahedral symmetry, and Jahn–Teller or crystal field distortions are frequent.

Distortions from cubic symmetry are usually tetragonal, trigonal or rhombic. Tetragonal distortions arise from alternation of the charge distribution along the *z*-axis, maintaining a fourfold axis. Trigonal distortions maintain the threefold axis, but rhombic distortions may be envisaged as destroying the fourfold axis of tetragonal symmetry by altering the charge distribution along the *x*- or the *y*-axis. This is illustrated in Figure 5.6.

In low-symmetry molecules, the electronic energy levels are usually singly degenerate and consequently are not split by spin–orbit coupling. Thus *g*-values are close to 2. However, deviations of *g*-values from 2 (Δg)

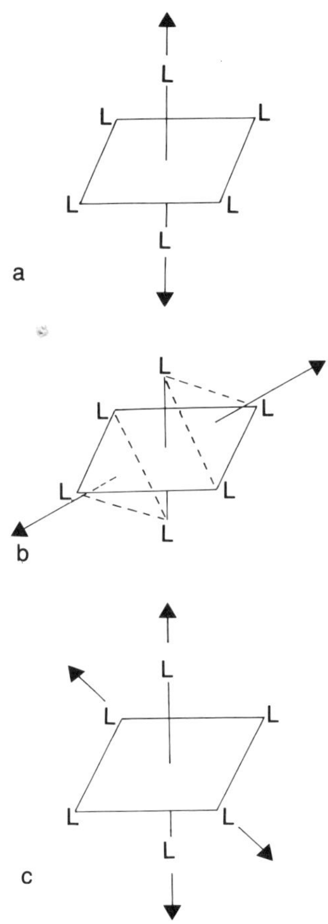

Figure 5.6 Illustrations of the distortions to an octahedron with ligands, L, that generate (a) tetragonal, (b) trigonal and (c) rhombic symmetries.

are produced by a mixing of the ground state orbitals with electronic excited states, the magnitude of Δg being inversely proportional to the energy separations between those states.

For the case of a single electron in a non-degenerate d-orbital, the g-value along any direction x, y or z is given by the expression:

$$g = g_0 - \frac{n\lambda}{\Delta},$$

where $g_0 = 2.0023$, λ is the spin–orbit coupling constant (Table 5.2) and Δ the energy separation between the orbital containing the unpaired electron and the orbital with which it may mix by spin–orbit coupling. The + or − sign in Table 5.2 corresponds to mixing with an empty or filled orbital, respectively. The coefficient n may be deduced from the 'magic pentagon' (Figure 5.7), n being the number on the line linking the orbitals under consideration. As an example, consider the case of a d[9] ion of near octahedral symmetry, but with a tetragonal distortion such as to elongate the charges along the z-axis. The symmetry is D_{4h} and the unpaired electron is in the $d_{x^2-y^2}$ orbital. It follows that

$$g_{\parallel} = g_0 + 8\lambda/\Delta E(d_{x^2-y^2} - d_{xy})$$

$$g_{\perp} = g_0 + 2\lambda/\Delta E(d_{x^2-y^2} - d_{xz,yz}).$$

If, however, there was a compression of charges, then the unpaired electron would be in d_{z^2} and

$$g_{\parallel} = g_0$$

$$g_{\perp} = g_0 + 6\lambda/\Delta E(d_{z^2} - d_{xz,yz}).$$

These equations are useful only if $\frac{\lambda}{\Delta}$ is small (i.e. ≤ 0.1). Further quadratic terms need to be used when $\frac{\lambda}{\Delta}$ is larger.

Table 5.2 Free-ion values for the spin–orbit coupling parameter, λ, for selected transition metal ions

Ion	*Number of d electrons*	*Weak octahedral ground term*		*Strong octahedral ground term*		*Tetrahedral ground term*	
V^{4+}	1	$^2T_{2g}$	248	$^2T_{2g}$	248	^2E	248
Cr^{3+}	3	$^4A_{2g}$	91	$^4A_{2g}$	91	$^4T_1$	91
Mn^{2+}	5	$^6A_{1g}$	–	$^2T_{2g}$	−300	$^6A_1$	–
Fe^{3+}	5	$^6A_{1g}$	–	$^2T_{2g}$	−460	$^6A_1$	–
Co^{2+}	7	$^4T_{1g}$	−172	^2E_g	−515	$^4A_2$	−172
Cu^{2+}	9	^2E_g	−830	^2E_g	−830	$^2T_2$	−830

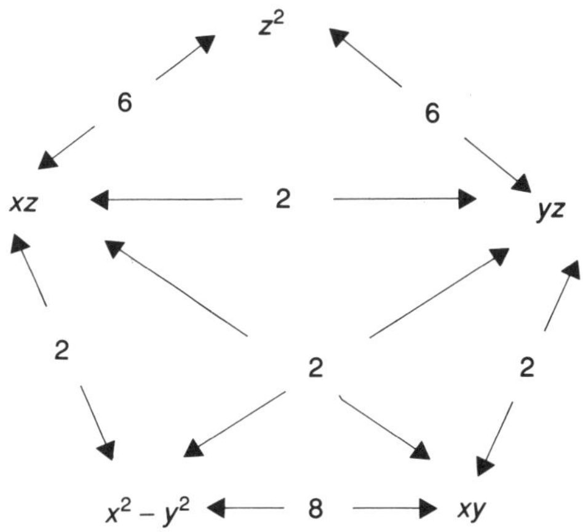

Figure 5.7 Coefficient, *n*, for the spin–orbit coupling between d-orbitals.

5.3 THE CASE OF TWO OR MORE UNPAIRED ELECTRONS: FINE STRUCTURE

5.3.1 Background

In ions with more than one unpaired electron there are interactions between the individual electron magnetic moments and the magnetic fields generated by the other electron(s). In this case, however, the ground state levels may not be degenerate in the absence of a magnetic field, i.e. spin–orbit coupling between various terms in the ground state may produce a zero-field splitting. When such a splitting is present the levels are no longer equally spaced in the external magnetic field and the various magnetic resonance transitions no longer coincide. A series of lines is thus obtained, which is commonly called the fine structure (Figure 5.8).

The variation of the energy levels with the magnetic field does not follow a simple relationship except when the magnetic field is parallel or perpendicular to the principal axis of the splitting, which is conventionally the *z*-axis. In many cases, if there is a metal ion with $I > 0$, then hyperfine structure may also be detected (section 5.4).

There are two important theorems which determine whether a spectrum is likely to be seen for various numbers of electrons in an ion. These are:

- Kramer's theorem, which states that any system containing an odd number of electrons will show at least twofold degeneracy in the absence of a magnetic field (these are often referred to as Kramer's doublets);

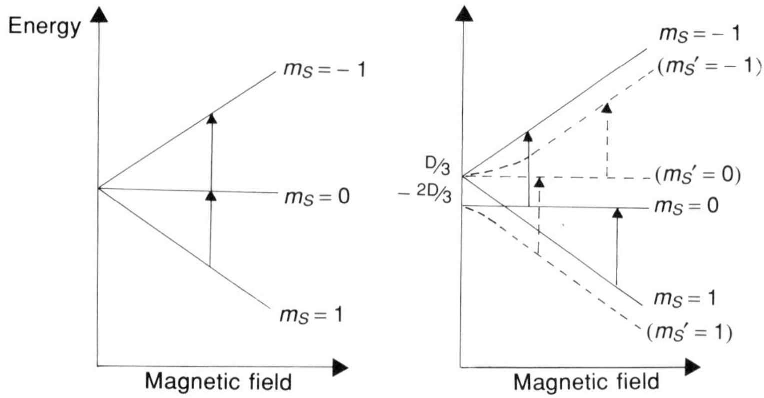

Figure 5.8 Diagrammatic representation of the splitting of the energy levels of a d^2 ion in a zero field and the origin of the fine structure in the EPR spectrum.

- Jahn–Teller theorem, which states that in any orbitally degenerate ground state there will be a distortion to remove the degeneracy, except in linear molecules and in systems having Kramer's doublets.

It follows from Kramer's theorem that in ions having an even number of electrons, the splitting of energy levels will be complete and all ground state levels will be non-degenerate. Because in most cases the separation between levels is very large ($> 1\ cm^{-1}$), the observation of paramagnetic resonance will not be expected. In paramagnetic ions having an odd number of electrons, levels showing Kramer's degeneracy will be present. Paramagnetic resonance is then expected, because in an external magnetic field the Kramer's doublets are split and transitions between their components are induced.

5.3.2 The nature of spectra

In the high-field limit, i.e. when $g\mu_B B$ is much greater than any zero-field splitting, transitions are of the type $m_S \rightarrow (m_S + 1)$. Thus a spectrum is obtained which arises from transitions between $2S + 1$ levels whose values of m_S differ by unity; the energies of the levels are given approximately by the relationship:

$$E = g\mu_e B m_S + \frac{1}{2} D(3\cos^2\theta - 1)\,[m_S{}^2 - S(S + 1)/3],$$

where θ is the angle between B and the z-axis. The intensities of the various lines vary according to the rule that, for the transition $m_S \rightarrow (m_S + 1)$, the intensity is proportional to $(S + m_S + 1)\ (S - m_S)$. Thus, for example, in the $S = \frac{5}{2}$ case, five lines are observed with relative intensities 5 : 8 : 9 : 8 : 5.

Interpretation of EPR spectra from ions with more than one unpaired electron is often fraught with difficulties because only part of the expected fine structure may be resolved. However, the zero-field splitting parameters D and E may be calculated. Parameter D is the principal value and is proportional to the square of the spin–orbit coupling constant, λ, and inversely proportional to δ, the separation between the ground state and the next higher state with which it can mix by spin–orbit coupling. Thus for any one ion, δ will become larger and D smaller with increasing distortion (tetragonal). The parameter E is zero if the ion is axially symmetric; its magnitude is a measure of the extent of further asymmetric distortion.

Because of its widespread occurrence in minerals and the complexity of the principles that determine its spectral characteristics, Fe^{3+} will be considered as a separate case in the following section.

5.3.3 EPR of Fe^{3+}

The Fe^{3+} (and also Mn^{2+}) ion contains five 3d electrons, which, in the presence of crystal fields, can exist in states with 1, 3 or 5 unpaired electrons. In minerals in which oxygen atoms are coordinated to the iron, it invariably occurs in the high-spin (^6S) state (Chapter 3). Consequently the other states will not be considered further. The sextet nature of the free-ion ground state remains in the presence of crystal fields of various symmetries, but spin–orbit coupling does lead to a splitting of the ground state energy levels. These are split further in the presence of a magnetic field as a result of interaction with the electron spin angular momentum (Figure 5.9). The right-hand side of Figure 5.9 illustrates the five allowable EPR transitions, but these are not always observed

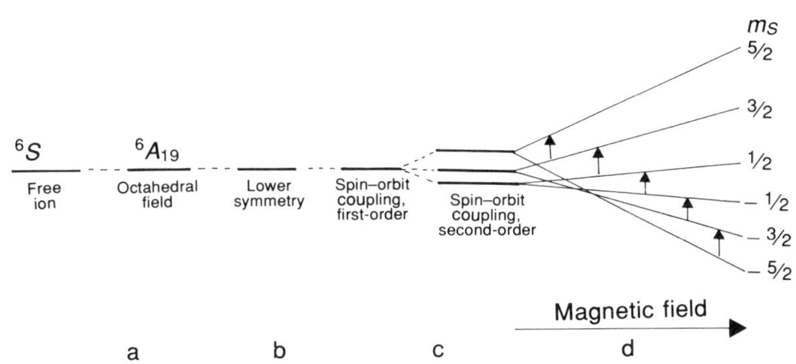

Figure 5.9 Diagrammatic representation of the lowest electronic energy state for high-spin Fe^{3+} in a crystal field of (a) O_h symmetry, (b) D_{4h} symmetry, with the effects of (c) spin–orbit coupling and (d) an external magnetic field.

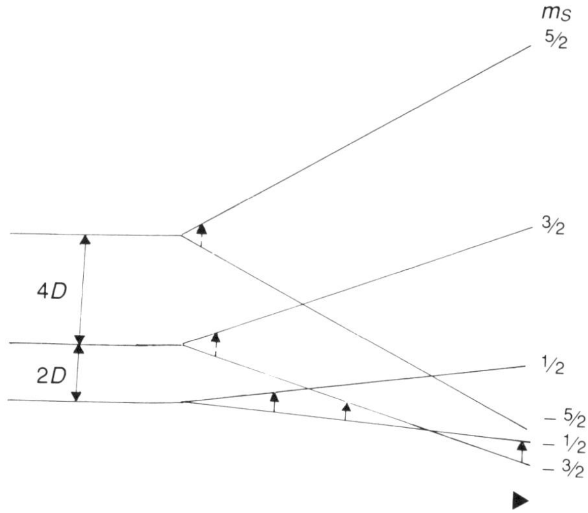

Figure 5.10 Diagrammatic representation of the EPR transitions for Fe^{3+} in an environment with a large zero-field splitting.

because of the relative magnitudes of the zero field and electronic Zeeman terms. When the zero-field splitting, D, is large, the transitions will be as illustrated in Figure 5.10, with those for $m_S > 1$ being forbidden, and consequently weak.

The zero-field splitting may be anisotropic (i.e. varies with orientation) and is conventionally described by two terms, D and E, where the axis system is defined so that $E \leqslant D/3$. The positions of the various resonance lines have been calculated by Aasa (1970) and are illustrated in Figure 5.11 for $E=0$ and Figure 5.12 for $E=D/3$ and $E=D/4$ for a range of values of $h\nu/D$. In these plots g is always close to 2.0, but a new parameter, g', is used to define the apparent g-value that is obtained from experimental peak positions. It is clear from Figures 5.11 and 5.12 that it is only when D is small that the resonance peaks are centred on $g = 2$, the true g-value.

In a powder sample, where all crystallite orientations are present, an EPR spectrum consists of a superimposition of peaks covering the range of positions between the x, y and z axes. When these are spread over a large energy range the peak height is very small, so that a powder spectrum is dominated by those transitions that are isotropic or near isotropic in character. The use of the Aasa diagrams to interpret the EPR spectral features will be illustrated by considering a major feature in the spectrum of the kaolinite in Figure 5.18(a). This spectrum was recorded at X-band frequencies (i.e. *c.* 9 GHz microwave frequency); hence the position of the free spin g-value is *c.* 330 mT. The main Fe^{3+} feature consists of three peaks centred at *c.* 155 mT, i.e. g' is *c.* 4.26 and g/g' is *c.* 0.47. Inspection of Figures 5.11 and 5.12 shows that a feature in this

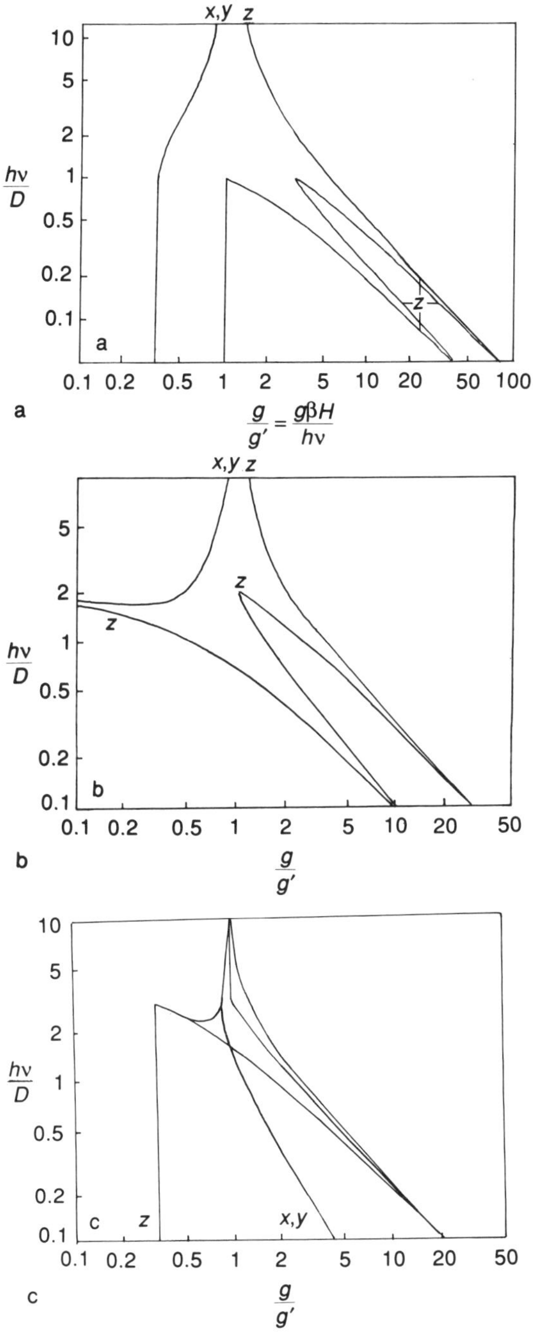

Figure 5.11 Variation as a function of the zero-field splitting, D, of the positions of the (a) $m_S(-\frac{5}{2}) \rightarrow m_S(-\frac{3}{2})$, (b) $m_S(-\frac{3}{2}) \rightarrow m_S(-\frac{1}{2})$, (c) $m_S(-\frac{1}{2}) \rightarrow m_S(\frac{1}{2})$, (d) $m_S(\frac{1}{2}) \rightarrow m_S(\frac{3}{2})$, and (e) $m_S(\frac{3}{2}) \rightarrow m_S(\frac{5}{2})$ transitions in the EPR spectra of Fe^{3+} for $E = 0$. (Adapted from Aasa, 1970.)

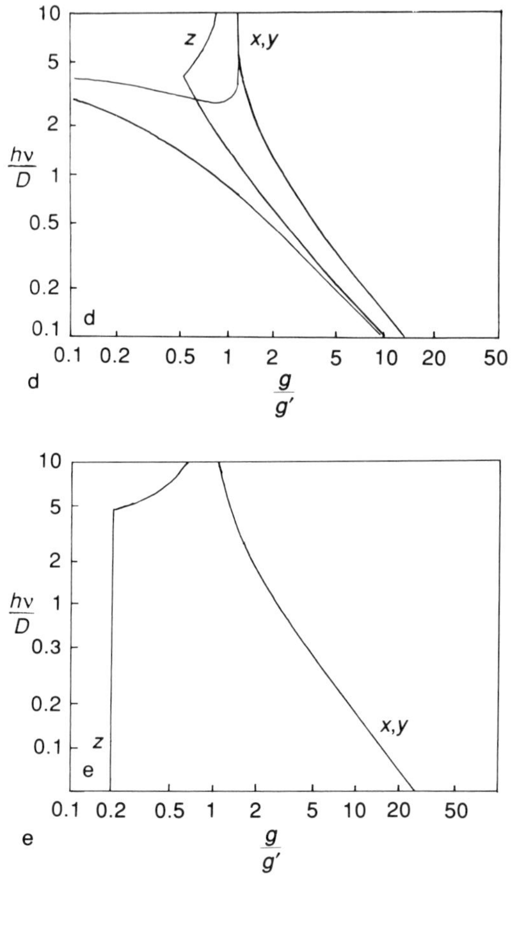

Figure 5.11 (contd).

position corresponds to the $m_s(-\frac{1}{2})\rightarrow m_s(+\frac{1}{2})$ transition for $E=D/3$ and large $D(h\nu/D<$c. 1.5). With slightly smaller values of E this feature splits into three components, and an examination of Figure 5.12(d), in which $E/D=0.25$, shows that it predicts transitions in positions that are very close to those observed in the experimental spectrum.

Peaks with g-values close to 4.27 are very common features of the EPR spectra of minerals and Fe^{3+} complexes with organic ligands. It must be stressed that this is not because of any inherent preference of Fe^{3+} for

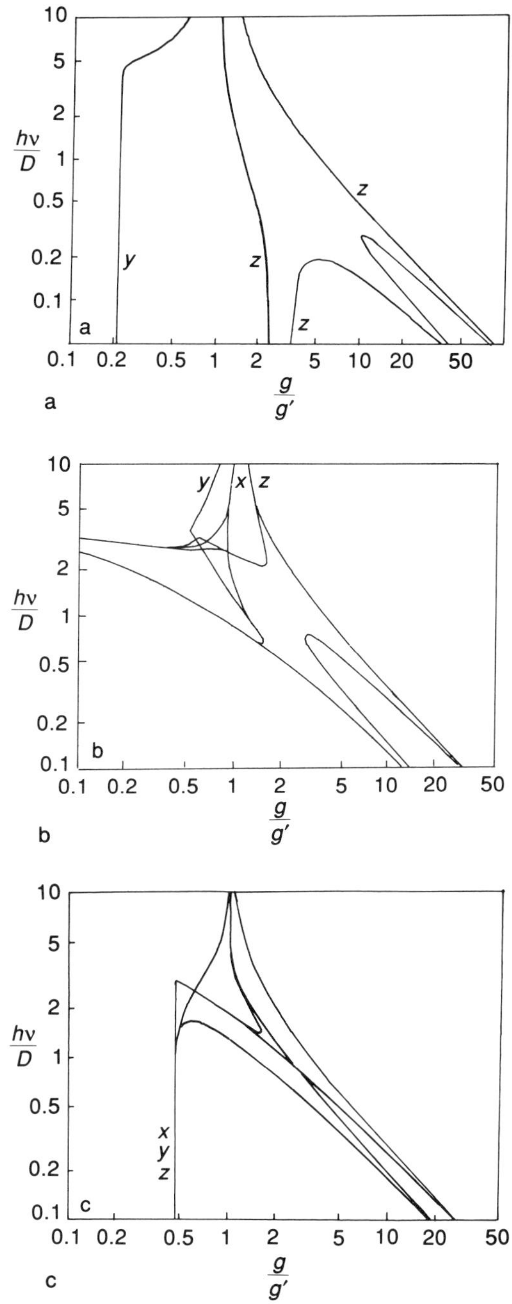

Figure 5.12 Variation as a function of the zero-field splitting, D, of the positions of the (a) $m_S(-\frac{5}{2}) \rightarrow m_S(-\frac{3}{2})$ and $m_S(\frac{3}{2}) \rightarrow m_S(\frac{5}{2})$, (b) $m_S(-\frac{3}{2}) \rightarrow m_S(-\frac{1}{2})$ and $m_S(\frac{1}{2}) \rightarrow m_S(\frac{3}{2})$, and (c) $m_S(-\frac{1}{2}) \rightarrow m_S(\frac{1}{2})$ transitions in the EPR spectra of Fe^{3+} for $E = D/3$, and (d) for the $m_S(-\frac{1}{2}) \rightarrow m_S(\frac{1}{2})$ transition for $E = D/4$. (Adapted from Aasa, 1970.)

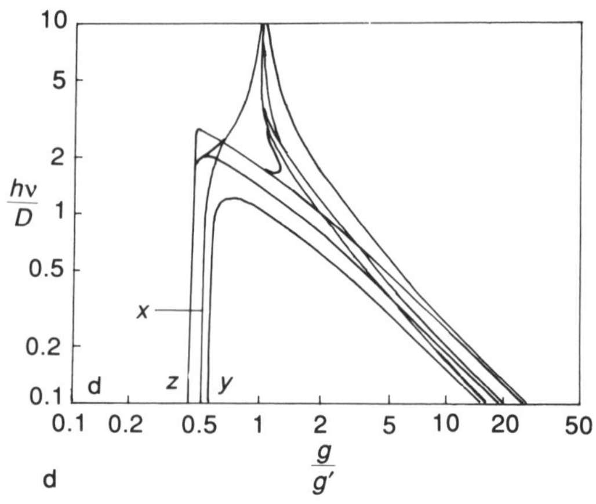

Figure 5.12 (contd).

environments with rhombic or near rhombic symmetry ($E/D \approx 0.3$), but because of the low level of anisotropy in the $m_S(-\frac{1}{2}) \rightarrow m_S(+\frac{1}{2})$ transition in this symmetry. Most other transitions in Fe^{3+} have high levels of anisotropy and hence extremely large powder spectral linewidths and small amplitudes. Two other situations that are revealed by examination of Figures 5.11 and 5.12 as giving powder spectral components with significant intensities are (1) small D, which produces a spectrum centred on $g = 2$, and (2) large D and zero E, which produces a spectrum with $g_{\perp} = 6$ and $g_{\parallel} = 2$.

The discussion so far has assumed that the iron is in a magnetically dilute environment. When magnetic interactions occur between neighbouring ions, dipolar splittings proportional to $1/r^3$ are produced, where r is the electron–electron separation. If a number of comparatively small values of r exist within a mineral specimen, then an envelope of lines rather than a single transition will be seen in the EPR spectrum. A diagrammatic representation of various types of EPR spectrum from Fe^{3+} that are encountered in aluminosilicate minerals is given in Figure 5.13.

5.3.4 EPR spectra of magnetically ordered materials

This section is included because of the importance of iron oxide phases in clay mineralogy. Apart from the individual oxide and oxyhydroxide minerals, for which EPR spectroscopy can add very little to our understanding of their chemical or physical properties, many naturally occurring clays have iron oxide phases associated with their surfaces. It is

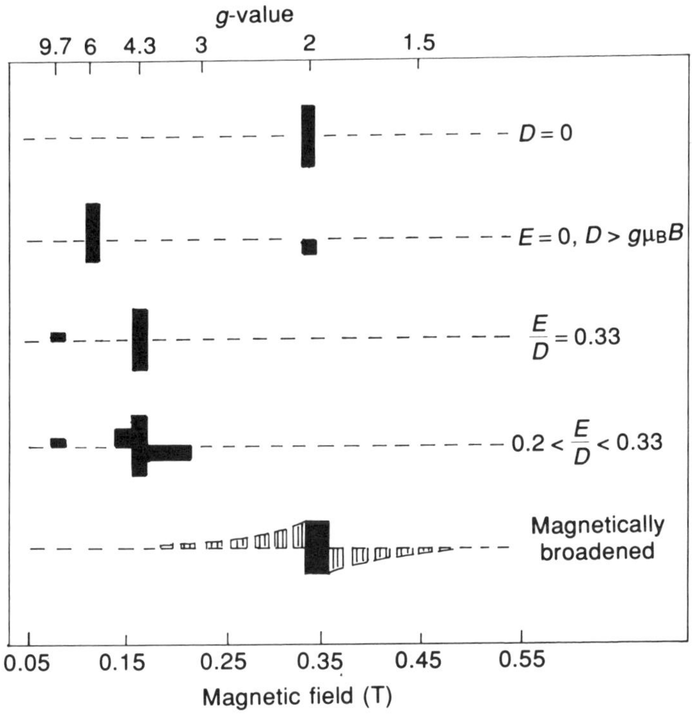

Figure 5.13 Diagrammatic representation of the various types of EPR spectrum for Fe^{3+} in the high-spin state.

important, therefore, to have an elementary understanding of the effects that these phases might have on the EPR spectra of these clays.

The interactions between the magnetic moments of paramagnetic ions in close proximity to one another result in large dipolar couplings. This usually has the effect of producing very broad spectra, especially if there is any non-uniformity in the spatial arrangement of these ions. In completely magnetically ordered minerals, however, the electron paramagnetic resonance is not produced by individual ions, but instead results from the interaction of the collective magnetic moments of magnetic domains with the magnetic field. With antiferromagnetic coupling there is a cancellation of the magnetic moments of individual ions and EPR is obtained from the few uncompensated magnetic moments that exist on the surfaces of particles. Thus antiferromagnetic ordering is usually accompanied by a reduction in spectral intensity over that for the corresponding paramagnetic phase. With ferrimagnetism the individual magnetic domains contain very large magnetic moments and consequently

Table 5.3 Magnetic properties of nuclei

Isotope	*Natural Abundance (%)*	*Magnetic moment (nuclear magnetons)*	*Spin (I)*
^{1}H	99.9844	2.79268	$\frac{1}{2}$
^{2}H	1.56×10^{-2}	0.857386	1
^{13}C	1.108	0.70220	$\frac{1}{2}$
^{14}N	99.635	0.40358	1
^{15}N	0.365	− 0.28304	$\frac{1}{2}$
^{17}O	3.7×10^{-2}	− 1.8930	$\frac{5}{2}$
^{19}F	100.00	2.6273	$\frac{1}{2}$
^{23}Na	100.00	2.2161	$\frac{3}{2}$
^{25}Mg	10.05	− 0.85471	$\frac{5}{2}$
^{27}Al	100.00	3.6385	$\frac{5}{2}$
^{29}Si	4.70	− 0.55477	$\frac{1}{2}$
^{31}P	100.00	1.1305	$\frac{1}{2}$
^{33}S	0.74	0.64274	$\frac{3}{2}$
^{35}Cl	75.4	0.82091	$\frac{3}{2}$
^{37}Cl	24.6	0.68330	$\frac{3}{2}$
^{39}K	93.08	0.39094	$\frac{3}{2}$
^{47}Ti	7.75	− 0.78711	$\frac{5}{2}$
^{49}Ti	5.51	− 1.1022	$\frac{7}{2}$
^{51}V	100.00	5.1392	$\frac{7}{2}$
^{53}Cr	9.54	− 0.47354	$\frac{3}{2}$
^{55}Mn	100.00	3.4611	$\frac{5}{2}$
^{57}Fe	2.245	0.0903	$\frac{1}{2}$
^{59}Co	100.00	4.6388	$\frac{7}{2}$
^{63}Cu	69.09	2.2206	$\frac{3}{2}$
^{65}Cu	30.91	2.3790	$\frac{3}{2}$

give rise to huge interactions with the external magnetic field. The resulting EPR spectra are very intense, but are easily saturated.

5.4 NUCLEAR HYPERFINE COUPLING

5.4.1 Introduction

As described in Chapter 4, nuclei have the property of spin, designated by the nuclear spin quantum number, I. Nuclei with even numbers of protons and neutrons have $I = 0$; nuclei with odd numbers of protons and neutrons have $I =$ an integer; nuclei with an even number of protons and an odd number of neutrons (or vice versa) have half integral nuclear spin, i.e. $I = \frac{1}{2}, \frac{3}{2}, \frac{5}{2}$, etc. The magnetic moment of the nucleus is given by $g_N \mu_N I$, where the nuclear magneton, μ_N, has a value of 5.0505×10^{-27} J/T.

Table 5.3 gives the magnetic properties and natural abundances of nuclei of relevance to EPR work on clay minerals. In the presence of a magnetic field, the nuclear spin states are quantized and the component m_I of the nuclear spin vector may take up the following values,

$$I, (I-1), (I-2), (I-3) \ldots -I.$$

In an external magnetic field the electron spin and nuclear spins interact and all combinations of electron and nuclear spins are allowed. Thus for $S = \frac{1}{2}$ and $I = \frac{1}{2}$,

$$m_S = -\frac{1}{2} \qquad m_I = \pm\frac{1}{2}$$

$$m_S = +\frac{1}{2} \qquad m_I = \pm\frac{1}{2}.$$

Only transitions between the same m_I states are allowed and the selection rule for EPR transitions is

$$\Delta m_S = \pm 1 \quad \text{and} \quad \Delta m_I = 0.$$

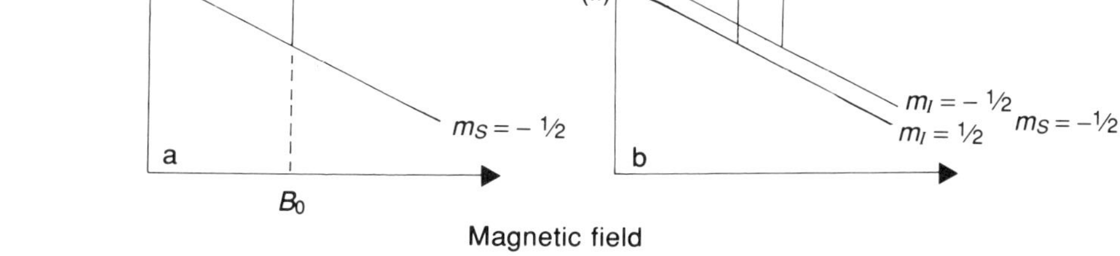

Figure 5.14 Effect of an external magnetic field on the energy levels of an electron interacting with a nucleus with $I = \frac{1}{2}$, showing the two transitions. Note that at zero field the energy at (i) is $g\beta A/4$ and that at (ii) is $-3g\beta A/4$.

At constant frequency the transitions are as shown in Figure 5.14. Interactions of an electron spin with a nucleus of spin I yields $2I+1$ transitions, which are, in the high-field approximation, equally separated with a spacing, A, which is known as the hyperfine coupling constant. Hyperfine couplings are reported in the literature in several different units, e.g. tesla, gauss, MHz or cm^{-1}. The relationship between the different units is given in Table 5.4.

Table 5.4 Relationship between units used in EPR spectroscopy

	T	*gauss*	*MHz*	*cm^{-1}*
T	1	10^4	1.4×10^4g	0.4669g
gauss	10^{-4}	1	1.400 g	4.669×10^{-5}g
MHz	7.1449×10^{-3}/g	71.449/g	1	3.3356×10^{-3}
cm^{-1}	2.1420/g	2.1420×10^4/g	2.9979×10^{-4}	1

5.4.2 Components of the hyperfine coupling tensor

The magnitude of nuclear hyperfine coupling is determined by the magnetic field at the electron produced by its interaction with the nucleus and depends strongly on the properties of the orbital in which the unpaired electron is located. The s-orbitals have high electron density at the nucleus and consequently produce large hyperfine couplings. Since these orbitals are symmetrical, the magnitude of the coupling is independent of direction and is called the isotropic hyperfine coupling (A_{iso}) or the Fermi contact interaction.

Isotropic hyperfine coupling can also be seen with ions in which the unpaired electron density is entirely in p- or d-orbitals. This arises because of an exchange interaction between the unpaired d-electron and the paired s-electrons in the core ($1s^2$, $2s^2$, $3s^2$, etc.) orbitals, which is spin-orientation dependent. In other words, there is a polarization of the ns^2 configuration which produces a magnetic field at the nucleus. The unpaired electron may be considered as attracting electrons of the same spin and repelling those of opposite spin, with the result that the nucleus feels the influence of one of the s-electrons more than the other, i.e. there is apparent unpaired spin density at the nucleus. This is of opposite sign to that of the unpaired d-electron.

In p- or d-orbitals, there is in addition a dipolar interaction between the unpaired electron and the nucleus, which is dependent upon both the direction of the orbital with respect to the applied magnetic field and the electron–nucleus separation. This interaction is called the anisotropic hyperfine coupling (A_{aniso}) and may be resolved into unique directions (x, y and z). When integrated over all space, the magnitude of the anisotropic hyperfine coupling is zero, i.e. $A_x + A_y + A_z = 0$.

Because any orbital may be considered as a hybrid of suitable combinations of s-, p- and d-orbitals, so also may a hyperfine coupling be divided into one contribution from p- and d-orbitals and one from s-orbitals. In a fluid, anisotropic coupling is averaged to zero and the only observed interaction is the isotropic coupling, A_{iso}. However, in a solid matrix the observed hyperfine splitting is the sum of anisotropic and isotropic interactions.

5.4.3 Superhyperfine coupling

Superhyperfine coupling may sometimes be observed when a metal ion is bound to atoms whose nuclei have non-zero magnetic moments. If the unpaired electron is located entirely in an atomic orbital on the metal and ligands are held entirely by crystal-field forces, the only interaction between a ligand nucleus and the unpaired electron is dipolar. This is inversely proportional to the mean cube of the distance of the electron from the ligand nucleus. As was the case for the dipolar interaction between the electron and its own nucleus, this interaction is anisotropic and vanishes when the molecule is tumbling rapidly. However, it is usual for a 'small amount' of the unpaired electron to be delocalized on to the ligands as a result of the formation of molecular orbitals. Therefore, ligand superhyperfine structure has both isotropic and anisotropic components, but the axis system of the latter is generally different from that of the metal orbital.

5.5 EXPERIMENTAL PROCEDURES

5.5.1 EPR spectrometers

5.5.1.1 Basic spectrometer design

A simplified block diagram of a continuous-wave EPR spectrometer is shown in Figure 5.2. Electromagnetic radiation either from a monochromatic klystron or a Gunn diode passes along waveguides to the three-port circulator, from which it is directed to the microwave reflection cavity and to the crystal detector. In the cavity, stationary waves are set up so as to generate the most intense region of the microwave magnetic field at the sample being investigated. A homogeneous magnetic field at the sample is produced by an electromagnet. The field is swept through the desired range by the power supply and is modulated at high frequency by modulation coils, which give a phase sensitive signal at the recorder.

5.5.1.2 *Frequency of operation*

Electron paramagnetic resonance spectrometers can operate at any of the microwave frequencies that have been developed for radar. The most commonly used is X-band (*c.* 9 GHz), which requires a magnetic field in the region of 0.30–0.35 T for EPR signals with *g*-values around that of a free electron. Other frequencies that are used are Q-band (*c.* 35 GHz), K-band (*c.* 24 GHz), S-band (*c.* 4 GHz) and L-band (*c.* 1 GHz).

5.5.1.3 *Cavity design*

There are several different types of resonator available for use with modern EPR spectrometers (e.g. one leading manufacturer of EPR equipment lists more than 40 different cavities in five frequency bands in their current catalogue). The selection of the most appropriate cavity for any particular experiment is governed by the requirements of that experiment. The most popular general purpose cavity is the rectangular design, shown diagrammatically in Figure 5.15. Coupling to the microwave bridge is achieved by a variable iris aperture and often this type of cavity is fitted with a grid for horizontal irradiation of the sample. A modification of this cavity used for accurate quantitative measurements is the double rectangular cavity, which contains two chambers and permits accurate comparison between samples and a reference standard. Cylind-

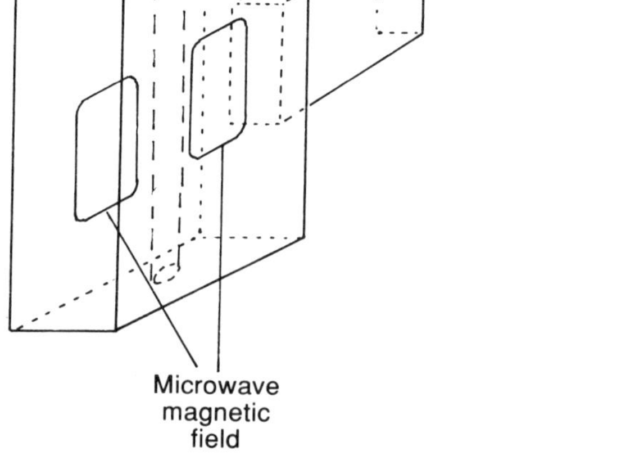

Figure 5.15 Basic design of a rectangular cavity for an EPR spectrometer.

rical cavities are used for samples with dielectric loss (especially those with high water content). A dielectric cavity has been developed recently and offers potentially greater sensitivity for such samples, although currently at the expense of a background signal from the cavity itself. For low-frequency spectrometers, split-ring resonators are becoming more common because of their vastly increased sensitivity for specimens with high dielectric loss.

5.5.2 Spectrometer operation

5.5.2.1 Microwave bridge tuning

The first step in acquiring an EPR spectrum is tuning the microwave bridge. This is done by 'wobbling' the microwave frequency with a low frequency, which displays a curve on a VDU screen with a dip when the bridge is tuned. A klystron is a broadband oscillator that produces a high power, low noise output and is easy to tune. It does, however, have a finite life and is costly to replace. The Gunn diode, in contrast, is a solid-state device and as such has a much longer life. It is, however, inherently more noisy than a klystron, especially at high microwave power levels, and also more difficult to tune.

5.5.2.2 Microwave power

The intensity of an EPR spectrum is proportional to the square root of the microwave power in the absence of saturation. Saturation occurs when the thermal relaxation processes are inadequate to maintain the Boltzmann equilibrium between the spin levels. It results in a broadening of peaks, which also become flattened in the region of maximum absorption. Optimum power levels are normally determined experimentally for each individual specimen, but as a general rule of thumb, the EPR spectra of transition metal ions saturate less readily than those of organic or inorganic free radicals, although some defect centres in mineral structures are particularly susceptible to saturation. Also, saturation effects are more pronounced at low temperatures. They may, however, be put to good use in the discrimination between components with different saturation characteristics by recording spectra at different power levels.

5.5.2.3 Modulation frequency

With many spectrometers a limited range of modulation frequencies is available. In general, for clay mineral specimens, spectra are recorded at as high a modulation frequency as possible (usually 100 kHz), low modulation frequencies being of value only for the resolution of very small splittings between peaks in free radicals with narrow linewidths.

5.5.2.4 *Modulation amplitude*

The amplitude of the field modulation is usually selected to be as high as possible without broadening the spectral peaks. Values have to be determined experimentally for individual specimens because of the huge variation in linewidths for different types of paramagnetic ion. In general, transition metal ions have linewidths of the order of a few tenths of a millitesla or greater, but free radicals and crystal defect centres may have much smaller linewidths.

5.5.3 Measurement of spectral parameters

Measurement of g-values requires accurate measurement of both the magnetic field and frequency at the appropriate positions in a spectrum. Unfortunately, few spectrometers are equipped with reliable frequency counters and gaussmeters and consequently parameters derived from an uncritical use of instrumental values may contain considerable errors. The use of standard materials whose parameters have been determined independently may partially alleviate this problem, especially for signals centred around $g = 2.0$. However, at the present time authenticated standards for signals remote from $g = 2$ are not commercially available. It is also unfortunate that few workers with clay mineral samples have taken the trouble to optimize their spectral parameters and many published numbers, particularly those of g-values, are not accurate.

The separations between adjacent peaks in a hyperfine split spectrum are equal only if the high-field resonance condition is satisfied (i.e. the magnitude of $g\mu_B B$ is very much greater than the hyperfine coupling). This will often not be the case for the spectra of transition metal ions, such as Cu^{2+}, Mn^{2+} or VO^{2+}, obtained with spectrometers operating at X-band or lower frequencies. Here line separations are generally unequal as a result of second-order contributions to the hyperfine structure. In fluid solution these are shifted to the low-field by an amount equal to

Table 5.5 Coefficient (x) for the second order correction to the isotropic hyperfine splitting

I								x							
0								0							
$\frac{1}{2}$							1		1						
1						2		4		2					
$\frac{3}{2}$					3		7		7		3				
2				4		10		12		10		4			
$\frac{5}{2}$			5		13		17		17		13		5		
3		6		16		22		24		22		16		6	
$\frac{7}{2}$	7		19		27		31		31		27		19		7

$xA^2/4B$, where the coefficient x is given in Table 5.5. This effect is illustrated in Figure 5.16 for a VO^{2+} ion. (Note the progressive increase in peak separations with increasing magnetic field.) The hyperfine splitting can be derived from the separation between the centremost peaks, but the correct g-value is not obtained from the centre of the experimental spectrum.

In anisotropic spectra the situation is more complicated and usually it is only possible to derive reliable spectral parameters by performing spectral simulations. Although computer programs have been widely available for several years for simulation of anisotropic spectra from $S = \frac{1}{2}$ ions, the time taken to simulate a spectrum generally exceeds by far the time taken to generate it. Thus few papers in the clay-minerals literature contain computer simulations to justify either the interpretation of the spectra or for the refinement of spectral parameters. The situation is more difficult for ions with $S > \frac{1}{2}$, as is illustrated in section 5.3.3. for Fe^{3+}.

5.5.4 Determination of the alignment of paramagnetic ions in clay-mineral structures

The small particle size of clay minerals precludes the use of single-crystal specimens to determine the orientations of constituent paramagnetic

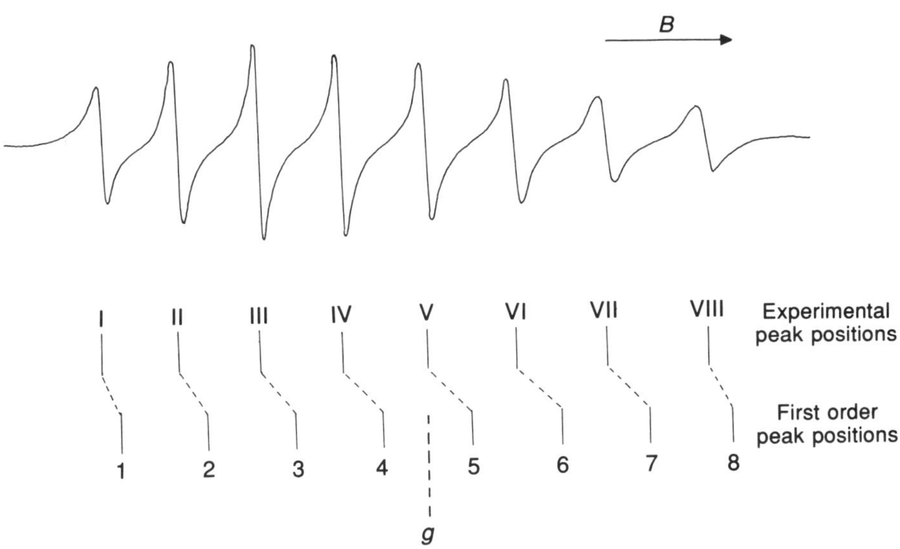

Figure 5.16 Electron paramagnetic resonance spectrum of the solvated VO^{2+} ion, showing the progressive increase in peak separations with increasing magnetic field and the relationship between experimental peak positions and those calculated to first order.

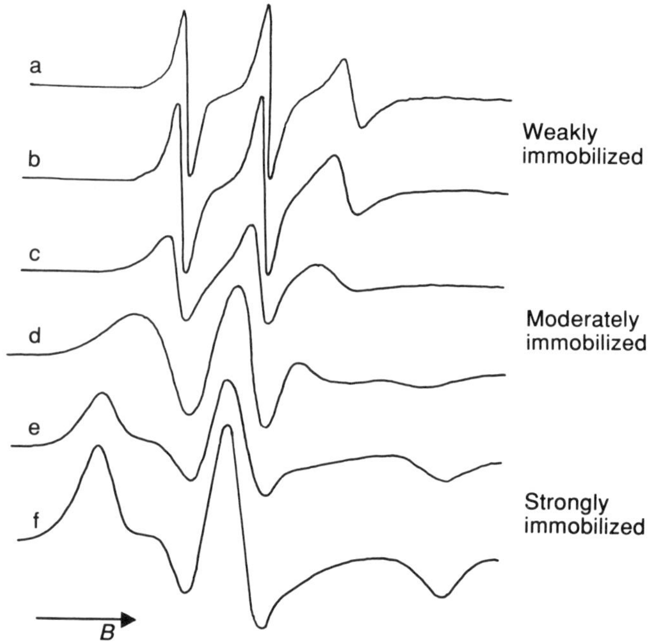

Figure 5.17 Variation of nitroxide spin-label spectral line-shapes with motional correlation times: (a) 15 ns, (b) 40 ns, (c) 70 ns, (d) 200 ns, (e) 700 ns and (f) 1200 ns.

components relative to the crystal axes. However, by use of oriented films it is sometimes possible to use the shapes of the EPR spectral envelopes to derive information on the alignment of component paramagnetic ions, especially if the principal molecular axes are aligned either perpendicular to or in the plane of the silicate sheet. It is more difficult to obtain information on intermediate angles and computer simulations are usually required to justify anything other than 'guesstimates', especially if allowance is to be made for non-perfect stacking of individual particles.

For molecules that undergo hindered motion, spectra are obtained with shapes intermediate between those of fluid solutions and rigid powders. Examples for a nitroxide radical with different rotational correlation times are given in Figure 5.17. Such spectra can be used to probe the effects of clay swelling on the mobilities of interlayer ions.

5.6 APPLICATIONS

5.6.1 Introduction

As mentioned at the beginning of this chapter, EPR spectroscopy is able to provide information on paramagnetic species that are present in minerals. However, where extensive dipolar interactions occur in structures with high levels of paramagnetic components, the technique is less discriminating and, of course, it provides no information at all on the major diamagnetic constituents. The strength of the EPR technique lies in its ability to characterize minor structural or surface components and, consequently, in this section some practical applications of EPR spectroscopy to clay mineralogy will be summarized. No attempt is made to provide a comprehensive coverage of the literature, and the selected examples merely illustrate the range of uses of the technique in clays research. The examples include the characterization of low-level substituents in aluminosilicate minerals, which may be either transition metal ions or defect centres (trapped electrons or electron 'holes'), the identification of the presence of iron oxides or oxyhydroxides associated with mineral specimens, the characterization of surface adsorbed ions and complexes, and the elucidation of surface reactions.

5.6.2 Paramagnetic components of kaolins

The EPR spectra of natural kaolins vary markedly according to their origins, typical spectra being presented in Figure 5.18. There have been extensive studies of both natural (Angel and Hall, 1973; Meads and Malden, 1975; Herbillon *et al.*, 1976; Mestagh, Vielvoye and Herbillon, 1980; Brindley *et al.*, 1986; Muller *et al.*, 1990) and synthetic specimens (Angel, Jones and Hall, 1974; Jones, Angel and Hall, 1974; Angel *et al.*, 1977; Cuttler, 1980, 1981) and the main spectral characteristics are now reasonably well understood. These results will be summarized in the next three subsections.

5.6.2.1 Low-field resonances from Fe^{3+} substituting for Al^{3+} in kaolinite

The characteristic EPR signal from kaolinites at relatively low magnetic fields consists of four main resonance lines with g-values of 4.9, 4.3, 3.7 and 3.5. All of these lines can be attributed to spin resonance arising from the central of three Kramer's doublets, whose energies are separated by crystal fields of rhombic symmetry. These observations require the existence of at least two sites or centres, both having rhombic (or near rhombic) symmetry. The peak with $g = 4.3$, which is generally referred to as Centre I, although it is also known as the I-line, is independent of the other three, and is associated with a site having $E/D \approx 0.33$ (section 5.3). It is observed in measurements at both X- and Q-band frequencies, which means that $D \geqslant 0.8\ cm^{-1}$ (Jones, Angel and Hall, 1974; Meads and Malden, 1975). The remaining peaks, termed Centre II or the E-lines, arise from a site having $E/D \approx 0.22$. The peaks with g-values of 3.7 and 3.5 are normally not resolved in spectra recorded at ambient temperatures, but may be separated

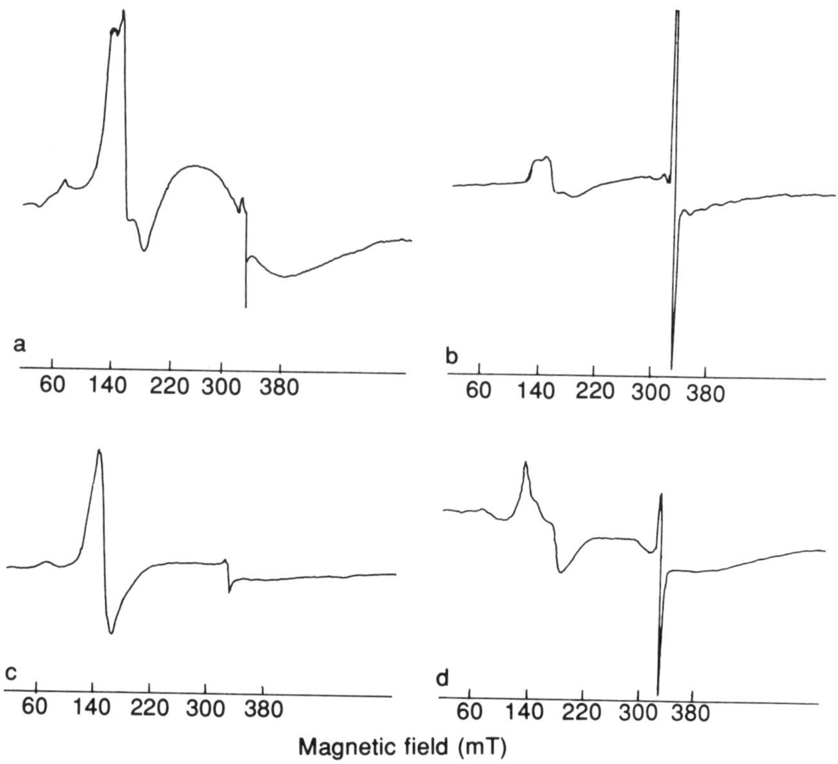

Figure 5.18 Electron paramagnetic resonance spectra of four natural kaolin samples from different localities. (a) Mount Crawford, Adelaide, Australia, Fe_2O_3 1.38%, Hinckley Index 0.65. (b) Washington County, Georgia, USA, Fe_2O_3 0.34%, Hinckley Index 0.9 (c) Pugu, Tanzania, Fe_2O_3 1.43%, Hinckley Index 0.07. (d) La Frontina Mine, Mexico, Fe_2O_3 0.62%, Hinckley Index 1.7. (Reproduced from Meads and Malden, 1975.)

at 77 K. The presence of both peaks is required on theoretical grounds and is confirmed by spectral simulations (Figure 5.19).

Both absolute and relative intensities of Centres I and II vary from sample to sample. The intensity of Centre I is enhanced in kaolinites of low crystallinity (Herbillon *et al.*, 1976; Mestagh, Vielvoye and Herbillon, 1980; Komusinski, Stoch and Dudiel, 1981; Brindley *et al.*, 1986). A similar enhancement of the intensity of Centre I has been observed with a kaolinite that was intercalated with dimethylsulphoxide (Lipsicas *et al.*, 1986). Centre I is, therefore, associated with stacking disorder and/or the disruption of interlayer hydrogen bonding. This component is well-developed in the poorly ordered Fe-substituted kaolinites typical of Oxisols. The lines associated with Centre II appear to be associated with more highly crystalline kaolinites and are lost when specimens are dehydroxylated by heating, as is the broad feature with $g = 2.0$. The resulting

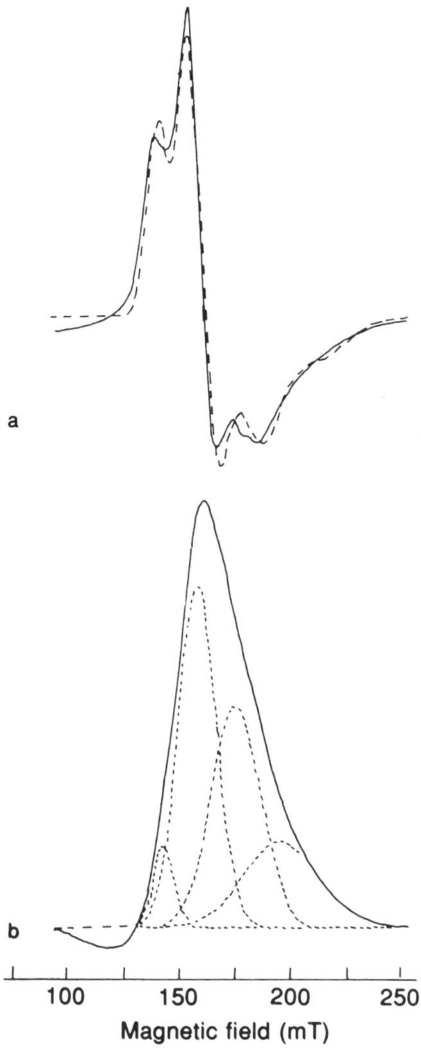

Figure 5.19 (a) Electron paramagnetic resonance first derivative spectrum of a Georgia kaolinite. The dashed line is the derivative of the sum of the Gaussian components delineated in (b). (b) The corresponding absorption spectrum obtained by integrating (a). (Reproduced from Brindley *et al.*, 1986.)

spectrum contains a single resonance peak with $g = 4.3$, which is similar in shape to that reported by Loveridge and Parks (1971) for Fe^{3+} in aluminosilicate glasses.

Spectra obtained at Q-band frequencies indicate the existence of three distinct octahedral Fe^{3+} sites in kaolinite. There appears to be two types of Centre II, with different symmetry properties, corresponding to parameters $D = 0.48$ cm^{-1}, $E/D = 0.234$ and $D = 0.32$ cm^{-1}, $E/D = 0.207$. These have been interpreted as corresponding to sites of Fe^{3+}-for-Al^{3+} substitution, which are associated with different orientations of adjacent surface OH groups (Jones, Angel and Hall, 1974; Meads and Malden, 1975).

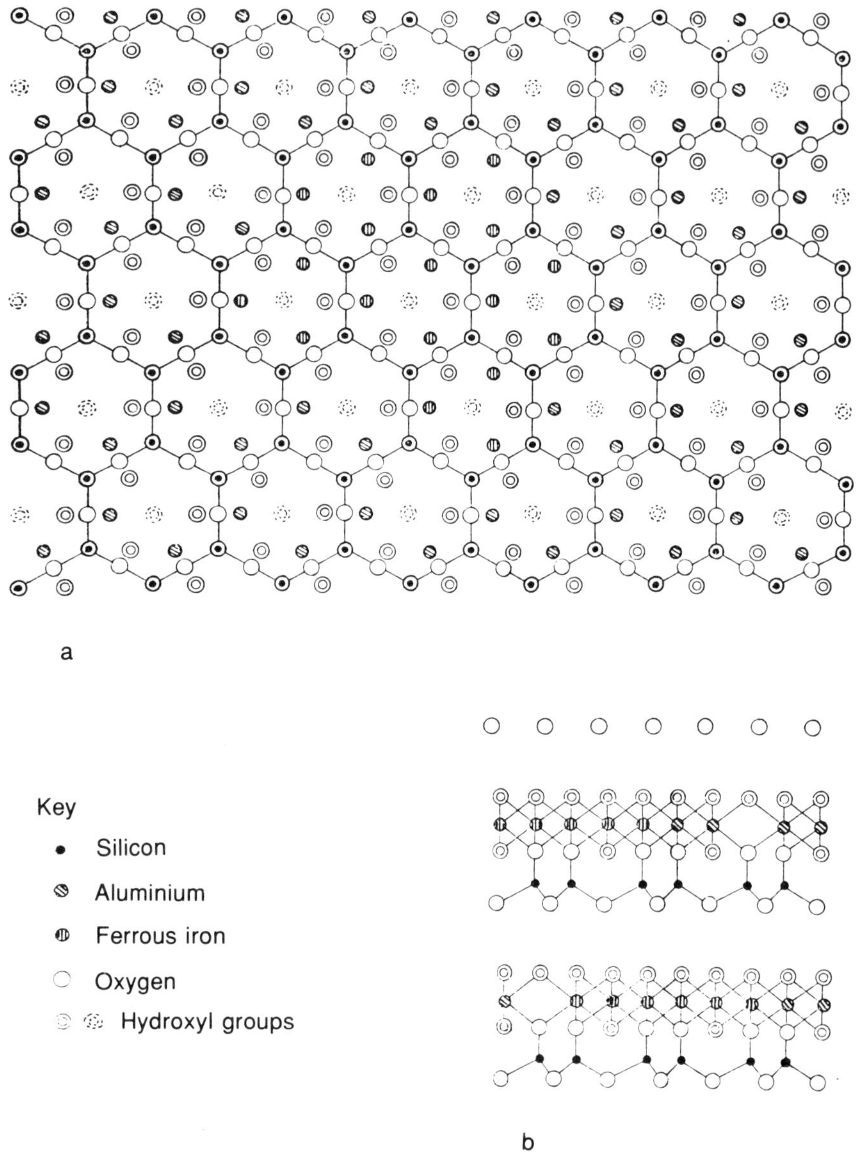

Figure 5.20 Structure of kaolinite containing trioctahedral Fe^{2+} domains shown as (a) a projection on the x–y plane, and (b) a projection on the y–z plane. (Reproduced from Cuttler, 1980.)

5.6.2.2 *Resonances with $g = 2$ associated with stable paramagnetic defect centres in kaolinites*

Excluding the broad lines associated with Fe^{3+} ions in magnetically ordered or superexchange environments, the dominant feature of the EPR spectra of natural kaolins in the region around $g = 2$ is an axially symmetric centre with $g_{\parallel} = 2.05$ and $g_{\perp} = 2.00$ and no hyperfine structure. This feature, which is commonly referred to as Centre A, has been shown by computer modelling (Swartz *et al.*, 1979) to have its principal axis aligned at *c.* 35° to the kaolinite *c**-axis. In addition there is a further feature with hyperfine structure from interactions of unpaired electrons with ^{27}Al nuclei (Angel and Hall, 1973; Angel, Jones and Hall, 1974; Jones, Angel and Hall, 1974; Meads and Malden, 1975), which has been designated as Centre B.

Studies on synthetic kaolinites doped with either Mg or Fe and X-irradiated have helped identify relationships between the observed features in the spectra of natural kaolinites. In the absence of irradiation, Mg-doping produced no EPR spectrum, whereas Fe^{3+}-doping generated a signal in the $g = 4.3$ region only. Irradiation resulted in the production of both Centres A and B (Angel, Richards and Jones, 1976). The intensity of Centre A increased with increasing crystallinity, but decreased with non-removable iron content (Meads and Malden, 1975; Mestagh, Vielvoye and Herbillon, 1980). Its concentration was, however, considerably less (10^{18}–10^{19} spins/g) than the typical Mg or Fe^{2+} contents from chemical analyses (Hall, 1980). After annealing at *c.* 200°C, X-irradiated Mg- or Fe^{2+}-doped kaolinites produced spectra equivalent to the Centres A and A′ present in natural samples (see below).

On the basis of combined EPR and Mössbauer spectroscopic data, Cuttler (1980, 1981) concluded that Centre A is essentially an electron hole associated with a 'trioctahedral domain', in which three divalent ions replace two aluminiums and a vacancy (Figure 5.20). Such a domain in an otherwise ordered structure generates an edge with a positive charge deficiency and another with an electron deficiency. The mechanism proposed for the formation of Centre A involves the ejection of an electron from the dioctahedral–trioctahedral interface where there is a positive charge deficiency, leaving a paramagnetic hole centred on an inner O atom bonded to Si (Figure 5.21(a)). The ejected electrons are presumably trapped elsewhere in the structure, although their EPR spectra have not been observed. (Note that trapped electron holes generally have $g > 2.0023$ and trapped electrons $g < 2.0023$). As the trioctahedral cell size is increased, the distance that ejected electrons have to migrate is also increased, a factor that favours the recombination reaction and a decreasing EPR signal with increasing Fe content. Also charge balance between normal dioctahedral zones and trioctahedral domains can be achieved by oxidation of some Fe^{2+} to Fe^{3+} (Figure 5.21(b)). This

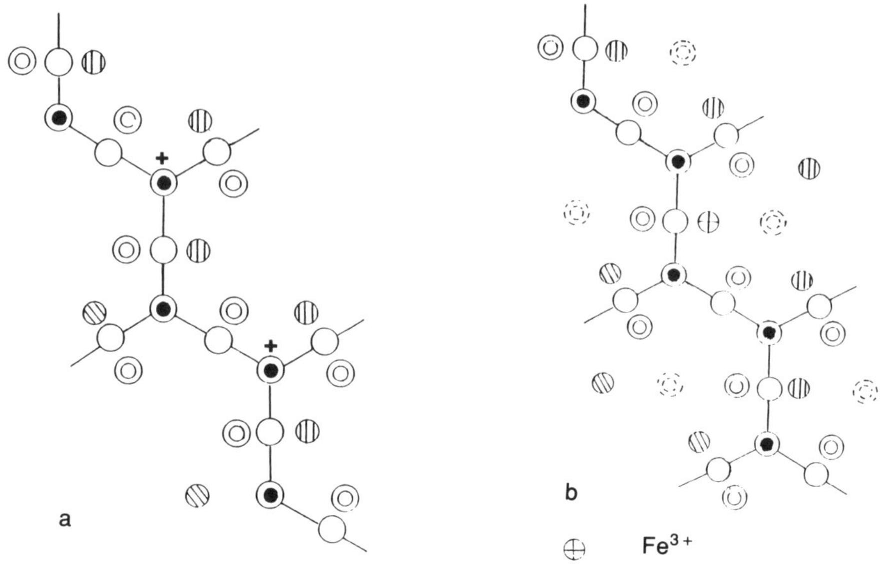

Figure 5.21 Diagrammatic representation of the sources of EPR signals in Fe^{2+} doped kaolinites. (a) creation of an electron hole on Si located on the edge of a dioctahedral–trioctahedral boundary (Centre A), and (b) oxidation of Fe^{2+} to Fe^{3+} (Centre I). (Reproduced from Cuttler, 1980.)

mechanism has been proposed for Centre I, which explains its association with crystal defects and disorder as well as its inverse relationship with Centre A. (Centres IIa and IIb can be attributed to sites of Fe^{3+} for Al substitutions remote from dioctahedral–trioctahedral interfaces.)

Recent detailed studies of the EPR spectra of kaolinites formed in a range of petrological environments have shown that there are in fact two distinct types of Centre A, termed A and A′ (Muller *et al.*, 1990; Ildefonse *et al.*, 1991). Centre A′ has a slightly less anisotropic g-tensor ($g_{\parallel} = 2.039$, $g_{\perp} = 2.009$) and slightly lower thermal stability than Centre A, being annealed at 350°C rather than 400°C. Both centres have been assigned to electron holes trapped on apical oxygens (Si–O^- centres), differing in their orientations with respect to the crystal axes of the kaolinite structure. Electron paramagnetic resonance spectra of Centres A and A′ in naturally irradiated kaolinites are illustrated in Figure 5.22.

The B centres resemble defects in irradiated Al-substituted quartz (O'Brien and Pryce, 1955) and may be assigned to electron holes located at Al–O–Si or Al–O–Al groups, which arise from tetrahedral Al-for-Si substitutions. These centres exhibit hyperfine structure, more easily seen at 77 K than at ambient temperature, as a result of interactions between

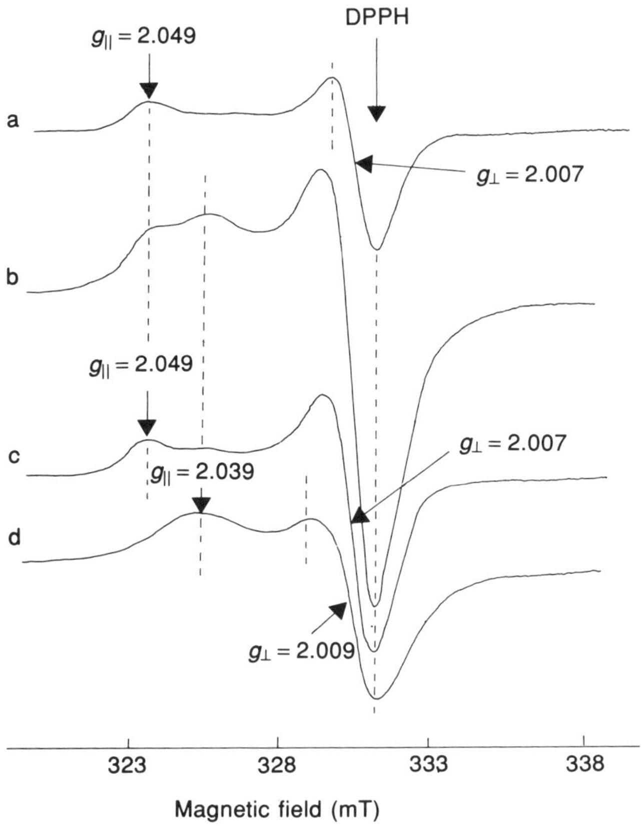

Figure 5.22 Comparative EPR spectra of hydrothermal kaolinites from an altered volcanic tuff. (a) kaolinite formed at some distance from uranium-bearing minerals; (b) kaolinite coexisting with uranium-bearing minerals; (c) sample (b) after annealing at 350°C; (d) difference spectrum obtained by subtracting spectrum (c) from spectrum (b). (Reproduced from Muller, Ildefonse and Calas, 1990.)

an unpaired electron and one or two ^{27}Al nuclei ($I = \frac{5}{2}$). They are of relatively limited thermal stability, being annealed at 200°C, and are usually of low intensity in natural kaolins.

Although the existence of Centres A, A′ and B depends on the presence of appropriate isomorphous substituent atoms, they are not observed until the minerals have been exposed to high-energy radiation. The nature of the geochemical processes involved have been elucidated recently by Muller and co-workers (Muller and Calas, 1989; Muller *et al.*, 1990; Ildefonse *et al.*, 1991). They investigated the relationship between the concentrations of Centres A and A′ in kaolinites formed in various

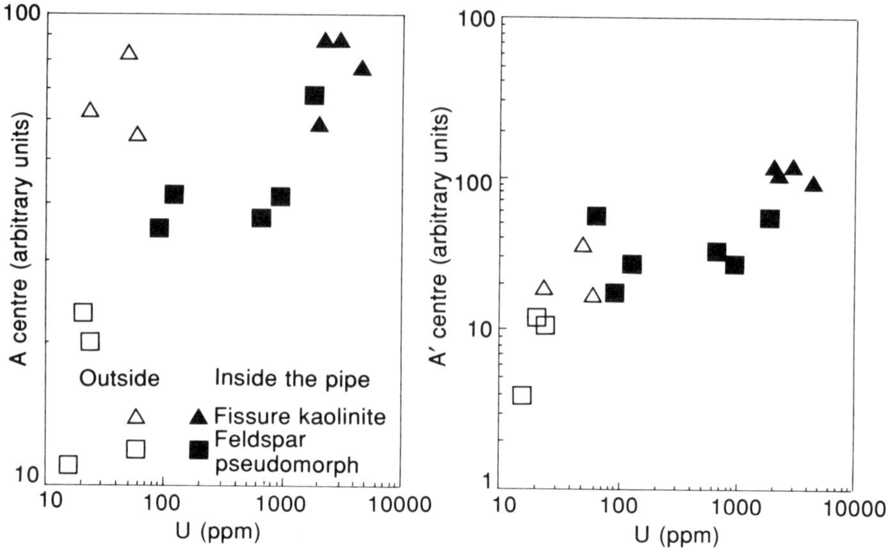

Figure 5.23 Relationship of concentration of A and A′ centres to uranium concentrations for hydrothermal kaolinites from various locations within an altered volcanic tuff. (Reproduced from Ildefonse *et al.*, 1991.)

natural petrological environments, including two tropical laterite weathering systems, one iron-rich and one iron-poor, and a hydrothermally altered tuff associated with a uranium deposit. There was a higher concentration of defect centres in the tuffaceous kaolinites. In both types of environment, however, there was a close association between defect centre concentration and the accumulated radiation dose associated with gamma-emission from isotopes in the uranium decay series. Figure 5.23 illustrates the relationship between the intensities of the A and A′ signals and uranium concentration of the hydrothermal tuff environment. In this case, uranium mineralization was restricted to a breccia pipe and kaolinite was formed either in fissures or as feldspar pseudomorphs. The intensities of the Centre A and Centre A′ peaks were also correlated with the total iron content in the iron-rich laterite. This was attributed to adsorption of U and Th by iron oxide gels, which are precursors to kaolinite formation.

The marked sensitivity of kaolinites to the level of external radioactivity suggests that they may have possible uses as dosimeters for assessing the performance of radioactive waste repositories (Ildefonse *et al.*, 1991). Also, if details of the dose levels and decay rates are known, the EPR spectra of defect centres can be used as an approximate dating technique (see Grün, 1989).

The stable defect centres in kaolinite have been shown to be associated with dehydration-induced luminescence phenomena and the ability of kaolinite surfaces to effect polymerization of amines and amino acids (Coyne, Lahav and Lawless, 1981; Coyne, Sweeney and Hovatter, 1983; Coyne, Mariner and Rice, 1991). These defect centres thus appear to function as centres of energy storage and transduction, and it has been suggested that they may be associated with prebiotic chemical evolution (Coyne, 1985; Coyne, McKeever and Blake, 1990).

5.6.2.3 EPR spectra of kaolin polytypes

The various kaolin polytypes show differences in their iron signals near $g = 4$, with kaolinite, dickite, nacrite and halloysite all having different numbers of peaks (Komusinski, Stoch and Dudiel, 1981; Chaikum and Carr, 1987; Nagasawa and Noro, 1987). There are also considerable variations in the broad signals around $g = 2$, although in many cases their intensities can be reduced considerably by magnetic separation. The removable components, therefore, correspond to iron-rich impurity phases mechanically admixed with the kaolin, whereas the remaining signal corresponds to superexchange coupling between Fe–O–Fe groupings that are chemically bonded to the mineral surface.

In a study of a range of natural kaolins of American and British origin, Angel and Vincent (1978) arrived at the conclusion that the iron oxide phases in the former were of hematitic or goethitic character, whereas they resembled lepidocrocite in the latter. In a study of acid dissolution of tropical soil kaolins, Herbillon *et al.* (1976) found two types of non-structural Fe, the major component dissolving at the same rate as Al and a minor component (responsible for part of the $g = 2.0$ signal) that was resistant to the acid treatment. The use of Mössbauer spectroscopy in the identification of iron-rich impurity phases associated with kaolins and other aluminosilicate minerals is described in Chapter 3.

5.6.3 Paramagnetic sites in smectites and illites

Most smectites give EPR spectra with resonances from Fe^{3+} ions in the $g = 4.3$ and $g = 2.0$ regions. Oliver, Vedrine and Pezerat (1975) have attributed the $g = 4.3$ resonances to a combination of two distinct octahedral and two tetrahedral sites, the former being assumed to arise from the different (*cis* and *trans*) arrangements of OH groups in the $FeO_4(OH)_2$ entity. However, the weak higher field peak near $g = 4$ has been shown to be influenced by the nature and position of exchangeable cations (McBride, Pinnavaia and Mortland, 1975a). The intensity of the weak signal decreased on drying Na- or K-exchanged clays, but was unaffected in Li or Ca clays (Figure 5.24). This indicates that the more easily dehydrated Na^+ and K^+ ions can perturb the local symmetry of the

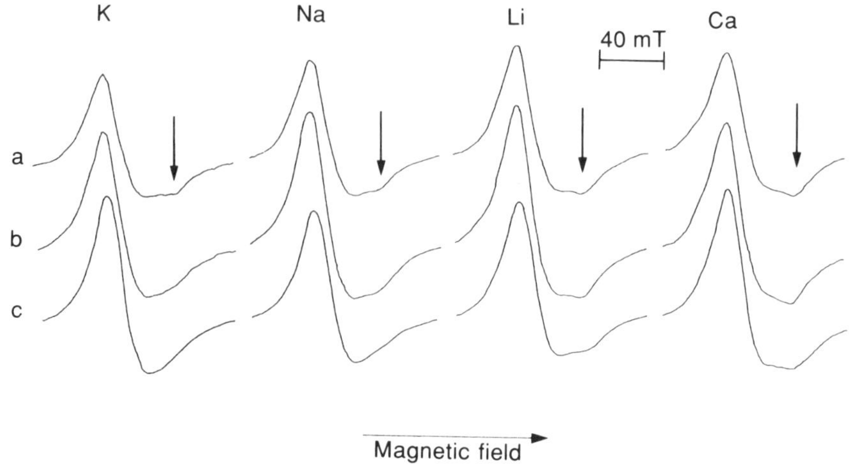

Figure 5.24 The $g = 4.3$ region of the Fe^{3+} EPR spectra of montmorillonite, with different exchange cations after equilibration at different relative humidities; (a) 93%, (b) 45% and (c) 0%. Arrows indicate the weak, higher field Fe^{3+} resonance. (Reproduced from McBride, Pinnavaia and Mortland, 1975a.)

Fe^{3+} octahedron if they can approach close enough. On this basis McBride, Pinnavaia and Mortland (1975a) suggested that the difference between the two octahedral Fe^{3+} sites may lie in the nature of the adjacent octahedral cation (divalent or trivalent). The origin of the difference remains somewhat speculative, especially as spectra may contain strong features associated with interactions between surface-adsorbed and substituted iron (e.g. Crăciun and Meghea, 1985). This problem is exacerbated further by the fact that smectites are often not single-phase specimens (Goodman, Nadeau and Chadwick, 1988).

A relatively limited amount of work has been performed on illites. For example, Michael and McWhinnie (1989) have used EPR and Mössbauer spectroscopy to follow changes in the local environments of ferrous and ferric ions in illite and montmorillonite following thermal decomposition and the formation of various higher temperature glassy and crystalline phases (e.g. spinels, mullite and hematite). However, as mentioned above for smectites, such work on natural illite specimens is often complicated by the fact that the iron is present initially in more than one phase (Goodman and Nadeau, 1988).

Evidence from EPR spectra for the substitution of other ions, such as Cu^{2+} and V^{4+}, for Al in the octahedral sheets of smectites and illite–smectites has been presented recently (Goodman and Nadeau, 1990; Mosser *et al.*, 1990). It is in the identification of such low-level substitutions that the EPR technique is able to make unique contributions to mineral characterization.

5.6.4 EPR spectra of interlayer cations

In this section we review applications of the EPR technique to the characterization of paramagnetic cations in the interlayer spaces of smectite clays. Early studies were performed with common transition metal cations, such as Cu^{2+} or Mn^{2+}, present either as homoionic exchange species or as dilute dopants (e.g. Mn^{2+} in a Ca-montmorillonite). In recent years there has been considerable interest in smectites having adsorbed polynuclear inorganic cations (e.g. Al_{13} or Zr_4 oxyhydroxides). These materials, which are known as 'pillared' clays, possess large c-axis spacings that are stable to relatively high temperatures, and have many catalytic applications (Pinnavaia, 1983). Following the initial characterization of the structure and bonding of copper ions in montmorillonite, an increasing number of studies have involved investigations of the behaviour of transition metal and rare-earth cations on pillared clays. In the following five subsections a variety of studies are reviewed and are organized according to the nature of the paramagnetic probe.

5.6.4.1 *Hydrated copper ions*

Electron paramagnetic resonance spectroscopy has proved to be a powerful technique for characterizing the coordination environment and orientation of hydrated metal ions in the interlayer spaces of expanding layer clay minerals. This is nicely illustrated by the Cu^{2+}–hectorite system (Clementz, Pinnavaia and Mortland, 1973). Under air-dried conditions, where the interlayer spacing corresponds to a monolayer of intercalated water (d_{001} of 12.4 Å), the EPR spectrum of Cu^{2+} has axial symmetry. It is characterized by parameters $g_{\parallel} = 2.34$, $g_{\perp} = 2.08$ and $A_{\parallel} = 165 \times 10^{-4}\,cm^{-1}$ (15.1 mT); $A_{\perp}$ being too small to be resolved. With oriented films of the clay the Cu^{2+} spectra showed orientation dependence: the $g_{\parallel}$ component corresponded to the situation where the magnetic field was perpendicular to the clay sheets, whereas the $g_{\perp}$ component was seen when the magnetic field was in the plane of the clay sheets (Figure 5.25). Thus the symmetry axis of the planar $Cu(H_2O)_4^{2+}$ complex is aligned perpendicular to the plane of the silicate sheets. Similar results have been obtained for hydrated Cu^{2+} ions in adsorbed monolayers of water in other clays such as montmorillonite, saponite and vermiculite.

In vermiculite, with a double layer of adsorbed water ($d_{001} = 14.24$ Å), the anisotropic Cu^{2+} EPR spectrum was found to be independent of orientation of the clay. The parameters $g_{\parallel} = 2.38$ and $g_{\perp} = 2.16$ are consistent with a copper (II) hexaquo complex, and hence its principal symmetry axis must be aligned at roughly 45° to the plane of the silicate sheets.

In contrast, when Cu^{2+} was doped at the 5% level into Mg-hectorite, which also had a double layer of adsorbed water ($d_{001} = 15.0$ Å), an

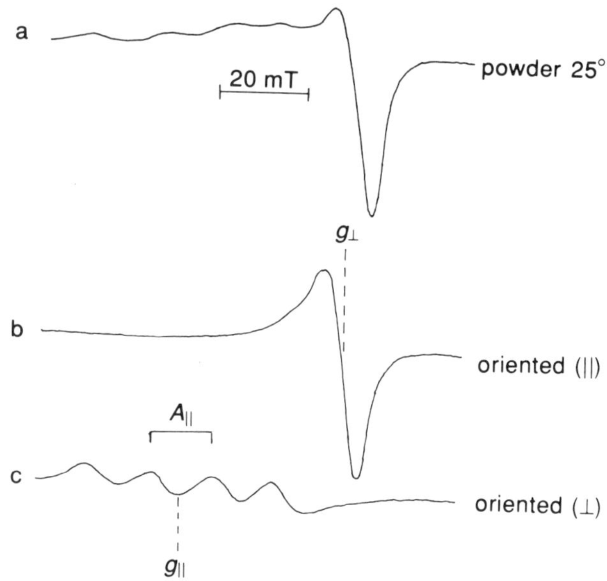

Figure 5.25 Electron paramagnetic resonance spectra at 300 K of air-dried Cu^{2+}–hectorite; (a) randomly oriented powder, (b) oriented film with the magnetic field in the basal plane, and (c) oriented film with the magnetic field perpendicular to the basal plane. (Reproduced from Clementz, Pinnavaia and Mortland, 1973.)

orientation-dependent spectrum was obtained with the principal axis perpendicular to the plane of the sheet. This result suggests a three-layer water structure with some keying of water molecules into the ditrigonal cavities between silicate oxygen atoms. Analysis of the *g*-values (McBride, 1976a) suggested that the copper was more covalently bound in hectorite than in vermiculite, the larger tetrahedral charge in the latter resulting in stronger hydrogen bonding between the surface oxygens and the interlayer water.

In fully hydrated clays ($d_{001} \approx 20$ Å) EPR spectra consist of a single isotropic peak, which can arise either because of rapid molecular tumbling or from a dynamic Jahn–Teller effect (i.e. a rapid interchange in the direction of orientation of the axial distortion from octahedral symmetry).

Copper has also been used to probe the swelling properties of reduced-charge montmorillonites (Clementz, Mortland and Pinnavaia, 1973). In collapsed layers, copper is coordinated to four water molecules, whereas expanded layers exhibit isotropic spectra as described above. Progressive charge reduction increases the tendency for the layers to remain collapsed even after prolonged equilibration at 100% relative humidity.

Electron paramagnetic resonance spectroscopy has been used to follow the migration of copper into the ditrigonal cavities on the siloxane surface, which occurs on heat treatment of Cu^{2+}-doped Ca-montmorillonite. This localization of the Cu^{2+} ions results in a reduction in spectral intensity, which is only partly reversible on rehydration (McBride and Mortland, 1974). In a mixed Na, Cu^{2+}-montmorillonite, McBride (1976a) has used EPR to demonstrate that in specimens with high sodium : copper ratios the ion exchange is not homogeneous and that there is selectivity of the surface for Cu^{2+}. In mixed copper tetraalkylammonium montmorillonites the copper ions were less fully hydrated than in the homoionic case (McBride and Mortland, 1975) and consequently less able to form complexes with other ligands.

The hydration state of Cu^{2+} in mixed Cu^{2+}–hexadecylpyridinium montmorillonites has been studied extensively under various controlled relative humidities, p/p_0, by Dyrek, Klapyta and Sojka (1983). The symmetry of the hydrated Cu^{2+} ion was found to increase with decreasing values of p/p_0. At a given p/p_0 the degree of hydration of the copper and the degree of covalency of the Cu–O bond was greater than in homoionic Cu^{2+}-montmorillonite.

Hydrolysis and thermal modification of Cu^{2+} ions in hectorite has also been investigated (McBride, 1982). Some hydrolysis occurs at ambient temperature even at low pH, but at 110°C extensive hydrolysis occurs, especially at high pH. Both air-drying and heating at 110°C promote the formation of the square planar $Cu(H_2O)_4^{2+}$ complex. At higher temperatures some or all of these ligands are lost and direct coordination of copper to the silicate oxygens occurs. This produces a concomitant reduction in the intensity and symmetry of the Cu^{2+} EPR signal. Guitierrez Le Brun and Gaite (1980) have performed computer simulations of the line-shapes for axially symmetric hydrated Cu^{2+} in dried films of Cu^{2+}–hectorite. They found that there was considerable misorientation of the platelets and that this could be represented by a distribution function of breadth *c.* 20°.

Both EPR and electron spin echo envelope modulation (ESEEM) have been used by Brown and Kevan (1988a) to investigate the hydration sphere of copper ions adsorbed on montmorillonite. (The ESEEM technique is not discussed in this Chapter and the reader is referred to Kevan (1981) for further details of the technique.) Brown and Kevan (1988a) showed that $Cu(H_2O)_n$ species exist, where $n = 6$, 4 or 1, depending on the degree of hydration or prior heat treatment of the clay. In the latter two cases the Cu^{2+} cations were either partially or largely coordinated to surface oxygen atoms of the clay. Similar studies on montmorillonites solvated with primary alcohols (Brown and Kevan, 1988b) showed that in this case the Cu^{2+} ions were coordinated primarily to four solvent molecules in a square planar configuration. Recently, Comets, Luca and Kevan (1992) have extended the range of surface studies to include synthetic hectorites and beidellite.

Braddell *et al.* (1987) and Kukkadapu and Kevan (1988, 1990) have attempted to characterize the location and bonding of Cu^{2+} ions introduced into alumina and zirconia pillared clays. Both mobile and immobile copper species have been observed, with the amounts in each form being dependent upon the experimental conditions.

5.6.4.2 *Hydrated vanadyl ions*

Fully hydrated VO^{2+}-hectorite gives an isotropic eight-line EPR spectrum which is similar to that obtained with VO^{2+} solutions and also to that of a full hydrated 1 : 1 VO^{2+} : Mg hectorite (Figure 5.26(a); McBride, 1979a). The slightly broader lines that were observed with the clay specimens indicate a somewhat restricted mobility for the complex in the interlayer spaces of the clay. On drying, the VO^{2+}-hectorite produced an anisotropic spectrum, which was dependent on the orientation of the clay, the $g_{\parallel}$ feature being observable only when the hectorite film was aligned perpendicular ($\perp$) to the magnetic field (Figure 5.26(b)). At low vanadium levels the spectra were dominated by an anisotropic component, even in fully wetted specimens. This spectrum showed little or no dependence on the orientation of the clay and was assigned by McBride (1979a) to the hydrolysis product $VO(OH)_2(H_2O)_3$. A similar investigation of solvated VO^{2+} in hydroxyaluminium pillared clays has been reported by Braddell, Barklie and Doff (1990).

5.6.4.3 *Hydrated Mn^{2+} ions*

Hydrated Mn^{2+} ions exchanged on clays also have lines that are broader than for the ion in aqueous solution (McBride, Pinnavaia and Mortland, 1975b). These linewidths increased progressively with decreasing levels of hydration and have been interpreted in terms of two broadening mechanisms: (1) dipolar spin–spin relaxation between neighbouring Mn^{2+} ions, and (2) relaxation processes involving ion–solvent molecule collisions. In the case of homoionic Mn^{2+} clays the dipolar interactions are dominant and provide information on the mean Mn–Mn distances in the various minerals. The anisotropy of the linewidths has been interpreted (McBride, Pinnavaia and Mortland, 1975b) in terms of an interaction between the Mn^{2+} and structural Fe^{3+} ions. In hectorite, which contains very little Fe^{3+}, dipolar broadening was minimized by doping a Mg-exchanged specimen with 5% Mn. The increase in linewidths with decreasing hydration in this sample gave a direct measure of the increase in orientational correlation time of the hydrated ions due to the decreased frequency of the Mn^{2+}-ion–water molecular collisions. In the fully hydrated Mn^{-}/Mg-hectorite ($d_{001} \approx 21$ Å), the linewidths indicated a relaxation time that was only 30% longer than for Mn^{2+} in aqueous solution, and indicated a liquid-like interlamellar environment. De-

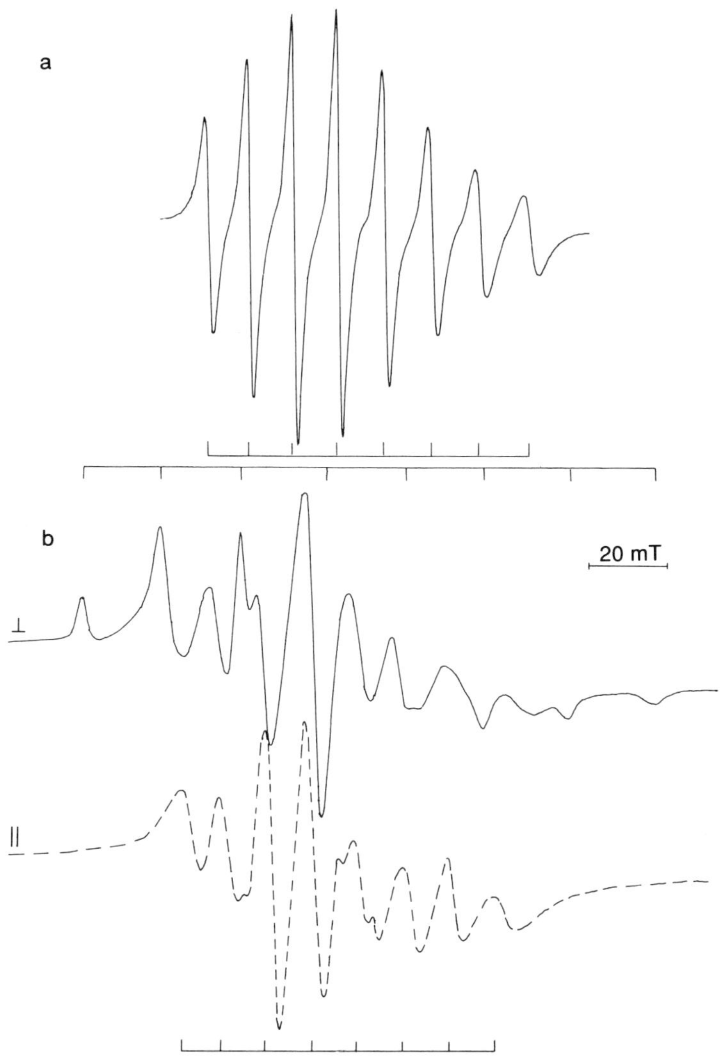

Figure 5.26 Electron paramagnetic resonance spectra at 300 K of Mg^{2+}-hectorite with 50% mol ratio VO^{2+}; (a) fully hydrated, and (b) air-dried film oriented parallel (||) and perpendicular (⊥) to the magnetic field. Positions of peaks are indicated by stick diagrams. (Adapted from McBride, 1979a.)

hydration at 200°C resulted in a crystalline rather than a solution-like EPR spectrum. This result has been attributed to migration of Mn into hexagonal cavities in the silicate structure.

Similar results, suggesting an overall reduction in the mobility of exchanged ions compared to those in bulk solution, have been reported for clays in which the water molecules coordinated to the Mn were replaced by organic molecules such as pyridine (Tarasevich and Ovcharenko, 1973). Duczmal and Buckmaster (1987) have shown that the change of mobility in aqueous complexes of Mn, Ca-exchanged montmorillonite at 373 K is essentially a liquid to liquid–crystal phase transition.

5.6.4.4 Other cations

Luca and Cardile (1989) have used a combination of EPR and Mössbauer spectroscopy to monitor the interactions of hydrated ferric ions with montmorillonite and hectorite (laponite) surfaces. Differences observed as a function of heat-treatment and resolvation could be explained in terms of the different orientations of structural OH groups with respect to the ditrigonal cavities in the dioctahedral and trioctahedral clays.

Brown, Luca and Kevan (1991) have investigated the solvation environment of Ag^0 atoms (produced by radiolytic reduction of Ag^+) in Na-exchanged and Al_{13}-pillared montmorillonites. In the fully hydrated sodium clay, silver had a coordination number of 4, as in bulk solution, but in other cases it was less strongly hydrated. The ^{27}Al spin echo envelope modulation indicated that silver atoms interacted with the pillars in the pillared clay, but not with the aluminium in the tetrahedral sheets of the clay. Similar studies have also been reported by Carrado *et al.* (1986) and Kukkadapu, Luca and Kevan (1992) on the interactions of Cr^{3+} and Pd^{2+} with pillared clays.

5.6.4.5 Dynamics of spin labels

Extensive EPR work has been carried out by McBride (1976b, 1977a, b, 1979b) on the dynamics of cationic spin labels in the interlamellar regions of smectites. When the spin label TEMPO$^+$ (TEMPO ≡ 2,2,6,6,-tetramethyl-1-piperidinyloxy-) was present at up to 10% of the cation exchange capacity (CEC) of hectorite exchanged with various mono- and divalent cations, rotational correlation times were in the range $1–5 \times 10^{-9}$ s, compared with 5×10^{-11} s for aqueous solution. Correlation times were also dependent on the nature of the solvent used to equilibrate the clays and in some cases a broad background resonance was observed, which was attributed to some TEMPO$^+$ molecules being immobilized on the EPR time-scale (10^{-8} s). This effect was most pronounced for solvents with low polarity.

Studies with oriented films showed that the spectra had some anisotropy, even for fully hydrated hectorite. Thus molecular motion is non-random and the N–O bond axis has a preference for alignment perpendicular to the plane of the silicate sheet. McBride (1979b) also showed that there was some non-heterogeneity in the adsorption of the spin probes on hectorite. There was a tendency for aggregation in certain layers even at low adsorption levels, because of a higher affinity of the probes for the hydrated clay surfaces than for the solvent.

From the above discussion it is clear that the mobility of ions in the interlamellar spaces of expanding layer silicates are lower than in pure solutions. The extent of this reduction in mobility is inversely dependent on the level of hydration of the clays. However, the interlayer regions are clearly solution-like in character and only under conditions of severe dehydration do the EPR spectra indicate a quasi-crystalline character. In such cases either the cations become embedded in the ditrigonal cavities on the silicate surface or hydrolysis of the exchange ions occurs, with the formation of polymeric oxyhydroxides.

5.6.5 Surface complexes and reactions

Electron paramagnetic resonance spectroscopy is a very powerful technique for the characterization of the nature of the bonding in paramagnetic transition metal complexes. Not surprisingly, therefore, it has been used extensively in characterizing the reactions between exchangeable cations at clay surfaces and organic adsorbate molecules. At its simplest, ion–adsorbate complex formation may be inferred from changes in spectral parameters. It is, however, often possible to identify the nature of the coordination environment of the paramagnetic metal centre or the nature of any organic radical species that might be formed.

A good example of the use of EPR spectroscopy is in the study of complexes of Cu^{2+} with ethylenediamine (en) (Velghe *et al.*, 1977). In solution the predominant species is the ion $Cu(en)^{2+}$ which has EPR parameters $g_{\parallel} = 2.261$, $g_{\perp} = 2.063$, $A_{\parallel} = 152 \times 10^{-4}\,cm^{-1}$ (14.4 mT) and $A_{\perp} = 13 \times 10^{-4}\,cm^{-1}$ (1.3 mT). However, on the surface of montmorillonite a reaction occurs to produce in addition the bis complex $Cu(en)_2^{2+}$, which has parameters $g_{\parallel} = 2.181$, $g_{\perp} = 2.030$, $A_{\parallel} = 204 \times 10^{-4}\,cm^{-1}$ (20.0 mT) and $A_{\perp} = 7 \times 10^{-4}\,cm^{-1}$ (0.7 mT), with its symmetry axis aligned at approximately 45° to the silicate sheets. With 1,10-phenanthroline, a tris complex $Cu(phen)_3^{2+}$ is also produced on reaction with montmorillonite (Berkheiser and Mortland, 1977). On heating to 475 K this complex was found to disproportionate to $Cu(phen)_2^{2+}$ and the free ligand. Work on the EPR spectra of these and other complexes has been extended by Schoonheydt (1978) and Schoonheydt *et al.* (1979), who calculated the in- and out-of-plane bonding parameters. The out-of-plane π-bonding was found to be slightly greater on clay surfaces than in pure solutions. The effect was small, however, with the π-bonding being weak in both cases.

In the reaction of Cu-montmorillonite with aromatic molecules, two different types of complex can be produced (Figure 5.27). Type I corresponds to a Cu^{2+} complex, whereas the Type II radical has the unpaired electron delocalized over the benzene ring. The EPR spectra of Type II benzene complexes consist of a sharp isotropic resonance near $g = 2.0$,

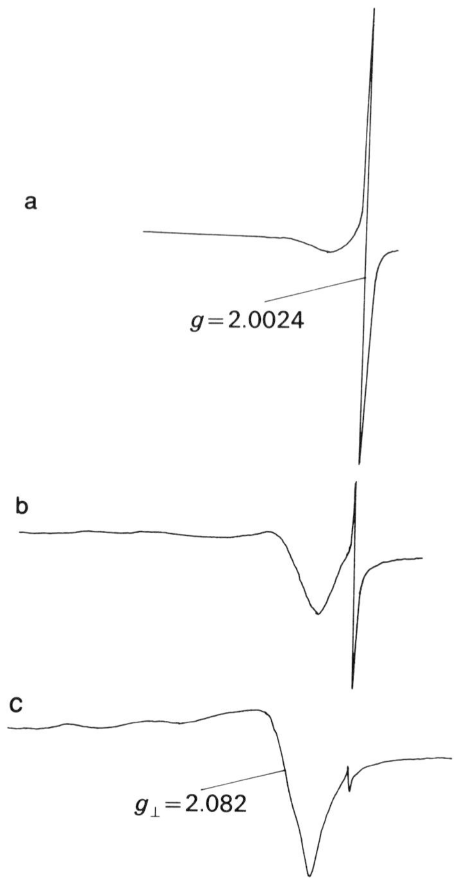

Figure 5.27 Electron paramagnetic resonance spectra of Types I and II Cu(II)–benzene complexes in montmorillonite; (a) Type II complex prepared at room temperature via P_2O_5, (b) mixed complex prepared at 80°C via azeotropic dehydration, and (c) Type I complex prepared by hydration of the complex that produced spectrum (b). (Reproduced from Rupert, 1973.)

which has been assigned to the radical $C_6H_6^{+\cdot}$. The absence of any Cu^{2+} resonance was interpreted as demonstrating a reduction to Cu^+. Dehydration of this radical complex resulted in polymerization to polyparaphenylene (Mortland and Halloran, 1976; Stoessel, Guth and Wey, 1977). A summary of the overall reaction scheme for the montmorillonite–copper–benzene system is presented in Figure 5.28.

Many other arenes undergo similar reactions on smectite surfaces with Cu^{2+} or other transition metal ions, such as VO^{2+}, Fe^{3+}, Ru^{3+} or Pd^{2+}, as the exchange cations. However, with vanadium and iron the polymeriza-

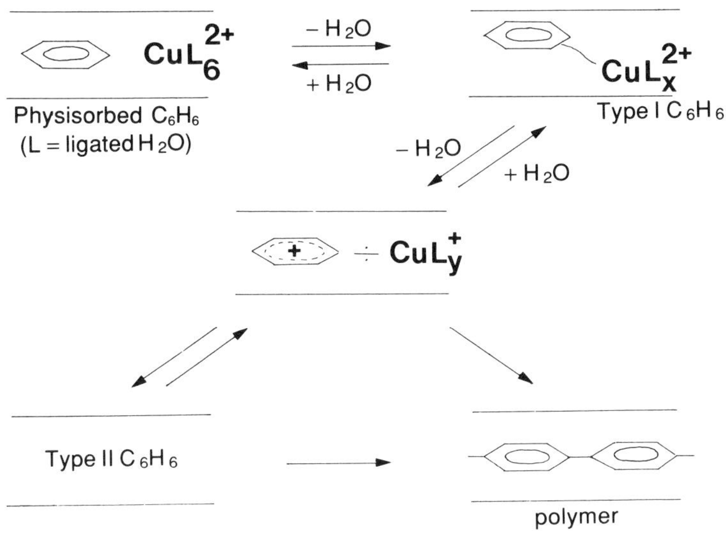

Figure 5.28 Reaction scheme for the montmorillonite–copper–benzene system. (Reproduced from Pinnavaia, 1981.)

tion reactions only proceed with freshly prepared films, presumably because of slow hydrolysis of the metal ions to produce inert oxyhydroxide species (Pinnavaia *et al.*, 1974). In other instances propagation of the polymer is limited, e.g. anisole produces only the dimer 4,4′-dimethoxybiphenyl, which itself can exist as a Type II radical complex (Fenn, Mortland and Pinnavaia, 1973; Johnston *et al.*, 1991). Other arene molecules, such as toluene, do not produce an observable Type II complex, although radical cation formation and subsequent polymerization do occur.

This general pattern of behaviour, in which modification and polymerization of adsorbed organic species is promoted by single electron transfer reactions and the formation of organic radical cation intermediates, extends to more complex environmentally significant ligands, such as chlorinated arenes and alkenes (Boyd and Mortland, 1986; Mortland and Boyd, 1987). Recent work has included the interactions of smectites with various types of herbicides (e.g. Pusino, Micera and Gessa, 1991), as well as a report of stereospecificity in solid-state intercalation of some *cis–trans* isomers into smectites (Ogawa *et al.*, 1992).

5.7 CONCLUSIONS

Although EPR spectroscopy is not one of the techniques for primary characterization of clay minerals it has been used extensively over the past two decades in investigations of minor structural substituents and in the interactions of clays with external cations, water and a wide variety of organic molecules. In conjunction with other complementary techniques EPR spectroscopy has proved to be an extremely powerful technique for addressing those problems that fall within its domain of applicability.

In deciding whether or not a particular problem is amenable to study by EPR spectroscopy, the researcher should always bear in mind the three principal aspects of the method, which can be summarized under the headings (1) selectivity, (2) sensitivity and (3) scale of investigation.

Firstly, EPR is highly selective in terms of the chemical species it can detect; they must contain unpaired electrons. Even then, other criteria must be satisfied before resonant absorption of microwave energy by the spin system can be observed experimentally. Electron paramagnetic resonance active species include transition metal and rare-earth metal ions, present either as individual substituents in clay mineral structures, as exchangeable cations, or as surface-adsorbed species; organic or inorganic free radicals and paramagnetic centres associated with crystal defects. However, if EPR active species are present in clay minerals or other materials, either naturally or as a result of treatment in the laboratory, they are detectable to very low levels (down to parts per billion under ideal circumstances). Finally, EPR spectroscopy is a technique whose scale of observation is atomic; it gives information about the immediate atomic or structural environments of the unpaired electron spins, but it is insensitive to major diamagnetic components of aluminosilicate minerals, except when, as in the case of Al in kaolinites, a non-zero nuclear spin can interact with unpaired electron spins to produce characteristic multiline hyperfine structures in the spectra.

Under the right conditions, EPR spectroscopy can, therefore, provide considerable detail about the local chemical and crystalline environments of unpaired electrons.

One particular strength of EPR spectroscopy is its ability to characterize redox or electron transfer reactions involving the formation or destruction of unpaired electron spins. Electron paramagnetic resonance has been applied extensively to studies of the chemical environments of paramagnetic ions present at low levels in the interlamellar regions of natural or artifically pillared smectites, occasionally in conjunction with the electron spin echo envelope modulation ('pulsed EPR') technique (Kevan, 1981). A considerable quantity of detailed information has been obtained on the nature of the bonding and solvation of a wide variety of probe cations present at the interlayer surfaces of smectites having a range of chemical compositions. Much of this work has been motivated by the potential applicability of these systems as highly selective heterogeneous catalysts.

Although EPR spectroscopy has probably provided less direct information relating to the paragenesis and transformation of clays in natural environments, an exception to this is the recent progress in relating the EPR spectra of stable defect centres in kaolinites to details of natural irradiation over geological or shorter time periods. This work holds out some promise for the use of kaolinites as natural dosimeters to monitor the performance of radioactive waste repositories (Ildefonse *et al.*, 1991). Over the past 25 years our understanding of the nature, origin and geochemical significance of the stable defect centres in kaolinites has evolved to the point where practical applications may be envisaged. This is a rewarding example of the advances which become possible when there is synergy between physics, chemistry, mineralogy and geology.

ACKNOWLEDGEMENTS

This work was carried out in part with funding provided by the Scottish Office Agriculture and Fisheries Department, for which acknowledgement is gratefully given. Figures 5.11, 5.12, 5.18, 5.19, 5.20, 5.21, 5.22, 5.23, 5.24, 5.25, 5.26, 5.27 and 5.28 are reproduced with permission from the copyright holders:
5.11 and 5.12 Aasa (1970) *Journal of Chemical Physics*, **52**, 3919–3930.
5.18 Meads and Malden (1975) *Clay Minerals*, **10**, 313–345.
5.19 Brindley *et al.* (1986) *Clays and Clay Minerals*, **34**, 239–249.
5.20 and 5.21 Cuttler (1980) *Clay Minerals*, **15**, 429–443.
5.22 Muller *et al.* (1990) *Clays and Clay Minerals*, **38**, 600–608.
5.23 Ildefonse *et al.* (1991) *Materials Research Society Symposium Proceedings*, **212**, 749–756.
5.24 McBride *et al.* (1975a) *Clays and Clay Minerals*, **23**, 162.
5.25 Clementz *et al.* (1973) *Journal of Physical Chemistry*, **77**, 196–200.
5.26 McBride (1979a) *Clays and Clay Minerals*, **27**, 97–104.
5.27 Rupert (1973) *Journal of Physical Chemistry*, **77**, 784–790.
5.28 Pinnavaia (1981) Elsevier.

REFERENCES

Aasa, R. (1970) Powder line shapes in electron paramagnetic resonance spectra of high-spin ferric complexes. *Journal of Chemical Physics*, **52**, 3919–3930.

Angel, B. R. and Hall, P. L. (1973) Electron spin resonance studies of kaolins. *Proceedings, Fourth International Clay Conference, Madrid 1972*, pp. 47–60.

Angel, B. R. and Vincent, W. E. J. (1978) Electron spin resonance studies of iron oxides associated with the surface of kaolins. *Clays and Clay Minerals*, **26**, 263–272.

Angel, B. R., Jones, J. P. E. and Hall, P. L. (1974) Electron spin resonance studies of doped synthetic kaolinites. I. *Clay Minerals*, **10**, 247–256.

Angel, B. R., Richards, K. and Jones, J. P. E. (1976) The synthesis, morphology and general properties of kaolinites specifically doped with metallic ions and defects generated by irradiation. *Proceedings, Fifth International Clay Conference, Mexico City 1975*, pp. 297–304.

Angel, B. R., Cuttler, A. H., Richards, K. and Vincent, W. E. J. (1977) Synthetic kaolinites doped with Fe^{2+} and Fe^{3+} ions. *Clays and Clay Minerals*, **25**, 381–383.

Berkheiser, V. E. and Mortland, M. M. (1977) Hectorite complexes with copper (II) and iron (II)–1, 10-phenanthroline chelates. *Clays and Clay Minerals*, **25**, 105–112.

Boyd, S. A. and Mortland, M. M. (1986) Radical formation and polymerization of chlorophenols and chloroanisole on copper (II) smectite. *Environmental Science and Technology*, **20**, 1056–1058.

Braddell, O., Barklie, R. C. and Doff, D. H. (1990) EPR of VO^{2+} ions in Al-pillared montmorillonite. *Clay Minerals*, **25**, 15–25.

Braddell, O., Barklie, R. C., Doff, D. H. *et al.* (1987) EPR of Cu^{2+} ions in pillared clays. *Zeitschrift fuer Physikalische Chemie*, **151**, 157–164.

Brindley, G. W. *et al.* (1986) On the relationship between order–disorder and other clay characteristics in kaolinites and dickites. *Clays and Clay Minerals*, **34**, 239–249.

Brown, D. R. and Kevan, L. (1988a) Aqueous coordination and location of exchangeable Cu^{2+} cations in montmorillonite clay studied by electron spin resonance and electron spin echo modulation. *Journal of the American Chemical Society*, **110**, 2743–2748.

Brown, D. R. and Kevan, L. (1988b) Solvation of exchangeable Cu^{2+} cations by primary alcohols in montmorillonite clay studied by electron spin resonance and electron spin echo modulation spectroscopies. *Journal of Physical Chemistry*, **92**, 1971–1974.

Brown, D. R., Luca, V. and Kevan, L. (1991) Electron paramagnetic resonance and electron spin echo modulation analysis of the silver atom environment in γ-irradiated silver-exchanged sodium montmorillonite. *Journal of the Chemical Society, Faraday Transactions*, **87**, 2749–2754.

Carrado, K. A., Suib, S. L., Skoularikis, N. D. and Coughlin, R. W. (1986) Chromium(III)-doped pillared clays (PILCs). *Inorganic Chemistry*, **25**, 4217–4221.

Chaikum, N. and Carr, R. M. (1987) Electron spin resonance studies of halloysites. *Clay Minerals*, **22**, 287–296.

Clementz, D. M., Mortland, M. M. and Pinnavaia, T. J. (1973) Properties of reduced charge montmorillonites: hydrated ions as a spectroscopic probe. *Clays and Clay Minerals*, **22**, 49–57.

Clementz, D. M., Pinnavaia, T. J. and Mortland, M. M. (1973) Stereochemistry of hydrated copper (II) ions on interlamellar surfaces of layer silicate: an electron spin resonance study. *Journal of Physical Chemistry*, **77**, 196–200.

Comets, J. M., Luca, V. and Kevan, L. (1992) Solvation of Cu(II) in Cu(II)-exchanged synthetic fluorohectorite, synthetic hydroxyhectorite, synthetic beidellite and montmorillonite studied by electron spin resonance. *Journal of Physical Chemistry*, **96**, 2645–2652.

Coyne, L. (1985) A possible energetic role of mineral surfaces in chemical evolution. *Origins of Life*, **15**, 161–206.

Coyne, L., Lahav, N. and Lawless, J. G. (1981) Dehydration induced luminescence in clay minerals. *Nature*, **292**, 819–821.

Coyne, L. M., McKeever, S. W. S. and Blake, D. F. (1990) Spectroscopic Characterization of Minerals and their Surfaces. *American Chemical Society Symposium Series*, **415**, 480 pp.

Coyne, L. M., Mariner, R. and Rice, A. (1991) Air oxidation of hydrazine. 1. Reaction kinetics on natural kaolinites, halloysites and model substituent layers with varying iron- and titanium-oxide and O^--center contents. *Langmuir*, **7**, 1660–1674.

Coyne, L., Sweeney, M. and Hovatter, W. (1983) Luminescence induced by dehydration of kaolin associated with electron-spin active centres and with surface activity for dehydration-polymerization of glycine. *Journal of Luminescence*, **28**, 395–409.

Crăciun, C. and Meghea, A. (1985) Electron spin resonance studies of montmorillonites. *Clay Minerals*, **20**, 281–290.

Cuttler, A. H. (1980) The behaviour of synthetic ^{57}Fe doped kaolin: Mössbauer and electron paramagnetic resonance studies. *Clay Minerals*, **15**, 429–443.

Cuttler, A. H. (1981) Further studies of a ferrous iron doped synthetic kaolinite: dosimetry and X-ray induced effects. *Clay Minerals*, **16**, 69–80.

Duczmal, T. and Buckmaster, H. A. (1987) Dynamics of $Mn(H_2O)_6^{2+}$ complex in Ca montmorillonite in the temperature interval typical for *in situ* recovery processes. *Applied Clay Science*, **2**, 273–289.

Dyrek, K., Klapyta, Z. and Sojka, Z. (1983) Hydration state of Cu^{2+} in mixed Cu^{2+} hexadecylpyridinium montmorillonite by electron spin resonance. *Clays and Clay Minerals*, **31**, 223–229.

Fenn, D., Mortland, M. M. and Pinnavaia, T. J. (1973) The chemisorption of anisole on Cu(II) hectorite. *Clays and Clay Minerals*, **21**, 315–322.

Goodman, B. A. and Nadeau, P. H. (1988) Identification of oxide impurity phases and distribution of structural iron in some diagenetic illitic clays as determined by Mössbauer spectroscopy. *Clay Minerals*, **23**, 301–308.

Goodman, B. A. and Nadeau, P. H. (1990) Electron paramagnetic resonance study of non-exchangeable vanadium (IV) in rectorites. *Clay Minerals*, **25**, 283–287.

Goodman, B. A., Nadeau, P. H. and Chadwick, J. (1988) Evidence for the multiphase nature of bentonites from Mössbauer and EPR spectroscopy. *Clay Minerals*, **23**, 147–159.

Grün, R. (1989) Electron spin resonance (ESR) dating. *Quaternary International*, **1**, 65–110.

Guitierrez Le Brun, M. and Gaite, J. M. (1980) Computer simulation study of partially oriented samples. Application to the EPR spectrum of copper hectorite. *Journal of Magnetic Resonance*, **40**, 105–113.

Hall, P. L. (1980) The application of electron spin resonance spectroscopy to studies of clay minerals: I. Isomorphous substitutions and external surface properties. *Clay Minerals*, **15**, 321–335.

Herbillon, A. J., Mestagh, M. M., Vielvoye, L. and Derouane, E. G. (1976) Iron in kaolinite with special reference to kaolinite from tropical soils. *Clay Minerals*, **11**, 201–219.

Ildefonse, P., Muller, J. P., Clozel, B. and Calas, G. (1991) Record of past contact between altered rocks and radioactive solutions through radiation-induced defects in kaolinite. *Materials Research Society Symposium Proceedings*, **212**, 749–756.

Johnston, C. T., Tipton, D., Stone, D. A. *et al.* (1991) Chemisorption of p-dimethoxybenzene on copper montmorillonite. *Langmuir*, **7**, 289–296.

Jones, J. P. E., Angel, B. R. and Hall, P. L. (1974) Electron spin resonance studies of doped synthetic kaolinites. II. *Clay Minerals*, **10**, 257–270.

Kevan, L. (1981) Electron spin echo studies of solvation structure. *Journal of Physical Chemistry*, **85**, 1628–1636.

Komusinski, J., Stoch, L. and Dudiel, S. M. (1981) Application of electron paramagnetic resonance and Mössbauer spectroscopy in the investigation of kaolin group minerals. *Clays and Clay Minerals*, **29**, 23–30.

Kukkadapu, R. K. and Kevan, L. (1988) Synthesis and electron spin resonance studies of copper-doped alumina-pillared montmorillonite clay. *Journal of Physical Chemistry*, **92**, 6073–6078.

Kukkadapu, R. K. and Kevan, L. (1990) Electron spin resonance and X-ray diffraction studies of copper (II) on Zr_4-pillared clay. *Journal of the Chemical Society, Faraday Transactions*, **86**, 691–696.

Kukkadapu, R. K., Luca, V. and Kevan, L. (1992) Electron spin resonance and electron spin echo modulation study of activated palladium (2+)-doped Al_{13} pillared clay: evidence for the migration of palladium cations from an Al_{13} pillar to the laponite layer with increasing activation temperature. *Journal of Physical Chemistry*, **96**, 415–421.

Lipsicas, M., Raythatha, R., Giese, R. F. Jr. and Costanzo, P. M. (1986) Molecular motions, surface interactions and stacking disorder in kaolinite intercalates. *Clays and Clay Minerals*, **34**, 635–644.

Loveridge, D. and Parks, S. (1971) Electron spin resonance of Fe^{3+}, Mn^{2+} and Cr^{3+} in glasses. *Physics and Chemistry of Glasses*, **12**, 19–27.

Luca, V. and Cardile, C. (1989) Cation migration in smectite minerals: electron spin resonance of exchanged Fe^{3+} probes. *Clays and Clay Minerals*, **37**, 325–332.

McBride, M. B. (1976a) Hydration structure of exchangeable Cu^{2+} in vermiculite and smectite. *Clays and Clay Minerals*, **24**, 211–212.

McBride, M. B. (1976b) Nitroxide spin probes on smectite surfaces: temperature and solvation effects on the mobility of exchange cations. *Journal of Physical Chemistry*, **80**, 196–203.

McBride, M. B. (1977a) Mobility and orientation of charged molecules at silicate surfaces. *Clay Minerals*, **12**, 273–277.

McBride, M. B. (1977b) Adsorbed molecules on solvated layer silicates: surface mobility and orientation from ESR studies. *Clays and Clay Minerals*, **25**, 6–13.

McBride, M. B. (1979a) Mobility and reactions of VO^{2+} on hydrated smectite surfaces. *Clays and Clay Minerals*, **27**, 91–96.

McBride, M. B. (1979b) Cationic spin probes on hectorite surfaces: demixing and mobility as a function of adsorption level. *Clays and Clay Minerals*, **27**, 97–104.

McBride, M. B. (1982) Hydrolysis and dehydration reactions of exchangeable Cu^{2+} on hectorite. *Clays and Clay Minerals*, **30**, 200–206.

McBride, M. B. and Mortland, M. M. (1974) Copper (II) interactions with montmorillonite: evidence from physical methods. *Soil Science Society of America, Proceedings*, **38**, 408–415.

McBride, M. B. and Mortland, M. M. (1975) Surface properties of mixed Cu (II)-tetraalkyl-ammonium montmorillonites. *Clay Minerals*, **10**, 357–368.

McBride, M. B., Mortland, M. M. and Pinnavaia, T. J. (1975a) Exchange ion positions in smectite: effects on electron spin resonance of structural iron. *Clays and Clay Minerals*, **23**, 162.

McBride, M. B. Pinnavaia, T. J. and Mortland, M. M. (1975b) Electron spin relaxation and the mobility of manganese (II) exchange ions in smectites. *American Mineralogist*, **60**, 66–72.

Meads, R. E. and Malden, P. J. (1975) Electron spin resonance in natural kaolinites containing Fe^{3+} and other transition metal ions. *Clay Minerals*, **10**, 313–345.

Mestagh, M. M., Vielvoye, L. and Herbillon, A. J. (1980) Iron in kaolinite: II. The relationship between kaolinite crystallinity and iron content. *Clay Minerals*, **15**, 1–13.

Michael, R. J. and McWhinnie, W. R. (1989) Mössbauer and ESR studies of the thermochemistry of illite and montmorillonite. *Polyhedron*, **8**, 2709–2718.

Mortland, M. M. and Boyd, S. A. (1987) Polymerization and dechlorination of chloroethenes on Cu(II) smectite via radical-cation intermediates. *Environmental Science and Technology*, **23**, 223–227.

Mortland, M. M. and Halloran, L. (1976) Polymerization of aromatic molecules on smectite. *Soil Science Society of America, Journal*, **40**, 367–370.

Mosser, C., Mestagh, M., Decarreau, M. and Herbillon, A. J. (1990) Spectroscopic (ESR, EXAFS) evidence of Cu for (Al–Mg) substitution in octahedral sheets of smectites. *Clay Minerals*, **25**, 271–282.

Muller, J. P. and Calas, G. (1989) Tracing kaolinites through their defect centres: kaolinite paragenesis in a laterite (Cameroun). *Economic Geology*, **84**, 694–707.

Muller, J. P., Ildefonse, P. and Calas, G. (1990) Paramagnetic defect centres in hydrothermal kaolinite from an altered tuff in the Nopal uranium deposit, Chihuahua, Mexico. *Clays and Clay Minerals*, **38**, 600–608.

Nagasawa, K. and Noro, H. (1987) An electron spin resonance study of halloysites. *Clay Science*, **6**, 261–268.

O'Brien, M. C. M. and Pryce, M. H. L. (1955) Paramagnetic resonance in irradiated diamond and quartz: an interpretation, in *Rept. Conf. Defects Cryst. Solids (Bristol, 1954)*, Physical Society, London, pp. 88–91.

Ogawa, M., Harata, M., Kurada, K. and Kato, C. (1992) Selective solid-state intercalation of *cis–trans* isomers into montmorillonite. *Chemistry Letters*, **1992**, 365–368.

Olivier, D., Vedrine, J. C. and Pezerat, H. (1975) Application de la RPE à la localisation du Fe^{3+} dans les smectites. *Bulletin du Groupe Français des Argiles*, **27**, 153–165.

Pinnavaia, T. J. (1981) Electron spin resonance studies in clay minerals, in *Advanced Techniques for Clay Mineral Analysis* (ed. J. J. Fripiat). Elsevier, Amsterdam, pp. 139–161.

Pinnavaia, T. J. (1983) Pillared clay catalysts. *Science*, **220**, 365–371.

Pinnavaia, T. J., Hall, P. L., Cady, S. S. and Mortland, M. M. (1974) Aromatic radical cation formation on the intracrystal surface of a transition-metal exchanged layer lattice silicate. *Journal of Physical Chemistry*, **78**, 994–999.

Pusino, P., Micera, A. and Gessa, C. (1991) Interaction of the herbicide Acifluorofen with montmorillonite: formation of insoluble Fe^{3+}, Al^{3+}, Cu^{2+} and Ca^{2+} complexes. *Clays and Clay Minerals*, **39**, 50–53.

Rupert, J. P. (1973). Electron spin resonance spectra of interlamellar copper (II)-arene complexes on montmorillonite. *Journal of Physical Chemistry*, **77**, 784–790.

Schoonheydt, R. A. (1978) Analysis of the electron paramagnetic resonance spectra of bis (ethylenediamine) copper (II) on the surfaces of zeolites X and Y and Camp Berteau montmorillonite. *Journal of Physical Chemistry*, **82**, 497–498.

Schoonheydt, R. A., Velghe, F., Baertz, R. and Utterhoeven, J. B. (1979) Complexes of diethylenetriamine (dien) and tetrethylenepentamine (tetren) with Cu (II) and Ni (III) on hectorite. *Clays and Clay Minerals*, **27**, 269–278.

Stoessel, F., Guth, L. J. and Wey, P. (1977) Polymerization of benzene to polyparaphenylene on copper (2+) montmorillonite. *Clay Minerals*, **12**, 255–259.

Swartz, J. C., Hoffman, B. M., Krizet, R. J. and Atmatzidis, D. K. (1979) A general procedure for simulating EPR spectra of partially oriented paramagnetic centres. *Journal of Magnetic Resonance*, **36**, 259–268.

Tarasevich, Y. I. and Ovcharenko, F. D. (1973) On the mechanism of interaction between nitrogenous organic substances and montmorillonite surface. *Proceedings, Fourth International Clay Conference, Madrid 1972*, pp. 627–636.

Velghe, F., Schoonheydt, R. A., Uytterhoeven, J. B. *et al.* (1977) Spectroscopic characterization and thermal stability of copper (II) ethylenediamine complexes on solid surfaces. II. Montmorillonite. *Journal of Physical Chemistry*, **81**, 1187–1194.

Wertz, J. E. and Bolton, J. R. (1986) *Electron Spin Resonance*. Chapman & Hall, New York, 497 pp.

CHAPTER 6

X-ray photoelectron spectroscopy

E. Paterson and R. Swaffield

6.1 INTRODUCTION

Surface reactions play a critical role in the behaviour and properties of clays and, as a result, many experimental and theoretical approaches have been developed for the physical and physico-chemical description of the interfacial regions involved in these reactions. These include measurements of the physical extent (specific surface area) and the topography of interfaces (porosity and surface roughness) using low-temperature gas adsorption methods, water vapour adsorption and mercury intrusion porosimetry (Gregg and Sing, 1982). The physico-chemical methods include the measurement of ion-exchange phenomena (Sposito, 1990) and the development of complex computer-based models of interfacial regions, whose properties are defined on the basis of a series of thermodynamic descriptors (Davis and Kent, 1990). Fewer methods are available for direct chemical and structural characterization of the clay surface, yet, in order to describe the mechanism of surface-controlled reactions on a molecular level and to validate mathematical models of interfacial reactions, direct measurements are required. Against this background, the development and application of highly surface-specific methods, such as photoelectron spectroscopy, has come at an opportune time.

Photoelectron spectroscopy comprises a group of techniques in which photoelectrons emitted from a sample, excited by electromagnetic radiation of a suitable wavelength, are separated on the basis of their kinetic energy, detected by a photomultiplier device and recorded in the form of electron yield against electron energy. In X-ray photoelectron spectro-

Clay Mineralogy: Spectroscopic and Chemical Determinative Methods.
Edited by M. J. Wilson. Published in 1994 by Chapman & Hall, London.
ISBN 0 412 53380 4.

scopy (XPS), the exciting radiation is an X-ray beam and the photoelectrons are emitted from the core and valence levels of the constituent atoms of the sample. Although originally conceived as an analytical technique, XPS can also give information on the 'chemical environment' of constituent atoms.

6.2 FUNDAMENTAL ASPECTS

6.2.1 Photoionization

Photoionization is one of a number of physical processes that can occur upon irradiation of a solid, liquid or gas. The incident radiation ejects an electron from the irradiated material and this can simply be formulated as

$$M + h\nu \rightarrow M^{+} + e^{-},$$

where M is the ground state species, $h\nu$ is the X-ray photon, M^{+} is the excited ionized species and e^{-} is the photoelectron.

Applying the principal of conservation of energy to the photoionization process yields the Einstein relationship:

$$KE = h\nu - BE$$

where KE is the kinetic energy of the photoelectron, BE is the binding energy of the electron, h is Planck's constant and ν is the wavenumber of incident radiation, hence $h\nu$ is the energy of incident beam. In practice, this fundamental equation must be modified to take account of experimental factors, such as the spectrometer work function, φ and/or sample charging (Defosse and Rouxhet, 1981). However, from the simple equation above, it can be seen that XPS peaks may be described either in terms of the experimentally measured parameter, the kinetic energy of the photoelectrons, or the derived parameter, the binding energy of the ejected electron. The two are simply related through the energy of the incident beam; thus, for Mg Kα radiation, $BE = 1253.6 - KE$, whereas for Al Kα the expression $BE = 1486.6 - KE$ may be used.

In chemical and mineralogical studies the spectrum is normally expressed in terms of electron yield against binding energy (Figure 6.1). No convention exists as to the correct manner to present curves but in the examples given here the peak intensity is given on the ordinate and the binding energy scale on the abscissa. The origin for both axes will be bottom left corner and hence BE will increase from left to right, although it should be noted that in some publications binding energy decreases from left to right.

6.2.2 Quantum mechanical aspects

Each electron in an atom is uniquely identified by a combination of four quantum numbers, the principal quantum number, n, the azimuthal

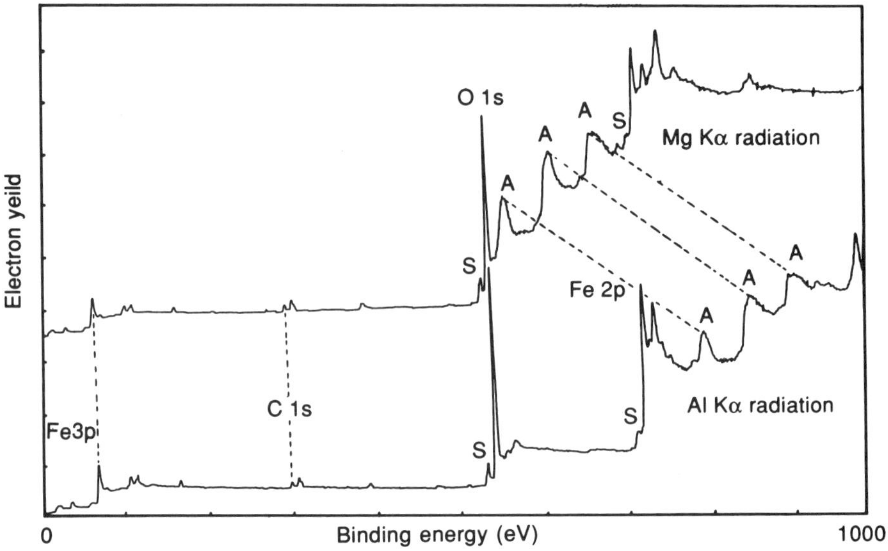

Figure 6.1 Spectrum for hematite (α-Fe_2O_3) showing XPS peaks (labelled with element and orbital), Auger peaks (marked A) and X-ray satellites (marked S). The dashed lines link corresponding Auger peaks for Al-Kα (lower curve) and Mg-Kα radiation (upper curve).

quantum number, l, the magnetic quantum number, m, and the spin quantum number, s. Each electron in an atom possesses a unique combination of these numbers, which then defines the energy level of the electron.

The principal quantum number, n, must be an integer and may be considered as representing the distance from the centre of the nucleus to the zone of electron density in which the electron is located. Thus $n=1$ corresponds to the shell closest to the nucleus and is equivalent to the K shell in conventional X-ray terminology.

The azimuthal quantum number, l, can take any integral value from 0 to $n-1$ and effectively defines the shape of the electron orbital. For a value of $l=0$ the orbital is spherically symmetrical about the nucleus whereas for other values different shapes and symmetries are involved.

The magnetic quantum number, m, can take any integral value from $-l$ to $+l$ and is related to the orientation of the orbital magnetic moment relative to that of any external field. In the absence of an external field it is degenerate, i.e. all electrons with the same values of n, l and s will have the same energy irrespective of the value of m.

Finally, there is the spin quantum number, s, which takes the value $+\frac{1}{2}$ or $-\frac{1}{2}$. Two electrons with the same n, l and m quantum numbers will be spin-paired in their orbital, i.e. they will spin in opposite directions, and one will have $s=+\frac{1}{2}$ while the other will have $s=-\frac{1}{2}$.

To take account of coupling between the orbital and spin magnetic moments, the azimuthal quantum number, l, and the spin quantum number, s, are combined to give the quantum number, j. In common with other spectroscopic techniques (Walker and Straw, 1961) this may be accomplished in two different ways but, for the purposes of this discussion, the details are unimportant, except in so far as they influence the nomenclature of spectral peaks. For naming XPS peaks the j–j coupling scheme is followed.

6.2.3 Spectral features

As predicted theoretically, each element can give rise to a number of different XPS peaks depending upon the core level involved. The relative intensities of the peaks for different core levels of the same element reflect differences in the probability of photoionization, the 'cross-section' of the process. These cross-sections may be obtained by experimental measurement or by quantum mechanical calculation, although quite wide divergence is observed in many cases.

In addition to the anticipated photopeaks other peaks may also be seen in the XPS spectrum (Figure 6.1). The peaks marked A arise from electrons produced by the Auger process, which forms the basis of another surface analytical technique known as Auger electron spectroscopy (AES) (Briggs and Seah, 1983). The electrons in this case are emitted as part of a relaxation process for the excited species produced after photoemission, in a similar manner to the fluorescent X-rays that form the basis of X-ray fluorescence (XRF) spectroscopy. The peaks can readily be distinguished from XPS peaks because of their greater linewidth and the fact that their **kinetic energy** is independent of the energy of the incident radiation because it depends only on the internal energy levels of the atomic orbitals involved in the process. Thus, on a **binding energy** scale, changing the X-ray source from Mg Kα to Al Kα will result in Auger peaks moving upscale by 233 eV (dashed lines in Figure 6.1).

The peaks marked S are X-ray satellite peaks and are the result of photoionization processes which involve X-rays of slightly different energy to the primary X-ray beam (section 6.3.1.1). They can be distinguished from other peaks because they are most easily seen in association with intense primary photopeaks and are displaced to lower binding energies by an amount that reflects the differences in energies of the X-ray beam components. The relative intensity of the satellite to primary photopeak is directly proportional to the intensity of the different components and thus X-ray satellites may be accounted for with ease by computer-based data systems. Satellites are not observed for Auger peaks because the X-ray photon is not directly involved in Auger emission.

6.3 EXPERIMENTAL ASPECTS

6.3.1 Instrumentation

The instrumentation required for XPS may be considered in three parts, a radiation source for excitation of photoelectrons, a method for separation of electrons according to their kinetic energy and a system for detecting and recording the emitted photoelectrons. Additional requirements are imposed on these component parts because the instrument operates under ultra-high vacuum (UHV) conditions and hence demands the use of UHV materials and fabrication techniques.

6.3.1.1 X-ray sources

The X-ray source used in XPS has to meet three main requirements:

- the beam should have an energy level sufficient to excite core-level photoemission from as wide a range of elements as possible;
- the beam should have a narrow energy spread in order to maximize energy resolution;
- the X-ray flux should be high in order to maximize photoelectron yield.

Examination of the energies and linewidths for various elements (Barrie, 1977) shows that the $K\alpha_{1,2}$ lines of Mg at 1253.6 eV and of Al at 1486.6 eV satisfy all of these criteria in that both have reasonably high energies, narrow linewidths < 1 eV and can be operated at fairly high power outputs. In practice, twin anode guns using both Mg and Al are widely used because the availability of two different energy sources facilitates the identification of Auger peaks in the spectrum.

Unlike conventional X-ray sources, the XPS gun has the anode at very high positive potential and the filament at earth. Thus, stray electrons from the filament will not strike the gun window or the surrounding metal case and thus generate X-rays of a different wavelength. In the twin-anode gun two semi-circular tungsten filaments are placed out of sight of the anodes, in order to minimize contamination by sputtering, and the appropriate filament–anode combination is used to generate X-rays of the chosen wavelength (Figure 6.2). A window of Al 2 μm thick can be used with Mg and Al anodes but a more X-ray transparent material, such as beryllium, must be used for softer radiations. The whole assembly is normally pumped separately from the spectrometer chamber in order to minimize contamination. Efficient production of X-rays is achieved using a variable accelerating voltage up to 15 kV with a stepped filament current to allow variation in X-ray flux. Guns with energy dissipations up to 1 kW are available.

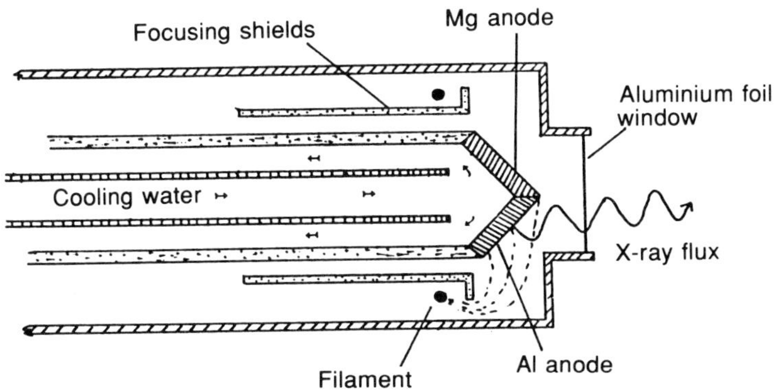

Figure 6.2 Schematic diagram of a twin-anode X-ray gun.

The X-ray spectrum produced in this way consists of a broad continuous background on which is superimposed the narrow characteristic lines, but for Al and Mg the principal Kα doublets have associated sets of weaker lines from K$\alpha_{3,4}$ and Kβ transitions. The window on the gun helps to filter the radiation but X-ray satellites are still observed on the spectrum and have to be accounted for, often with the help of the data system. Further improvement in linewidth is possible using a monochromator but, despite the improvements in overall energy resolution and signal-to-noise ratio by the removal of continuous radiation, the decrease in intensity associated with the process is considerable and for routine work monochromators are rarely used. More intense monochromatic radiation is available from synchrotron sources but this is very much a specialist application, where transport of the spectrometer to the synchrotron is justified!

6.3.1.2 Energy analysers

The energy resolution of XPS is a key factor in the use of the technique to yield structural or chemical information, although for qualitative analysis a lower resolution is acceptable. The width of a photopeak, normally expressed as the full width at half maximum (FWHM), is a complex function of the energy spread in the X-ray beam, the intrinsic width of the photopeak itself and the resolving power of the energy analyser. Of these three factors, those relating to the X-ray beam have already been discussed, whereas the second, the inherent width of the photopeak, is a function of the electron energy levels in the atom and is not amenable to experimental control. The final factor, the resolution of the electron analyser assembly, is a function of both design and mode of operation.

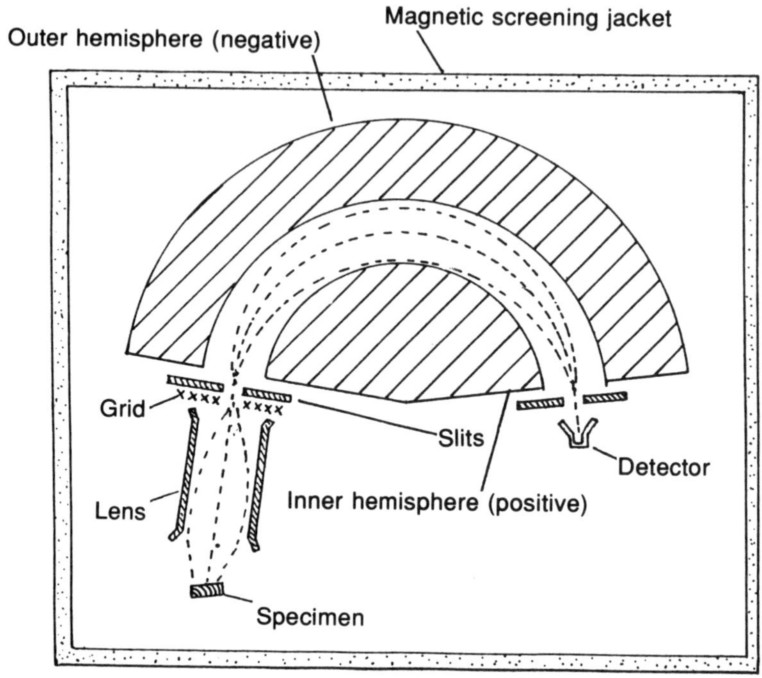

Figure 6.3 Schematic diagram of concentric hemispherical analyser assembly.

The purpose of the analyser assembly is to separate the negatively charged photoelectrons according to their kinetic energy. Two main types have emerged in the development of electron spectroscopy, the concentric hemispherical analyser (CHA) and the cylindrical mirror analyser (CMA) (Figure 6.3). For XPS work the CHA is the more commonly used. Two hemispherical surfaces are concentrically located on an insulating pedestal with, at one end, an entrance slit and a retarding–accelerating grid and at the other an exit slit and electron detector. An electrical potential, applied across the two hemispheres, sets up a series of equipotential surfaces and any electron with a suitable kinetic energy entering the analyser will travel along an equipotential, leave via the exit slit and strike the detector assembly. Clearly very precise geometry is required and in order to reduce manufacturing difficulties the velocities of the electrons are reduced before they enter the analyser chamber. This is achieved using a retarding–accelerating grid situated at the entrance to the analyser. The method by which the electrons are retarded gives rise to the two modes of operation, constant analyser energy (CAE) or constant retard ratio (CRR).

In order to understand the impact of these two methods of operation on the spectrum it is important to appreciate that resolution may be defined in two ways, absolute and relative. Absolute resolution is obtained from the full width at half maximum height for any given peak, whereas relative resolution is defined as the absolute resolution divided by the kinetic energy of the measured peak.

In the CRR mode all the energies are decreased by a constant ratio thus narrowing the energy range to be handled by the analyser. The potentials across the hemispheres in the analyser assembly are scanned through the range of interest, causing electrons of different energy to be focused on the detector at different times. In the CAE mode the hemispheres are held at a constant potential and scanning is achieved by varying the potentials on the retarding–accelerating grid, positioned before the entrance to the analyser. For most work CAE is the preferred mode of operation.

In order to maximize the count rate the entrance to the analyser should be as close to the sample as possible, but this can create many practical difficulties in an already congested area of the sample chamber. Fortunately, the problem can be overcome by using an electron transfer lens to transmit electrons emitted from the sample to the entrance of the analyser. In addition, by programming the potentials applied to the lens assembly the area analysed may be held constant throughout the analysis.

6.3.1.3 Detection and recording systems

In XPS, the detector has to measure an electric current of the order of 10^{-15} A and, with such low currents, an electron multiplier becomes necessary. Originally, discrete dynode type multipliers were used but these have now been replaced by the channel electron multiplier or channeltron. This unit is much more compact and is less susceptible to poisoning. The device consists of a glass tube with an inside surface coated with a resistive material of a high secondary emission coefficient. The end nearest the analyser exit slit is flared out to a conical funnel, about 10 mm in diameter, and an electrical contact is attached. At the far end, the anode is sealed with a metal contact coming through the glass. The whole tube is formed into a spiral. Any electron entering the tube generates many more electrons by collision with the tube wall and these in turn will be accelerated and strike the wall further in, creating an avalanche effect. The channeltron is an ideal detector for weak signals because it has a very low dark current and an amplification factor of around 100. At high count rates, however, saturation occurs, giving flat-topped peaks and poor resolution. High count rates will also degrade the tube coating and must be avoided. Several channeltron devices can be fused together in an array to give a multichannel detector system. On a spectrometer using this system the analyser exit slit is replaced with the array, which results in greater sensitivity because of the greater area seen by the detectors with the exit slits removed. The output from the channeltron can either be fed through a ratemeter to an X-Y recorder or via an analogue-to-digital converter to a microcomputer.

Data acquisition by a small computer opens up many possibilities for both instrument control and data handling. The computer can be used to control the spectrometer, and with carousel control and several energy regions scanned per sample, extended, unattended operation is possible. The computer will control all aspects of the spectrometer from pass-energy settings, through scan-width settings, to final X-ray gun switch-off, and multiple scanning with improved signal-to-noise ratios becomes possible. Once data acquisition is complete, the software will handle data storage and the important data manipulation. X-ray satellite subtraction, peak smoothing, peak deconvolution, background subtraction, spike editing, peak area determination and peak identification are some of the tasks that a well-designed data system must undertake (Sherwood, 1983). The final manipulated data can then be hardcopied to a printer or plotter. Clearly, there are significant advantages to be gained from using some form of computer control.

6.3.2 Energy calibration

Energy calibration is an important aspect of the practical application of XPS. The process may be conveniently considered as the determination and adjustment of the gradient and intercept of the linear relationship that should exist between indicated and actual binding energies. This principle has been used in the analysis of errors from an interlaboratory survey, which shows that an error of some 0.3 eV is associated with the determination of the zero point, an error of 0.05% is attributable to the instrument scaling or gradient and that random error, indicative of instrument precision, amounts to 0.1 eV (Antony, 1983). However, behind this deceptively simple description there are many theoretical and practical difficulties, even for conductors, and in the case of insulators these problems multiply rapidly (Evans, 1977).

A number of procedures for calibration have been summarized by Evans (1977) and the importance of 'traceability' of standards stressed by Antony (1983). The principal method is to use secondary standards of metals with well-defined XPS (and Auger) peaks with as wide a separation as possible between the peaks. In this regard, copper is particularly useful, but the ease with which surface oxides form requires the sample to be cleaned *in situ*, perhaps by Ar ion etching. This problem may be circumvented by using the noble metals, Au and Ag, although the energy separation of the two peaks used is not so great as with Cu. These standards can also be used in the calculation of the spectrometer work function, φ.

During the photoionization process, electrons are lost from the sample and this leads to a build-up of positive charge on the sample. The charge reaches a steady state when the rate of photoionization is equal to the rate of neutralization of positive holes, either by electron flux from the sample holder in the case of conductors or by bombardment with low-energy electrons present in the analyser chamber in the case of insulators. Any nett charge remaining will result in a reduction in the kinetic energy of emitted photoelectrons and an apparent increase in binding energy on the spectrum. Thus, corrective procedures are required for insulators, whereas for conductors it is important to ensure that conductivity is maintained by using a suitable sample-mounting procedure.

A variety of methods have been proposed to indicate whether the sample has acquired a charge (Evans, 1977). These fall into two groups:

- methods based on disturbing the steady state between positive hole production and electron transfer;
- methods based on estimation of the charge by comparing the peak positions for a particular element in the sample with literature values or with uncharged standards.

In the first group the rate of positive hole production may be reduced by reducing the incident X-ray flux, or the rate of neutralization through electron transfer from the vacuum may be enhanced by bathing the sample in slow, thermally generated electrons. In either case sample charging is indicated by movement of XPS peaks on the spectrum. In methods based on comparison of peak position a choice has to be made of which element to use for charge referencing (Swift, Shuttleworth and Seah, 1983). The most widely used choice is the C 1s peak at *c*. 285 eV, thought to arise from adventitious contamination of samples by hydrocarbon pump oil. As Swift (1982) has indicated there are a number of problems associated with this approach, but nevertheless the method is widely used. A similar method has been developed where, instead of an adventitious surface deposit, either a specific organic compound (Connor, 1977) or a noble metal, such as gold, is deposited on the surface, but the difficulty of ensuring that the coating is at electrical equilibrium with the sample remains. A similar difficulty is experienced with admixture of specific components, such as graphite.

For the examination of a group of related compounds, standardization of *BE* to that of an element common to the group can be of value. Clearly, it is important to ensure that the selected element shows little variation in its chemical environment across the group of samples and that it shows little variation in its binding-energy–chemical-environment relationship. For clay minerals Si is ubiquitous and presents two well-developed photoelectron peaks, the 2s singlet and 2p unresolved doublet at *c*. 154 and 103 eV. Seyama and Soma (1984) have used this approach in an XPS study of ion exchange in smectites, taking *BE* Si 2s as 153.4 eV in phyllosilicates. Another peak, ubiquitous to clays is the O 1s peak at *c.* 531 eV but this has not been widely used because of the range of chemical environments available for oxygen in clays, e.g. oxide ions, hydroxyl or even sorbed water.

6.3.3 Depth of analysis

One feature which distinguishes XPS as an analytical technique is its high degree of surface specificity. This does not arise from any properties of the incident X-ray beam, which will penetrate deeply into the sample, but arises from the properties of the emitted electrons.

Electron movement in solids is severely curtailed by collisions with constituent atoms, during which the electrons may lose energy in inelastic processes, to the extent of not emerging from the solid phase. This can severely attenuate the signal from deeper layers within the solid phase (Figure 6.4). In order to account for these factors in a semiquantitative way the concepts of attenuation length and escape depth have been applied (Hochella, 1988). The attenuation length, λ, is the distance from which $\frac{1}{e}$ photoelectrons produced at that depth escape from the surface of the solid without energy loss. From the exponential form of the relationship it can be shown that 95% of the signal originates within a depth of 3λ. Seah and Dench (1979) have produced a compilation, for different phases, of the inelastic mean free-path for electrons of various kinetic energies.

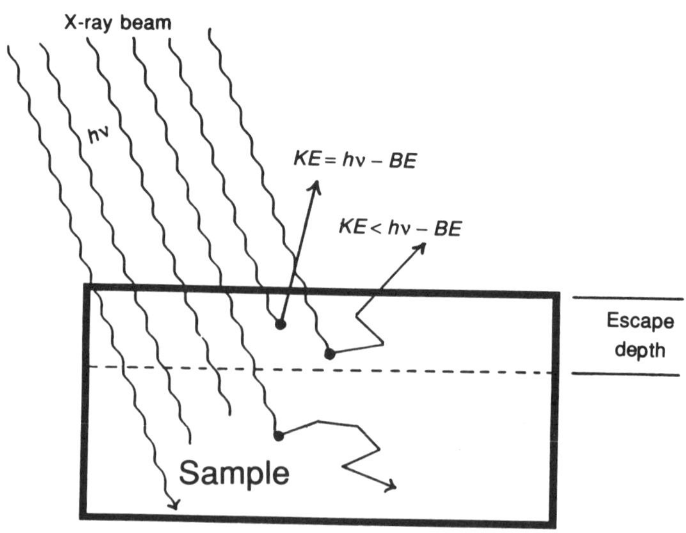

Figure 6.4 Attenuation of photoelectron yield at depth in a solid sample by inelastic collision.

In view of the shallow sampling depth, the presence of surface contaminants, either introduced before entry into the analyser chamber or accumulated within it, can present a major difficulty. These surface overlayers attenuate the signal arising from the sample but, even worse, the attenuation is not constant throughout the spectrum because it varies with the kinetic energy of the electron (Adams *et al.*, 1977). With modern equipment and suitable samples a vacuum of 10^{-9} to 10^{-10} mbar is routinely available which reduces contamination within the spectrometer. Calculation shows that in an atmosphere of oxygen at a pressure of 10^{-6} mbar, assuming a sticking probability of 1, a monolayer of oxide will form on a suitable metal in 1.4 s, whereas at 10^{-9} mbar it takes 1400 s (Fuggle, 1977). However, for freshly cleaved mineral surfaces contamination before entry into the analyser chamber is potentially a far greater source of error than contaminant build-up within the vacuum system. Stipp and Hochella (1991) have demonstrated clearly the rigour required to eliminate contaminant C on freshly cleaved calcite surfaces.

Contaminants are clearly best avoided but this is not always possible and 'cleaning' methods may have to be considered. The primary technique used is Ar ion sputtering, which can be carried out *in situ* in the spectrometer chamber. In the simplest case this technique involves bombardment of the sample surface with a beam of Ar ions, which then sputter the contaminant layer from the sample. Care must be exercised when using this technique in order to avoid sputtering the components of the sample itself. Indeed, more prolonged exposure to ion bombardment provides the basis of sputter depth profiling (Hofmann, 1983), which can be used for mineral analysis (Hochella *et al.*, 1988). However, the interpretation of depth profiles obtained in this way is not always easy because, as well as differential sputtering of various elements, mixing of components on an atomic scale can occur over a depth of some 1–20 nm (Seah, 1983).

Another form of depth analysis may be obtained by tilting the sample, altering the electron take-off angle and enhancing the surface specificity of the technique (Figure 6.5). This method has been widely used in the examination of metallic samples with oxide overlayers, where it has proved possible to calculate the thickness of such layers by simple geometry.

6.3.4 Spatial resolution

The spatial resolution of XPS is generally poor and the area analysed in most instruments is of the order of several square millimeters, although it is possible to reduce this by physically limiting the sample area from which photoelectrons are accepted. The use of a neutral excitation beam severely restricts the options available for small-area analysis by excluding the use of electromagnetic lenses to focus the excitation beam in the

a

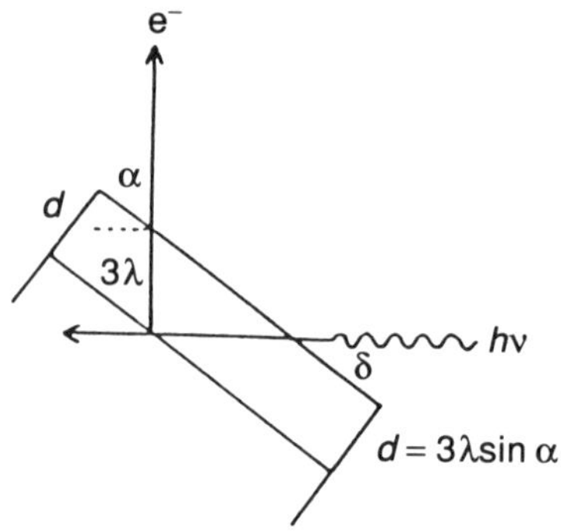

b

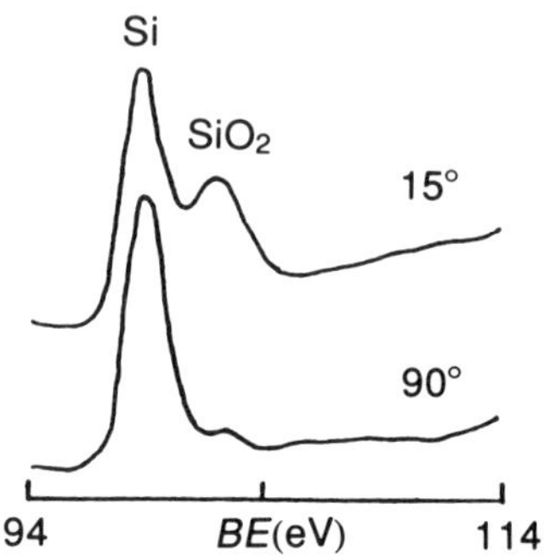

Figure 6.5 (a) Geometry of electron 'take-off' angle. (b) Si 2p photopeak showing enhancement of signal from the oxide overlayer as α decreases. (Reproduced from *Handbook of X-ray and Ultraviolet Photoelectron Spectroscopy*.)

same manner as that used in electron probe microanalysis (EPM). However, reduced-area analysis is possible, for example, on the VG Escalab instrument by reducing the entrance slits and the lens acceptance aperture. In this way, analysis on mineral single crystals some 1 mm in diameter have been carried out, although Hochella (1988) quotes analyses from 150 μm spots. The practical limitation on the area analysed is the requirement for a reasonable photoelectron count rate to provide a good signal-to-noise ratio. The X-ray flux is already maximized in most instruments by reducing the distance between gun-tip to sample and can be increased only by using higher intensity beams. In addition to these mechanical methods of reducing the analysed area, instruments are now available which will allow resolutions of only a few micrometres using a new input lens with a microfocused X-ray beam (Fisons Surface Science Division, 1992). Other ingenious methods have been devised in order to enhance spatial resolution, including the use of *in situ* generated X-rays produced by focusing a high-intensity electron beam on Al foil in close proximity to the sample in the analyser (Cazaux, Mouze and Perrin, 1982), but this method cannot be considered routine.

6.3.5 Sample preparation

X-ray photoelectron spectroscopy can be used to study gases, liquids and solids but, within a mineralogical context, only solids need be considered. Many methods of sample preparation have been proposed and tested with the objectives of presenting the surface in a suitable physical form for examination under ultra-high vacuum conditions, maintaining the chemical integrity of the sample by minimizing contamination and attempting to ensure reproducibility. Within this general context there is scope for considerable ingenuity in preparation and pretreatment techniques.

Although clay mineralogy is by definition concerned with finely particulate materials there are circumstances where larger samples of a size similar to the area of analysis are utilized. These samples may include macrocrystalline materials, such as mica flakes, or cleaved planes of single crystals, such as calcite, to be used as model surfaces. In addition, some of the preparative techniques described later produce coherent discs that have to be attached to the sample holder. Attachment may be achieved in a number of ways including double-sided sticky tape, mechanical attachment via lugs on the sample stub, or the use of an electrically conductive paste of colloidal silver or graphite. In the authors' experience the most satisfactory of these is the colloidal paste technique, although with appropriately designed sample holders the mechanical methods can also be of value.

For finely divided powders, a high-pressure die, similar to that used for infrared spectroscopy, can produce a coherent disc from a sample weight of some 50–100 mg that can be glued to the sample stub. There is, however, a risk of contamination using this method and some authors have recommended that the surface of the disc should be abraded with a steel blade after mounting (Adams *et al.*, 1977). The pressure exerted during disc formation will result in sample orientation, although the extent to which abrasion may randomize the analysed volume is uncertain. In addition, it has been shown that pressure-induced cation exchange may occur for some clay-mineral mixtures (Paterson and Clark, 1991). Despite these reservations the method is widely used.

A second group of methods, particularly appropriate for clay minerals, involves evaporation of a suspension placed on the sample stub. The liquid phase of the suspension may be either water or an organic liquid, such as acetone. For aqueous suspensions of clay minerals good adherence to the sample holder is frequently obtained, but even where this is not the case the procedure still produces a coherent film that can subsequently be glued to the sample stub. The surface prepared in this way will consist almost entirely of particles oriented with their basal planes parallel to the surface. For polydisperse samples the time taken for evaporation at room temperature will allow considerable particle size sorting and the analysed surface may not be representative of the whole. Whether this is a problem or an advantage clearly depends on the aims of the study. Another feature which should be noted is that any water soluble material will be deposited on the surface, possibly as small 'islands' of impurity (Bancroft, Brown and Fyfe, 1977). Acetone suspensions can be evaporated much more quickly than aqueous ones so that size fractionation is less of a problem. It has been shown that from acetone evaporation randomly oriented samples are obtained (Paterson, Bunch and Duthie, 1986). In both cases, the risk of contamination is always present and must be guarded against by the use of good laboratory practice.

Samples may also be mounted mechanically, either by pressing into indium foil or into a fine metal mesh. Indium, with a melting point of 121°C, is very malleable and flows around the particles of powder, holding them in place but it should be noted that indium is very toxic and must be handled with care. Noble metal mesh attached to the sample stub also has been used as a support for powdered samples.

6.4 INFORMATION CONTENT

The information content of the XPS spectrum may be considered in two parts, namely (1) elemental composition and (2) chemical speciation (Figure 6.6). X-ray photoelectron spectroscopy is basically a form of atomic spectroscopy and, as such, it has a clear and well-defined analytical role based on the positions and sizes of peaks within the spectrum. However, the exact energy levels of the core and valence electrons re-

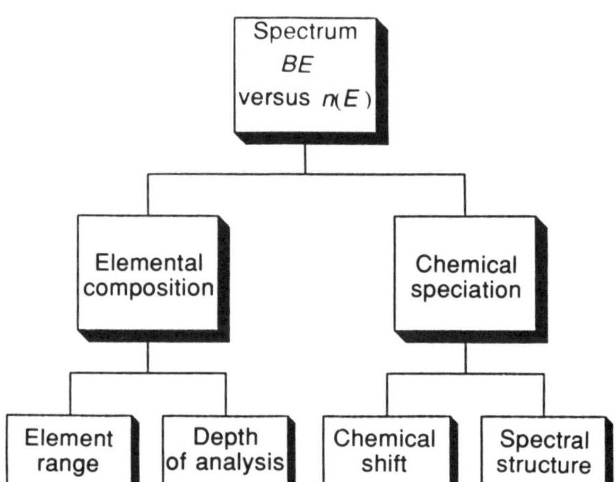

Figure 6.6 Information content of XPS spectra.

spond to their electronic environment and additional structural or chemical information may be obtained from binding energy shifts and spectral fine structure.

6.4.1 Analytical information

6.4.1.1 Qualitative analysis

The analytical capacity of XPS is readily apparent from the fact that all elements, with the exception of H and He, can be detected. Within this huge range the sensitivity for different elements varies considerably (Evans, Pritchard and Thomas, 1978) but, in this, the technique compares favourably with many others. The majority of elements contribute more than one XPS peak to the spectrum, to the extent that it may seem that overlapping of peaks will become a major difficulty, but in practice this is not the case. On the occasions when peak overlap does occur it can be readily identified and other peaks used for identification or quantification. High accuracy in binding-energy values is not required for peak identification and for charged samples internal referencing is normally sufficiently accurate for most purposes. However, the positive identification of every peak on the spectrum for some clay minerals can be a daunting task, partly because many of the collations or computer-based libraries of peak binding energies contain only the most intense peaks.

Few studies rely solely on the qualitative use of XPS but, as an indication of the surface sensitivity and utility of the technique, it is instructive to consider a simple example. Cation exchange is one of the most important properties exhibited by clay minerals and exchange reactions play an important role in both nutrient retention and pollutant immobilization. For swelling minerals exchange processes can take place between the layers of the phyllosilicate structure, but for non-swelling minerals, such as mica, chlorite, or kaolinite, exchange is restricted to the outermost surfaces of individual particles, on basal planes and edge sites. Cleavage of mica along the basal planes may be viewed as the exposure of a number of K^+ cations and/or vacant cation-exchange sites. Cation exchange is thus restricted to the outermost single atomic layer of the mineral and involves only a small proportion of the K^+ ions present and a very small absolute amount of exchangeable cation. Nevertheless, in these circumstances, introduced cations, such as Cs^+, can be readily detected (Figure 6.7). Clearly, XPS has considerable merit for the study of such reactions and the physical properties of mica, including the ease with which large flakes may be obtained and cleaved, suggest that this group of minerals may be useful as models in the study of the behaviour of negatively charged phyllosilicates in general.

The qualitative use of XPS can be clearly seen in a recent study of crack surfaces in large single crystals of olivine produced by volcanic activity

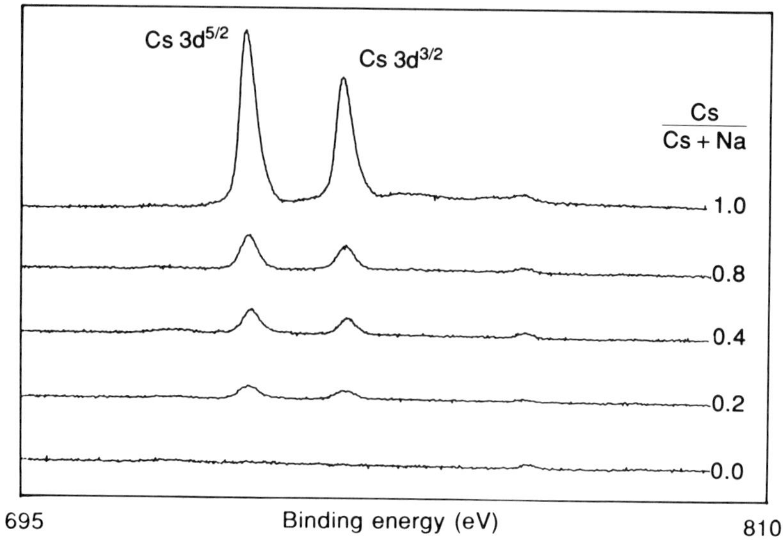

Figure 6.7 X-ray photoelectron spectroscopy of freshly cleaved mica flakes treated with mixed solutions of sodium and caesium chloride.

in Arizona and Hawaii (Tingle *et al.*, 1990). This study essentially confirmed and extended earlier XPS observations of the presence of carbon-containing coatings on crack surfaces within minerals formed in a volcanic environment (Mathez, 1987). The very high surface sensitivity of XPS enabled analytical data to be obtained on the 3–4-nm-thick carbonaceous coatings present in these minerals and shows that in addition to C they also contain O, and smaller amounts of N, Al and Na. Whether these coatings are biogenically or abiogenically produced remains speculative at this stage but there is little doubt that few, if any, analytical techniques could have supplanted XPS in these studies.

6.4.1.2 *Quantitative analysis*

In principle, quantification in XPS is relatively straightforward. Unlike other forms of spectroscopy, such as XRF or EPM, interelement effects do not have to be considered and, for a homogeneous solid, XPS peak areas are directly proportional to atomic concentrations. The various parameters involved in the proportionality constant, termed the atomic sensitivity factor by Hochella (1988), include the X-ray flux, the volume of sample analysed, the angle of tilt of the sample with respect to the analyser, the analyser detection efficiency and the probability of peak attenuation due to inelastic collisions. Thus, for spectra collected under identical experimental conditions and utilizing XPS peaks with energies within a few hundred electron volts of each other, quantitative analysis to within 10% should be readily obtained.

In practice, various approaches have been utilized, ranging in rigour and complexity from the simple method outlined above to a priori treatments of the relevant theory to calculate atomic sensitivities through quantum mechanical calculations of photoionization cross-sections. In addition, methods utilizing published data sets of photoionization cross-sections have also been used (Seah, 1980), and it is noteworthy that early work by Adams *et al.* (1977) using this approach resulted in an accuracy of 5% for the principle constituents of clay minerals. For a comprehensive treatment of considerations in quantitative analysis the reader is referred to Seah (1983).

Common to all of the methods of quantitative analysis is the requirement to determine peak areas. In computer-based data systems the software frequently allows a choice of method by which the area may be calculated, the principal difference between the methods being the treatment afforded to the baseline under the peak (Sherwood, 1983). The simplest method is to draw a straight line from the beginning of the peak to the end. The difficulty is, of course, deciding where the peak begins and ends, particularly for an intense peak with a stepped baseline arising from inelastically scattered electrons. Spectral fine structure adds to the difficulty. A more sophisticated approach is to calculate a step function between the baseline at the start of the peak and that at the end and subtract this from the experimentally determined curve. The approach adopted is dictated to a large extent by the available software, but it should be remembered that, for an empirical method, reproducibility is more important than mathematical elegance.

6.4.2 Chemical and structural information

The use of XPS to obtain information on the 'chemical environment' of atoms and ions has already been alluded to and the valency, coordination and other bonding information on specific elements can be obtained in favourable circumstances. It is, however, important to note two points at this stage:

- there is no a priori method of assessing the ease with which this information may be obtained for any particular element;
- the examples given here have been selected and it should not be assumed that similar success will be attained for every sample examined.

The information carriers include peak position in terms of binding energies, differences in energies between peaks, both XPS and Auger, and spectral fine structure.

6.4.2.1 *Binding energies*

As has been shown, the position of an XPS peak can be used to identify the element from which the electron was emitted. However, for a given core level and element, slight changes in binding energy (normally < 5 eV) may be observed. These changes can be related to the valency of the ion or to the polarity of the bonds and although complex physically based explanations may be advanced to account for these 'chemical shifts' it is possible to give a qualitative explanation based on simple electrostatic considerations. As the electron density around the atom or ion decreases the binding energy for photoemission from the remaining core levels tends to increase. The extent of the shift varies from element to element, with sulphur showing a particularly pronounced increase in the binding energy of the 2p doublet on increasing its formal oxidation number from -2 in sulphides to $+6$ in sulphates (Figure 6.8).

However, there are a number of problems associated with the use of binding-energy shifts to provide chemical information. The first of these is the occurrence of charging, because frequently the charge will exceed the shift. The second is that most XPS peaks are due to core-level transi-

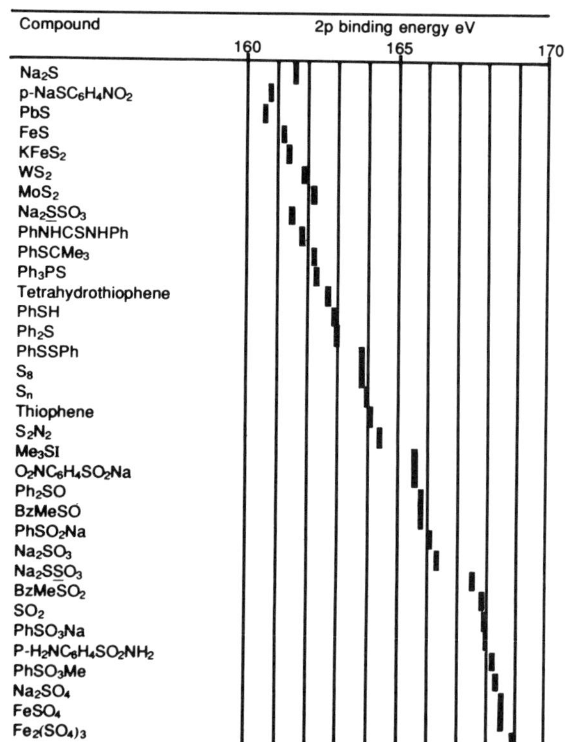

Figure 6.8 'Chemical shift' of S 2p doublet in a series of S-containing compounds. (Reproduced from Wagner *et al.*, 1979.)

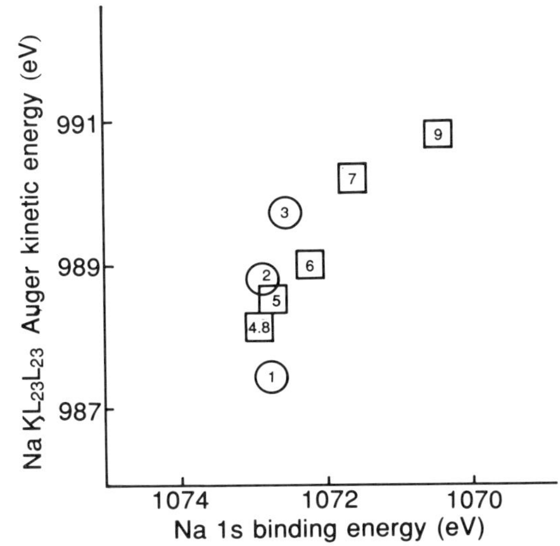

Figure 6.9 Chemical state plot for a series of Na-containing compounds: (1) NaF, (2) NaCl, (3) NaBr, (4) analcite, (5) Na-montmorillonite, (6) albite, (7) riebeckite, (8) tetrasilicic fluor-mica (exchangeable), (9) tetrasilicic fluor-mica (non-exchangeable). (Reproduced from Soma *et al.*, 1990.)

tions that may be little affected by changes in the population of valence bands. However, peaks arising from the Auger process may also be observed on XPS spectra and the chemical shifts associated with these peaks may be up to ten times greater than those for XPS peaks (Perry, Taylor and Wagner, 1990).

Wagner (1977) has suggested the use of the so-called Auger parameter in order to further characterize solid materials. This parameter is obtained by subtracting the kinetic energy of the XPS peak from that of the Auger peak. The parameter is independent of charging because both peaks are influenced equally. In some instances, instead of subtraction to obtain the Auger parameter, scatter plots (Figure 6.9) of the binding energy on the abscissa and the Auger peak kinetic energy on the ordinate have been used to show similarities in the chemical environment of particular elements in minerals (Wagner, 1977; Seyama and Soma, 1985; Hochella and Brown, 1988). It should be noted that Auger-peak shapes are frequently complex and peak position may be difficult to define.

6.4.2.2 Spectral fine structure

Additional information may be obtained by examination of the fine structure associated with XPS peaks. This arises from a number of different physical processes including spin–orbit splitting, multiplet splitting and shake-up satellites.

The vectorial combination of the orbital and spin contribution to the energy of the electron results in the creation of doublet peaks in the XPS spectrum for all orbitals where the azimuthal quantum number, l, is greater than zero. The magnitude of the splitting observed varies according to the shell and orbital involved (Orchard, 1977), but the influence of electronic configuration can result in second-order effects which can reveal information on, for example, valency.

Multiplet splitting can also be of considerable value in specific cases, particularly for transition metals. The process essentially arises from the difference in energy between the pair of electrons in closed core-level shells due to interactions with spin-parallel unpaired electrons in valence shells (Fadley and Shirley, 1970). The effect can be seen very clearly in the 3s photopeak of cations of the first transition series. This peak is normally a singlet due to the degeneracy of the spherically symmetrical s-orbital but where the 3d orbitals contain unpaired electrons the s-electron remaining after photoionization can have spin parallel or anti-parallel to the unpaired d-electrons and give rise to doubling of the photopeak. The process also occurs for other core levels and, as early as 1974, Gupta and Sen (1974) calculated theoretically the splitting expected for the 2p photopeaks for Fe. McIntyre and Zetaruk (1977) have observed much of the anticipated fine structure in iron oxides.

Other examples of spectral fine structure include shake-up and shake-off satellites (Brisk and Baker, 1975). In this type of configurational

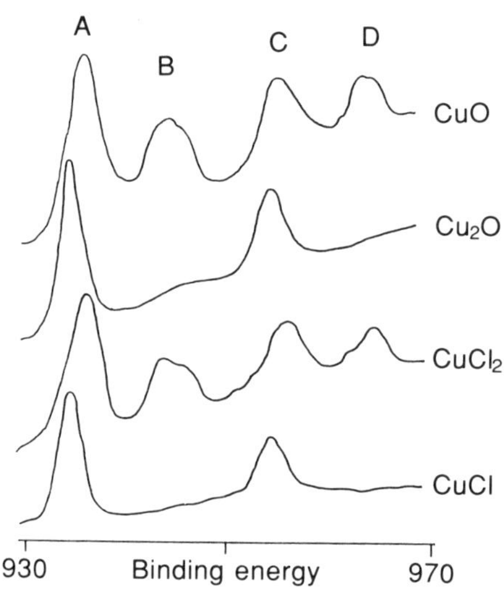

Figure 6.10 X-ray photoelectron spectroscopy peaks and satellites for Cu(I) and Cu(II) compounds: (A) Cu $2p^{3/2}$, (B) shake-up satellite, (C) Cu $2p^{1/2}$, (D) shake-up satellite.

interaction with valence band electrons a proportion of the photoelectrons associated with a particular photopeak lose energy due to the promotion of remaining valence electrons to higher levels. In other words, the ion remaining after photoionization is in an excited state but, unlike other mechanisms of energy loss, the system remains quantized and a satellite peak appears associated with the main photopeak. Perhaps the best-known example of this effect can be seen for the 2p photopeak of Cu(II), where intense satellites can be seen compared with the spectrum for Cu(I), where none are observed (Figure 6.10). The reasons for this may be seen by comparing the electron configurations of the Cu(II) and the Cu(I)ions. Cu(II), with intense satellite structure, has a $3d^9$ occupancy, so interaction between the 2p electrons and the unfilled d shell can occur, but Cu(I) with no satellite structure has $3d^{10}$ occupancy preventing interactions. Thus the valency of Cu may be readily discerned from the XPS spectrum but it should be noted that this is an unusually good example and should not be taken as typical for all transition metals.

6.5 APPLICATIONS IN CLAY MINERALOGY

The application of XPS to clay mineralogy *per se* has been sporadic, with only a small number of groups making significant contributions to the subject, although applications in mineralogy as a whole have increased considerably over the past decade. There are a number of reasons for the relatively low usage of the technique in clay mineralogy, including the cost of the instrumentation required, the range of skills demanded of the experimentalist, and the fact that the technique does not lend itself to routine use in the manner of many other techniques.

6.5.1 Dissolution of primary minerals

In order to understand the weathering reactions that are fundamental to the formation and properties of soils and sediments, quantitative XPS and other surface-specific spectroscopies have been used. The central problem addressed in these studies may be defined as the classical mechanistic problem in the dissolution of primary minerals (Casey and Bunker, 1990). Is the reaction diffusion controlled via a region depleted in cations or is it controlled by the surface reactivity of the original material? Despite the fact that initial studies by XPS (Petrovic, Berner and Goldhaber, 1976) did not support the suggestion of cation-depleted surface layers, work has continued and the application of a very wide range of surface analytical tools now suggests that a surface layer of the order of 100–200 Å in depth can form during dissolution of labradorite (Muir, Bancroft and Nesbit, 1989) under appropriate conditions. However, as Inskeep *et al.* (1991) have shown the analysis by XPS is complicated by the formation of etch pits and the preferential dissolution of sodic exsolution lamellae (Wilson and McHardy, 1980). In addition, it is thought likely that different silicates will behave differently. Using hydrothermal dissolution of albite in a flow system to minimize the likelihood of secondary precipitation on the surface, Hellman *et al.* (1990) have used XPS and Ar-ion sputtering to show that dissolution does indeed produce a surface layer depleted in Na, Al and O. However, the presence of Ba and Cl ions from the solution phase throughout this leached layer suggests that ion diffusion can occur readily and hence it was concluded that the dissolution reaction is not diffusion-limited. In addition to the quantitative aspects, XPS was also used to give some indication of the chemical environment of various components (section 6.4.2).

An interesting variant on this type of behaviour can be seen in the dissolution of ferromagnesian silicates (Schott and Berner, 1983). Again a cation-depleted surface layer was formed which, in the absence of dissolved O_2, did not inhibit diffusion. However, when O_2 is present two surface layers appear to form. The outermost of these is a hydrous oxide of Fe(III), probably goethite-like, which can be removed by ultrasonic agitation and does not inhibit diffusion, whereas the inner layer appears to consist of Fe(III) in a hydroxylated or magnesium–silicate matrix. In this case, diffusion is limiting and this is reflected in the dissolution kinetics. The XPS data clearly provide a method by which reaction mechanism can be elucidated. Similar studies may be found in the literature for many other mineral species, including chrysotile (Thomassin *et al.*, 1977), phosphate minerals (Thirioux *et al.*, 1990), as well as the surface chemistry of sulphide minerals (Perry, 1986).

6.5.2 Investigation of mineral structures

6.5.2.1 Binary oxides

From the pioneering efforts of Allen *et al.* (1974) and McIntyre and his co-workers (McIntyre and Cook, 1975; McIntyre and Zetaruk, 1977) the oxides of Fe and other metals of the first transition series provide excellent examples of the range of information that may be obtained by XPS. Binding-energy shifts have been used to distinguish Fe(II) and Fe(III) in magnetite as well as Fe(III) in the oxide, hematite, from that in the oxide hydroxide, goethite. In the latter case, the O 1s peak was examined in some detail and deconvoluted into three components thought to be due to oxide, hydroxide and chemisorbed water. In the example from the authors' laboratory (Figure 6.11) only two components have been considered, oxide and hydroxide. For synthetic goethite (α-FeOOH), the O 1s peak can be seen to be broad, almost flat-topped, whereas that for hematite (α-Fe_2O_3), produced by ignition of the goethite, has a much

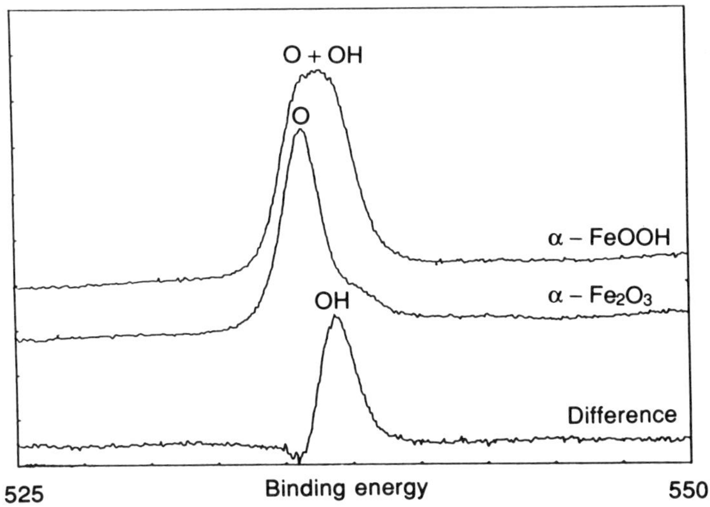

Figure 6.11 O 1s peaks for synthetic goethite before and after dehydroxylation to hematite with difference spectrum.

narrower peak at lower binding energy, with only a shoulder at higher binding energy. Subtraction of the hematite peak from that of goethite gives an indication of the proportion of the OH contribution of the O 1s peak in the goethite spectrum.

The hydrous oxides of manganese have considerable environmental significance as accumulators of heavy metals, and aspects of their formation and structure have been investigated by XPS. Much of this work is based on the early observation by Oku, Hirokawa and lkeda (1975) that multiplet splitting of the 3s photopeak in manganese–oxygen systems could yield valency information. Murray *et al.* (1985) have used a similar approach using standard minerals containing Mn(II), Mn(III) and Mn(IV) (Figure 6.12) and compared these with the oxidation product of Mn(II) in mildly alkaline solutions. Using data from the 3s multiplet splitting they followed the course of the oxidative hydrolysis from an initial precipitation of hausmannite (Mn_3O_4) to the final product γ-MnOOH. Changes in the O 1s photopeak consistent with hydroxylation were also observed.

In addition to information on Mn itself, Dillard and his co-workers have shown how XPS provides useful information on other constituents of the ferromanganese nodules. Thus Dillard, Crowther and Murray (1982) have shown that Co, Ti, Fe and Pb can be detected in selected nodules and, in a later publication (Dillard, Crowther and Calvert, 1984), that speciation is possible for these minor components. The observation of Co(III) in the nodules confirmed suggestions that Co(II) could undergo

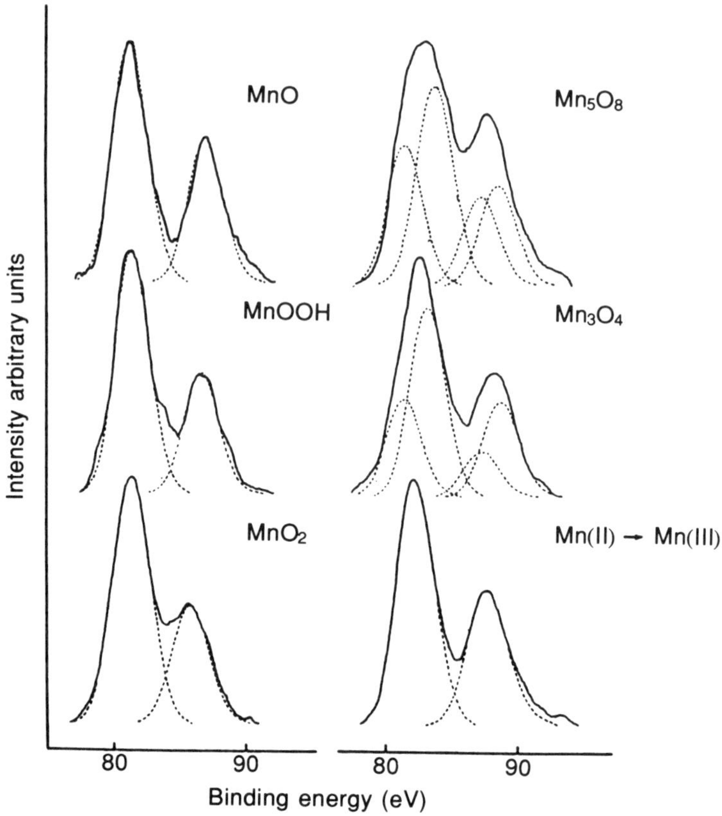

Figure 6.12 Multiplet splitting of Mn 3s photopeak for manganese oxides. (Reproduced from Murray *et al.*, 1985.)

oxidation on adsorption to manganese oxides, thus greatly influencing the reversibility of uptake.

6.5.2.2 Aluminosilicates

The structural environment of Fe within phyllosilicates was the subject of early work by Stucki, Roth and Baitinger (1976), who examined the spectra for oxidized and reduced forms of nontronite and biotite. They showed that Fe(II) and Fe(III) could be distinguished on the basis of the binding energy of the Fe 2p doublet. They also showed that the O 1s peak was also affected by the valency of the Fe, presumably through changes in charge density, although some of the changes in the overall spectrum were suggested to arise from changes in coordination. Experiments in the authors' laboratory suggest that great care must be exercised in studies of this type because of the possibility of radiation-induced change (Storp, 1985).

Investigation of Al and Si binding within aluminosilicates has proved to be less successful than has been the case with incorporated transition metal cations (Hochella, 1988; Perry, Taylor and Wagner, 1990). Early studies using a wide range of minerals of known coordination have shown that either small shifts or no shift at all were noted for the Al and Si 2p photopeaks. Measurement of the Auger parameter (Wagner, Gale and Raymond, 1979; West and Castle, 1982) proved to be more successful and clustered values were obtained for various structurally related mineral groups. Latterly, the wheel has turned full circle and the early work of Yin, Ghose and Adler (1971) on the O 1s line-shape has been developed in a number of more recent studies (quoted in Hochella, 1988) to show that the peak may be resolved into bridging and non-bridging oxygens. However, resolution is not obtained for all minerals (Hochella and Brown, 1988) and it must be said that, at this stage, XPS can play only a confirmatory role.

An interesting and potentially valuable extension to XPS, X-ray photoelectron diffraction, has been developed and applied to micas by Evans, Raftery and Thomas (1979). In this technique the photoelectron yield for the elements of interest from a single crystal of mica is measured as a function of angular rotation about a specific crystallographic axis. The diffraction effects encountered by the emitted photoelectrons as they travel through the sample modulate the electron yield and by comparing the angular yields for pairs of elements inferences may be drawn concerning the similarity of the structural sites occupied by the atom pairs. In this way Evans and Raftery (1980) have shown that the similarity in angular yields for Ti, Fe and Mg suggest that at low levels of substitution (1–2%) the Ti in micas is entirely located in the octahedral layer. A subsequent study of lepidolite has shown that Rb occupies similar sites to those occupied by K (Figure 6.13) and that there is no segregation of the cations in different interlayers despite their difference in size (Evans and Raftery, 1982).

6.5.3 Sorption and exchange reactions on mineral surfaces

6.5.3.1 Cation exchange on phyllosilicates

Cation exchange is one of the most important properties exhibited by clay minerals and some of the earliest applications of XPS to clay mineralogy concentrated on this aspect. The use of XPS to measure cation exchange capacities of beidellites (Adams and Evans, 1979) suggested that for the large cations, Ba, K and Pb, excess cation was present on the clay surface. This surface excess has been confirmed by the results of Seyama and Soma (1984) for Ba on montmorillonite, although the authors suggest that

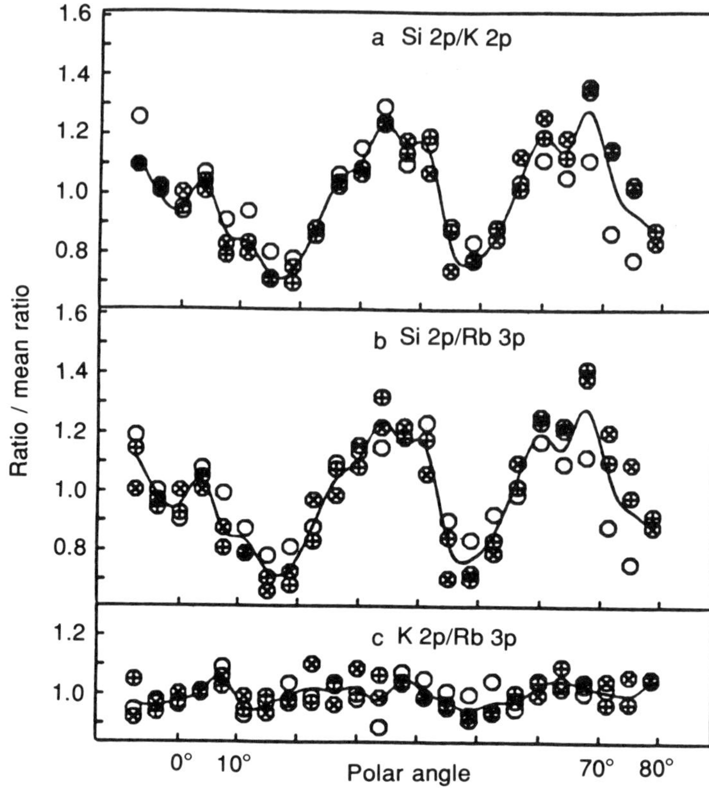

Figure 6.13 Photoelectron yield for Rb-containing mica showing angular variation of (a) peak area ratios for Si 2p/K 2p photopeaks, (b) peak area ratios for Si 2p/Rb 3p photopeaks and (c) the K 2p/Rb 3p photopeaks. The different symbols refer to replicative determinations at each angle. (Reproduced from Evans and Raftery, 1982.)

the result may be an experimental artefact caused by the attenuating effect of surface contamination on the high binding-energy Ba 3d peak area used for analysis. This ambiguity could be resolved by either using a contaminant overlayer correction term in the calculation (Adams *et al.*, 1977) or restricting the peaks used in quantification to those with similar binding energies. Clearly, further work is required in order to settle this question.

The nature of the exchanged species has also been studied using XPS. In an early study of the uptake of Cu(II) and Ni(II) by the clay minerals, illite, kaolinite and chlorite, Koppelman and Dillard (1977) have shown that whereas Ni(II) is taken up as the hexaquo cation, it is likely that Cu(II) is partially hydroxylated to produce the $CuOH^+$ species at the pH of saturation. Hydroxylated species were also observed for Co uptake on illite (Dillard *et al.*, 1981), but in this case uptake was measured over a wider range of pH. At pH 6.00 only Co(II) species were present, at pH > 7.8 precipitation of $Co(OH)_2$ occurred, and between these two values a mixture of various hydroxylated components was thought to be present. The affinity of Co and Ni for various mineral surfaces has been investigated by Davison, McWhinnie and Hooper (1991) using XPS to examine clay samples after saturation and washing an arbitrary number of times. Differences were observed which suggest that hectorite has a greater affinity for Co than montmorillonite whereas the reverse was the case for Ni. In addition, the lability to washing suggested that uptake took place on at least two types of site.

Examination of a synthetic Na fluor-mica (Soma *et al.*, 1990) shows very clearly that exchangeable and non-exchangeable Na may be differentiated both on the basis of the Na 1s photopeak and the Na KLL Auger peak (Figure 6.14). The binding energy of the 1s photopeak for non-exchangeable Na is seen to be some 2.5 eV lower than that of the exchangeable Na. The two components were well separated using an Auger plot, with the exchangeable Na occupying a position close to that of Na-montmorillonite whereas the non-exchangeable Na is closer to that for riebeckite, a hornblende mineral in which Na is eight coordinate, i.e. a position of high electro-negativity.

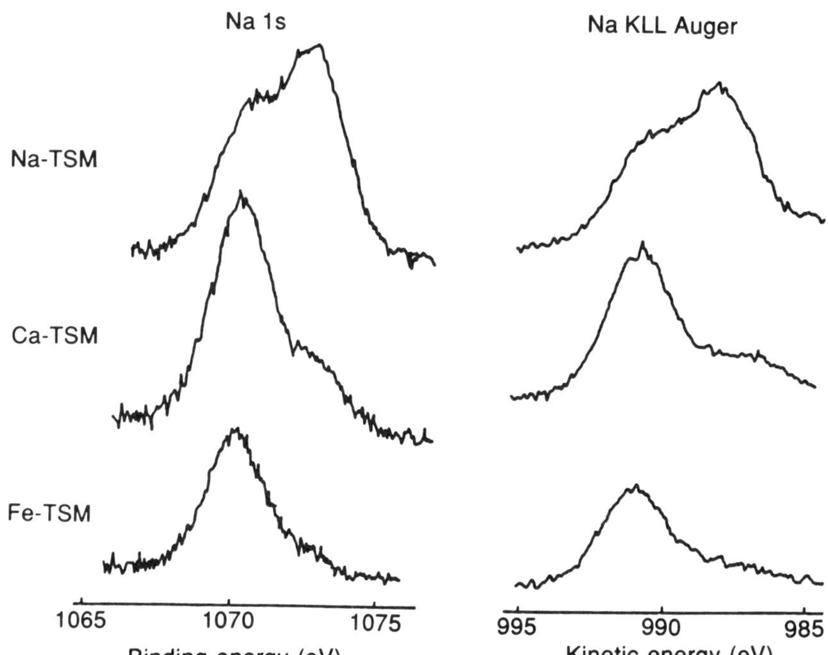

Figure 6.14 Na 1s and Na Auger spectra for various cation forms of tetrasilicic fluor-mica. (Reproduced from Soma *et al.*, 1990.)

6.5.3.2 *Adsorption on oxide surfaces*

It is generally recognised that oxides play an important role in sorption and exchange reactions in soils and sediments (Jenne, 1967) and considerable efforts have been made to describe and understand the reactions taking place. However, an important aspect of the application of XPS in this area is the extent to which artefacts are created by the UHV conditions in which the samples are examined. Harvey and Linton (1981) have concluded that in the case of ferrihydrite, only physically adsorbed water is removed, but it is interesting to note from their study that peak resolution of the O 1s photopeak for goethite suggests a 1:1 ratio of oxide to hydroxide species which, although correct for the bulk sample does not take account of the surface hydroxylation that occurs for well-crystallized synthetic goethite (Russell *et al.*, 1974). This problem remains to be resolved.

An interesting variant of the normal experimental procedures used to prepare samples for examination by XPS may be seen in the work of Brûlé *et al.* (1980) who synthesized manganese oxide directly on to the XPS stub instead of preparing the synthetic product by conventional means as a powder then depositing it on the sample holder. In this way they obtained a smooth coherent film, ideal for XPS investigation and, in addition, the film could survive immersion in aqueous solutions. In this way they obtained data on both the amount and kinetics of Ni(II), La(III) and Ba(II) uptake. Clearly, for multicomponent adsorption and ternary surface complexes, additional information may be obtained by angle-resolved experiments.

Adsorption and exchange in the natural environment rarely occurs from the one- or two-component solutions frequently used in the laboratory. In particular, the presence of silicate and of dissolved organic carbon in soil solutions and the consequent impact on adsorption is difficult to measure using conventional solution measurements. Early work by Alvarez *et al.* (1976) on the adsorption of silicate by gibbsite shows that only in the presence of Ca^{2+} does appreciable adsorption of silicate occur. Similarly, adsorption from solutions of calcium dihydrogen phosphate on to gibbsite results in the uptake of both Ca and P. Analagous behaviour may be anticipated for dissolved organic matter and XPS could have a valuable role in validating the sophisticated ternary surface complex computer models currently being developed (Schindler, 1990).

X-ray photoelectron spectroscopy has recently been used by Vempati *et al.* (1990) to study the nature of adsorbed silicate on synthetic ferrihydrite. The Si 2p and O 1s photopeaks were used to indicate that, at lower levels of adsorbed silicate, the binding energy values were consistent with low molecular weight silicate species, whereas at higher levels an XPS peak could be deconvoluted to indicate the presence of a Si-rich second phase. The binding energy of the Si 2p peak for silica–ferrihydrite coprecipitates was similar to that observed for aluminosilicates and may indicate the presence of Fe–O–Si linkages within the ferrihydrite structure.

6.6 CONCLUSIONS

X-ray photoelectron spectroscopy is a powerful technique, well suited to the examination of many processes of considerable importance in mineralogy, such as ion exchange, adsorption and dissolution. Advances in instrumentation and particularly in ultra-high vacuum technology has enabled the method to be utilized more widely than ever before. In addition the theoretical basis of the phenomena recorded is now fully developed and the technique may be considered as mature and beyond the speculative and phenomenonological investigations typical of new techniques. There is little doubt that from the examples now appearing in the literature, XPS can now play a very significant role in improving our understanding of interfacial chemistry, upon which so many natural processes ultimately depend.

ACKNOWLEDGEMENTS

Acknowledgement is given to the Scottish Office Agriculture and Fisheries Department for providing funding for this work. Figures 6.5, 6.8, 6.9, 6.12, 6.13 and 6.14 are reproduced with permission from the copyright holders:
6.5 Heyden, London.
6.8 *Analytical Chemistry*, **51**, 466–482.
6.9 *Clay Science*, **8**, 1–8.
6.12 *Geochimica et Cosmochimica Acta*, **49**, 463–470.
6.13 *Clay Minerals*, **17**, 443–452.
6.14 *Clay Science*, **8**, 1–8.

REFERENCES

Adams, J. M. and Evans, S. (1979) Exchange and selective surface uptake of cations by layered silicates using X-ray photoelectron spectroscopy (XPS). *Clay Minerals*, **27**, 248–252.

Adams, J. M., Evans, S., Reid, P. I. *et al.* (1977) Quantitative analysis of aluminosilicate and other solids by X-ray photoelectron spectroscopy. *Analytical Chemistry*, **49**, 2001–2008.

Allen, G. C., Curtis, M. T., Hooper, A. J. and Tucker, P. M. (1974) X-ray photoelectron spectroscopy of iron–oxygen systems. *Journal of the Chemical Society, Dalton Transactions*, **1974**, 1525–1530.

Alvarez, R., Fadley, C. S., Silva, J. A. and Uehara, G. (1976) A study of silicate adsorption on gibbsite ($Al(OH)_3$) by X-ray photoelectron spectroscopy (XPS). *Soil Science Society of America, Journal*, **40**, 615–617.

Antony, M. T. (1983) Spectrometer calibration, in *Practical Surface Analysis by Auger and X-ray Photoelectron Spectroscopy* (eds D. Briggs and M. P. Seah). Wiley, Chichester, pp. 429–435.

Bancroft, G. M., Brown, J. R. and Fyfe, W. S. (1977) Calibration studies for quantitative X-ray photoelectron spectroscopy of ions. *Analytical Chemistry*, **49**, 1044–1048.

Barrie, A. (1977) Instrumentation for electron spectroscopy, in *Handbook of X-ray and Photoelectron Spectroscopy* (ed. D. Briggs). Heyden, London, pp. 79–119.

Briggs, D. and Seah, M. P. (1983) *Practical Surface Analysis by Auger and X-ray Photoelectron Spectroscopy*. Wiley, Chichester.

Brisk, M. A. and Baker, A. D. (1975) Shake-up satellites in X-ray photoelectron spectroscopy. *Journal of Electron Spectroscopy and Related Phenomena*, **7**, 197–213.

Brûlé, D. G., Brown, J. R., Bancroft, G. M. and Fyfe, W. S. (1980) Cation adsorption by hydrous manganese dioxide: a semi-quantitative X-ray photoelectron spectroscopic (ESCA) study. *Chemical Geology*, **28**, 331–339.

Casey, W. H. and Bunker, B. (1990) Leaching of mineral and glass surfaces during dissolution, in *Reviews in Mineralogy*, Vol. **23**, *Mineral–Water Interface Geochemistry*, (eds M. F. Hochella and A. R. White). Mineralogical Society of America, Washington, DC, pp. 397–426.

Cazaux, J., Mouze, D. and Perrin, J. (1982) Scanning X-ray radiography: first tests in an electron spectrometer. *Journal of Applied Physics*, **53**, 3299–3302.

Connor, J. A. (1977) XPS studies of inorganic and organo-metallic compounds, in *Handbook of X-ray and Photoelectron Spectroscopy* (ed. D. Briggs). Heyden, London, pp. 183–209.

Davis, J. A. and Kent, D. B. (1990) Surface complexation modelling in aqueous geochemistry, in *Reviews in Mineralogy*, Vol. **23**, *Mineral–Water Interface Geochemistry*, (eds M. F. Hochella and A. R. White). Mineralogical Society of America, Washington, DC, pp. 177–260.

Davison, N., McWhinnie, W. R. and Hooper, A. (1991) X-ray photoelectron spectroscopic study of cobalt(II) and nickel(II) sorbed on hectorite and montmorillonite. *Clays and Clay Minerals*, **39**, 22–27.

Defosse, C. and Rouxhet, P. G. (1981) Introduction to X-ray photoelectron spectroscopy, in *Advanced Chemical Methods for Soil and Clay Mineral Research* (eds J. W. Stucki and W. L. Banwart). Reidel, Dordrecht, pp. 169–203.

Dillard, J. G., Crowther, D. L. and Murray, J. W. (1982) The oxidation state of cobalt and selected metals in Pacific ferromanganese nodules. *Geochimica et Cosmochimica Acta*, **46**, 755–759.

Dillard, J. G., Crowther, D. L. and Calvert, S. E. (1984) X-ray photoelectron spectroscopic study of ferromanganese nodules: chemical speciation for selected transition metals. *Geochimica et Cosmochimica Acta*, **48**, 1565–1569.

Dillard, J. G., Koppelman, M. H., Crowther, D. L. *et al.* (1981) X-ray photoelectron spectroscopy (XPS) studies on the chemical nature of metal ions adsorbed on clays and minerals, in *Adsorption from Aqueous Solutions* (ed. P. H. Tewari). Plenum, New York, pp. 227–240.

Evans, S. (1977) Energy calibration in photoelectron spectroscopy, in *Handbook of X-ray and Photoelectron Spectroscopy* (ed. D. Briggs). Heyden, London, pp. 121–151.

Evans, S. and Raftery, E. (1980) X-ray photoelectron studies of titanium in biotite and phlogopite. *Clay Minerals*, **15**, 209–218.

Evans, S. and Raftery, E. (1982) X-ray photoelectron diffraction studies of lepidolite. *Clay Minerals*, **17**, 443–452.

Evans, S., Pritchard, R. G. and Thomas, J. M. (1978) Relative differential subshell photo-ionisation cross-sections (Mg Ka) from lithium to uranium. *Journal of Electron Spectroscopy and Related Phenomena*, **14**, 341–358.

Evans, S., Raftery, E. and Thomas, J. M. (1979) Angular variations in core-level XPS peak intensity ratios from single crystal solids. *Surface Science*, **89**, 64–75.

Fadley, C. S. and Shirley, D. A. (1970) Multiplet splitting of metal-atom electron binding energies. *Physical Review A*, **2**, 1109–1120.

Fuggle, J. C. (1977) XPS in ultra-high vacuum conditions, in *Handbook of X-ray and Photoelectron Spectroscopy* (ed. D. Briggs). Heyden, London, pp. 273–312.

Gregg, S. J. and Sing, K. S. W. (1982) *Adsorption, Surface Area and Porosity*. Academic Press, London.

Gupta, R. P. and Sen, S. K. (1974) Calculation of multiplet structure of core p-vacancy levels. *Physical Review B*, **10**, 71–77.

Harvey, D. T. and Linton, R. W. (1981) Chemical characterization of hydrous ferric oxides by X-ray photoelectron spectroscopy. *Analytical Chemistry*, **53**, 1684–1688.

Hellman, R., Eggleston, C. M., Hochella, M. F. and Crerar, D. A. (1990) The formation of leached layers on albite surfaces during dissolution under hydrothermal conditions. *Geochimica et Cosmochimica Acta*, **54**, 1267–1281.

Hochella, M. F. (1988) Auger electron and X-ray photoelectron spectroscopies, in *Reviews in Mineralogy*, Vol. **18**, *Spectroscopic Methods in Mineralogy and Geology* (ed. F. C. Hawthorne). Mineralogical Society of America, Washington, DC, pp. 573–637.

Hochella, M. F. and Brown, G. E. (1988) Aspects of silicate surface and bulk structure analysis using X-ray photoelectron spectroscopy (XPS). *Geochimica et Cosmochimica Acta*, **52**, 1641–1648.

Hochella, M. F., Lindsay, J. R., Mossotti, V. G. and Eggleston, C. M. (1988) Sputter depth profiling in mineral-surface analysis. *American Mineralogist*, **73**, 1449–1456.

Hofmann, S. (1983) Depth profiling, in *Practical Surface Analysis by Auger and X-ray Photoelectron Spectroscopy* (eds D. Briggs and M. P. Seah). Wiley, Chichester, pp. 141–179.

Inskeep, W. P., Nater, E. A., Bloom, P. R. *et al.* (1991) Characterization of laboratory weathered labradorite surfaces using X-ray photoelectron spectroscopy and transmission electron microscopy. *Geochimica et Cosmochimica Acta*, **55**, 787–800.

Jenne, E. A. (1967) Controls on Mn, Fe, Co, Ni, Cu and Zn concentrations in soils and water: the significant role of hydrous Mn and Fe oxides. *Advances in Chemistry Series*, **73**, 337–387.

Koppelman, M. H. and Dillard, J. G. (1977) A study of the adsorption of Ni(II) and Cu(II) by clay minerals. *Clays and Clay Minerals*, **25**, 457–462.

McIntyre, N. S. and Cook, M. G. (1975) X-ray photoelectron studies on some oxides and hydroxides of cobalt, nickel, and copper. *Analytical Chemistry*, **47**, 2208–2213.

McIntyre, N. S. and Zetaruk, D. G. (1977) X-ray photoelectron spectroscopic studies of iron oxides. *Analytical Chemistry*, **49**, 1521–1529.

Mathez, E. A. (1987) Carbonaceous matter in mantle xenoliths: composition and relevance to the isotopes. *Geochimica et Cosmochimica Acta*, **51**, 2339–2347.

Muir, I. J., Bancroft, G. M. and Nesbit, H. W. (1989) Characteristics of altered labradorite surfaces by SIMS and XPS. *Geochimica et Cosmochimica Acta*, **50**, 1235–1241.

Murray, J. W., Dillard, J. G., Giovanoli, R. *et al.* (1985) Oxidation of Mn(II): initial mineralogy, oxidation state and ageing. *Geochimica et Cosmochimica Acta*, **49**, 463–470.

Oku, M., Hirokawa, K. and Ikeda, S. (1975) X-ray photoelectron spectroscopy of manganese–oxygen systems. *Journal of Electron Spectroscopy and Related Phenomena*, **7**, 465–473.

Orchard, A. F. (1977) Basic principles of photoelectron spectroscopy, in *Handbook of X-ray and Photoelectron Spectroscopy* (ed. D. Briggs). Heyden, London, pp. 1–77.

Paterson, E. and Clark, D. R. (1991) Pressure-induced cation exchange in bentonite/laponite mixtures. *Clay Minerals*, **26**, 371–375.

Paterson, E., Bunch, J. L. and Duthie, D. M. L. (1986) Preparation of randomly-oriented samples for X-ray diffractometry. *Clay Minerals*, **21**, 101–106.

Perry, D. L. (1986) Applications of surface techniques to chemical bonding studies of minerals. *American Chemical Society, Symposium Series*, **323**, 388–402.

Perry, D. L., Taylor, J. A. and Wagner, C. D. (1990) X-ray induced photoelectron and auger spectroscopy, in *Instrumental Surface Analysis of Geologic Materials* (ed. D. L. Perry). VCH Publishers, Mannheim, pp. 45–86.

Petrovic, R., Berner, R. A. and Goldhaber, M. B. (1976) Rate control in dissolution of alkali feldspar – I. Study of residual feldspar grains by X-ray photoelectron spectroscopy. *Geochimica et Cosmochimica Acta*, **40**, 537–548.

Russell, J. D., Parfitt, R. L., Fraser, A. R. and Farmer, V. C. (1974) Surface structures of gibbsite, goethite and phosphated goethite. *Nature*, **248**, 220–221.

Schindler, P. W. (1990) Co-adsorption of metal ions and organic ligands: formation of ternary surface complexes, in *Reviews in Mineralogy*, Vol. **23**, *Mineral–Water Interface Geochemistry* (eds M. F. Hochella and A. R. White). Mineralogical Society of America, Washington, DC, pp. 281–307.

Schott, J. and Berner, R. A. (1983) X-ray photoelectron studies of the mechanism of iron silicate dissolution during weathering. *Geochimica et Cosmochimica Acta*, **47**, 2233–2240.

Seah, M. P. (1980) The quantitative analysis of surfaces by XPS: a review. *Surface and Interface Analysis*, **2**, 222–239.

Seah, M. P. (1983) Quantification of AES and XPS, in *Practical Surface Analysis by Auger and X-ray Photoelectron Spectroscopy* (eds D. Briggs and M. P. Seah). Wiley, Chichester, pp. 181–216.

Seah, M. P. and Dench, W. A. (1979) Quantitative electron spectroscopy of solids: a standard data base for electron inelastic mean free paths in solids. *Surface and Interface Analysis*, **1**, 2–11.

Seyama, H. and Soma, M. (1984) X-ray photoelectron spectroscopic study of montmorillonite containing exchangeable divalent cations. *Journal of the Chemical Society, Faraday Transactions I*, **80**, 237–248.

Seyama, H. and Soma, M. (1985) Bonding-state characterization of the constituent elements of silicate minerals by X-ray photoelectron spectroscopy. *Journal of the Chemical Society, Faraday Transactions I*, **81**, 485–495.

Sherwood, P. M. A. (1983) Data analysis in X-ray photoelectron spectroscopy, in *Practical Surface Analysis by Auger and X-ray Photoelectron Spectroscopy* (eds D. Briggs and M. P. Seah). Wiley, Chichester, pp. 445–475.

Soma, M., Tanaka, A., Seyama, H. *et al.* (1990) Bonding states of sodium in tetrasilicic sodium fluor mica. *Clay Science*, **8**, 1–8.

Sposito, G. (1990) Molecular models of ion adsorption on mineral surfaces, in *Reviews in Mineralogy*, Vol. **23**, *Mineral–Water Interface Geochemistry* (eds M. F. Hochella and A. R. White), Mineralogical Society of America, Washington, DC, pp. 261–279.

Stipp, S. L. and Hochella, M. F. (1991) Structure and bonding environments at the calcite surface as observed with X-ray photoelectron spectroscopy (XPS) and low energy electron diffraction (LEED). *Geochimica et Cosmochimica Acta*, **55**, 1723–1736.

Storp, S. (1985) Radiation damage during surface analysis. *Spectrochimica Acta*, **40B**, 745–756.

Stucki, J. W., Roth, C. B. and Baitinger, W. E. (1976) Analysis of iron-bearing clay minerals by electron spectroscopy for chemical analysis (ESCA). *Clays and Clay Minerals*, **24**, 289–292.

Swift, P. (1982) Adventitious carbon – the panacea for energy referencing? *Surface and Interface Analysis*, **4**, 47–51.

Swift, P., Shuttleworth, D. and Seah, M. P. (1983) Static charge referencing techniques, in *Practical Surface Analysis by Auger and X-ray Photoelectron Spectroscopy* (eds D. Briggs and M. P. Seah). Wiley, Chichester, pp. 437–444.

Thirioux, L., Baillif, P., Touray, J. C. and Ildefonse, J. P. (1990) Surface reactions during fluorapatite dissolution–recrystallization in acid media (hydrochloric and citric acids). *Geochimica et Cosmochimica Acta*, **54**, 1969–1977.

Thomassin, J. H., Goni, J., Baillif, P. *et al.* (1977) An XPS study of the dissolution kinetics of chrysotile in 0.1 N oxalic acid at different temperatures. *Physics and Chemistry of Minerals*, **1**, 385–398.

Tingle, T. N., Hochella, M. F., Becker, C. H. and Malhotra, R. (1990) Organic compounds on crack surfaces in olivine from San Carlos, Arizona, and Hualalai Volcano, Hawaii. *Geochimica et Cosmochimica Acta*, **54**, 477–485.

Vempati, R. K., Loeppart, R. H., Dufner, D. C. and Cocke, D. L. (1990) X-ray photoelectron spectroscopy as a tool to differentiate silicon-bonding state in amorphous iron oxides. *Soil Science Society of America, Journal*, **54**, 695–698.

VG Scientific (1992) Escalab 220i reaches new 2 mm record. *Surface Science Inview*, Winter, 1992/93. Fisons Scientific Instruments, East Grinstead.

Wagner, C. D. (1977) The role of auger lines in photoelectron spectroscopy, in *Handbook of X-ray and Photoelectron Spectroscopy* (ed. D. Briggs). Heyden, London, pp. 249–272.

Wagner, C. D., Gale, L. H. and Raymond, R. H. (1979) Two-dimensional chemical state plots: a standardized set for use in identifying chemical states by X-ray photoelectron spectroscopy. *Analytical Chemistry*, **51**, 466–482.

Walker, S. and Straw, H. (1961) *Spectroscopy*, Vol. 1, *Atomic, Microwave and Radiofrequency Spectroscopy*. Science Paperbacks, London.

West, A. R. and Castle, J. E. (1982) The correlation between the Auger parameter with refractive index: an XPS study using Zr Lα radiation. *Surface and Interface Analysis*, **4**, 68–75.

Wilson, M. J. and McHardy, W. J. (1980) Experimental etching of microcline perthite and implications regarding natural weathering. *Journal of Microscopy*, **120**, 291–302.

Yin, L. I., Ghose, S. and Adler, I. (1971) Core binding energy differences between bridging and nonbridging oxygen atoms in a silicate chain. *Science*, **173**, 633–635.

CHAPTER 7

X-ray fluorescence spectroscopy and microanalysis

D. C. Bain, W. J. McHardy and E. E. Lachowski

7.1 INTRODUCTION

This chapter deals with the X-ray emission that can be generated by bombarding a specimen with high-energy X-radiation, which in turn is generated by bombarding an appropriate anode material with high-energy electrons. The basis for chemical analysis by X-ray fluorescence (XRF) was established nearly 80 years ago, with Moseley's study of the X-ray spectra of the elements (Moseley, 1913). The first commercial X-ray spectrometer was produced about 1948 but it was some ten years later that XRF spectrometry became commonly used as a reliable method of chemical analysis.

X-ray emission is also generated by direct bombardment with an electron beam and this is the basis for electron beam microanalysis, which evolved in the early 1950s when Castaing and Guinier (1950) described the first static beam electron microprobe. During the same years that Cosslett and Duncumb (1956) advanced the concept by using a scanning probe, Oatley and co-workers at Cambridge University (Oatley, Nixon and Pease, 1965) were developing a practical scanning electron microscope (SEM). In the 1960s several high-performance microprobes and SEMs became available, but initially the two instruments generally maintained separate places, with the so-called electron probe microanalyser used for X-ray analysis and the SEM for high-quality imaging. However, there is much less of a distinction between the two in modern

Clay Mineralogy: Spectroscopic and Chemical Determinative Methods.
Edited by M. J. Wilson. Published in 1994 by Chapman & Hall, London.
ISBN 0 412 53380 4.

instrumentation and either can provide a high-quality image or a quantitative analysis.

Because the generation of X-rays and their detection by either energy dispersive X-ray spectroscopy (EDS) or wavelength dispersive spectroscopy (WDS) is similar in principle to XRF and electron beam systems, those aspects will be covered in general terms in section 7.2. Section 7.3 will deal in more detail with those aspects particular to XRF analysis using WDS, section 7.4 with what can be achieved with an EDS system as an attachment to an SEM and section 7.5 with an EDS system on a transmission electron microscope (TEM), usually referred to as analytical electron microscopy (AEM).

7.2 GENERAL PRINCIPLES

7.2.1 Nature of X-rays

An X-ray is a high-frequency electromagnetic radiation of energy intermediate between the far ultraviolet and gamma-ray regions of the spectrum and arises from an electron transition between discrete orbital shells of an atom. The energy of an X-ray is conveniently measured in units of an electron volt, 1 eV being the increase in energy imparted to an electron after acceleration through a potential field of 1 V. However, like other electromagnetic radiations, X-rays possess a sinusoidal waveform so that they may be characterized by a wavelength where energy (E) and wavelength (λ) are related by the equation

$$E = h\nu = \frac{hc}{\lambda},$$

where h is the Planck constant, c the velocity of light in a vacuum and ν the frequency of the waveform in cycles per second. Putting the appropriate numerical values into this equation leads to the more convenient expression

$$E(\text{in keV}) = \frac{12.396}{\lambda},$$

where λ is the wavelength expressed in angstrom units (Å), which by tradition are still used to express wavelength in X-ray spectrometry.

7.2.2 X-ray generation

X-ray emission from a specimen is generated by bombarding it directly with an electron beam or with an X-ray beam which is itself generated by bombarding an appropriate anode with a high-energy electron beam in a sealed X-ray tube. If the bombarding electromagnetic energy is of

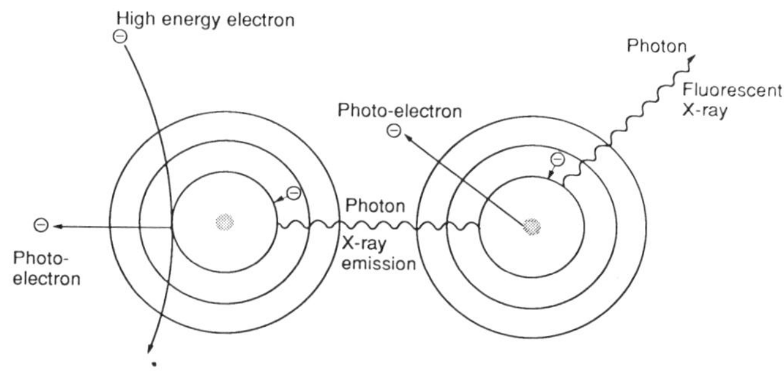

Figure 7.1 Schematic diagram of the atomic processes governing the emission of fluorescent X-radiation.

sufficient energy, an inner shell electron may be ejected and for the atom to return to its lowest energy state an outer shell electron (of greater energy) will fill the vacancy in the lower energy shell (Figure 7.1). In so doing an X-ray photon will be released with an energy equal to the difference between the energy of the electron in its initial and final states. A simplified mode of X-ray production according to the Bohr model of the atom is illustrated in Figure 7.2. The gap created by ejection of an electron from the K shell may be filled by an L shell electron, causing a Kα X-ray to be emitted; if filled by an M shell electron then Kβ radiation will be emitted. If an electron is ejected from an L shell then the resulting vacancy may be filled by an electron transition from an M shell, giving rise to Lα X-ray emission. Because all electrons in a given shell do not possess the same energy there will be specific variations with the α or β radiation; for example the Kα radiation is comprised of $K\alpha_1$ and $K\alpha_2$ X-rays, which are very close together in energy.

For an electron or X-ray beam to eject an inner shell electron from the atom, it must possess a minimum amount of energy. This is the binding

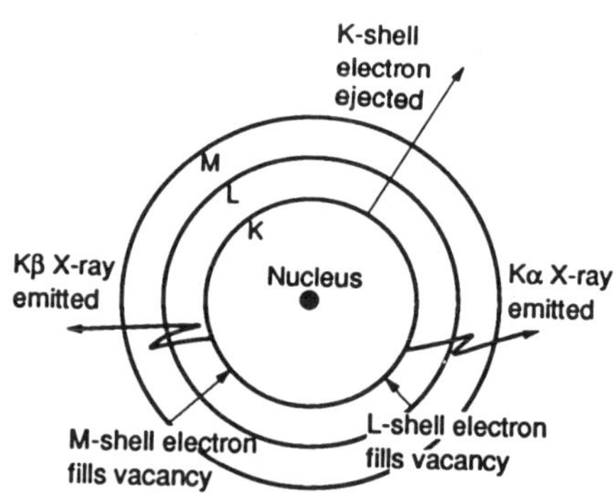

Figure 7.2 X-ray production according to the Bohr model of the atom.

energy for each electron in the atom. Because the K electron is closer to the nucleus its binding energy will be greater than that for an L shell electron, which in turn will be greater than that of an M shell electron. Therefore, if sufficient energy exists in the incident beam to excite K X-rays, L and M X-rays will also be excited if these shells are occupied, as they will be in high atomic number elements. The discrete and specific energy requirement necessary to excite any given X-ray line is called the absorption edge energy, or critical excitation energy and is always slightly greater than the corresponding X-ray emission line.

Just as the two electrons in the K shells have slightly different energy levels giving rise to $K\alpha_1$ and $K\alpha_2$ X-rays the electrons within the other shells have several slightly different energy levels, giving rise to complex fine structure within the various α and β lines. In practice many of the lines are weak in intensity and some are so close together that they are not resolved by some detection systems.

The energy of the characteristic radiation within a given series of lines varies monotonically with atomic number. This is Moseley's Law (Moseley, 1913) expressed by the following equation

$$E^{\frac{1}{2}} = C_1(Z - C_2),$$

where E is the energy of the emission line for a given X-ray series (e.g. $K\alpha$ lines), Z is the atomic number of the emitter, and C_1 and C_2 are constants. It follows from this equation that if the energy of a given K, L or M line is measured, then the atomic number of the element producing that line can be determined. It is the measurement of this energy and its intensity that is the basis for electron beam and X-ray fluorescence analysis.

In an electron beam instrument the range of spectral lines that can be available for analytical purposes depends on the maximum kilovoltage of the electron gun. Thus the K lines of silver will only be excited if more than 25.5 kV is available for the electron beam while the L lines of lead require at least 16 kV. Moreover, for the most efficient generation of spectra a voltage about three times the energy of the line is desirable. In most electron beam instruments analysis will be carried out using lower energy lines (e.g. below 20 keV). However, in transmission electron microscopy, where gun voltages of 100 kV or higher are available, the K lines of heavier elements may be used (section 7.3).

For XRF the primary excitation beam is generated by sealed high power X-ray tubes where a chromium, rhodium, scandium or gold anode is bombarded by a high energy, 40–100 kV, electron beam depending on the particular application.

7.2.3 X-ray measurement

As was pointed out earlier an X-ray can be described in terms of its energy or wavelength and thus it can be measured either by EDS or

WDS. As many electron columns today have an EDS detector and some have one or more wavelength spectrometers, both types of detector being used in XRF instrumentation, it is worth describing briefly the two systems and comparing them in terms of their ability to perform X-ray analysis.

7.2.3.1 *Wavelength dispersive spectrometers*

In the wavelength dispersive system, a crystal of known interplanar spacing, d, is used. A divergent beam of X-rays emanating from the specimen strikes the curved crystal and is diffracted according to the Bragg Law,

$$\lambda = 2d \sin \theta$$

where λ is the wavelength of the X-rays and θ the angle of incidence of the X-rays to the crystal. The curvature of the crystal converges the diffracted beam on to a detector, usually a gas-flow proportional counter. For any element there is a unique set of Bragg angles at which there will be X-rays detected by the counter. To detect a second wavelength, the angle of the crystal to the incident beam (i.e. the Bragg angle) must be changed. When the instrument is in use, the crystal is rotated on a motorized goniometer, while the counter sweeps through an arc, collecting X-rays and registering a signal as the Bragg condition is satisfied.

The great advantage that WD spectrometers have over energy dispersive spectrometers is their excellent resolving power. However, they are large and mechanically complex: their size means that the detector will be at some distance from the specimen, so necessitating lengthy analysis times. Each crystal in the spectrometer is only useful for a small range of wavelengths. This means that there must be an arrangement which allows selection from several crystals if the whole X-ray spectrum is to be investigated. Wavelength dispersive spectrometers are still the definitive form of XRF microprobe analysis and are used to calibrate standards, although their use has been supplanted in recent years by the introduction of energy dispersive detectors. However, computer control, which relieves the operator of most of the spectrometer manipulation, has revived interest in wavelength spectrometers as attachments to SEMs.

7.2.3.2 *Energy dispersive spectrometers*

The EDS detector is a small, solid-state device cooled by liquid nitrogen. It produces a voltage pulse which is proportional in energy to the energy of the incident photon. The pulses are sorted, measured and counted by a multichannel analyser system whose output is displayed as a spectrum on a cathode-ray tube. Energy dispersive spectroscopy systems are mechanically very simple and can be readily interfaced with most electron microscopes, both the scanning and transmission types. They display simultaneously most of the X-ray energies emitted from the specimen, which makes them ideal for a fast, preliminary analysis. The lower limit of the detectable energy level is set by the thin beryllium window, through which X-rays must pass to reach the detector. Normally, elements of atomic number 11 (Na), and greater, are readily detectable when using one of the K, L, or M lines. Detectors with an ultra-thin window or removable window are now available, where, it is claimed, the range of detection has been extended down to and including beryllium.

The EDS detector can be positioned very close to the specimen, collecting X-rays over a large solid angle, which allows short analysis times (*c.* 100 s) and small beam currents, both of which reduce the chance of specimen damage or volatilization of labile elements. In addition, the use of small beam currents means that X-ray microanalysis is compatible with the electron beam conditions that are required for reasonably high-image resolution on an SEM. The large collection angle means that the semiconductor detector is relatively insensitive to specimen height, and is therefore more appropriate than a WD system for the study of rough surfaces.

One serious disadvantage of EDS systems is their inferior spectral resolution, leading to severe peak overlap. Failure to account for overlaps can result in grossly inaccurate determinations. Fortunately, the output from an EDS detector readily lends itself to computer manipulation, and software is available to overcome this problem. The typical resolution of a modern semiconductor detector is *c.* 140 eV, but recent developments have been mainly in the degree of sophistication of the data reduction techniques.

A useful summary of the relative advantages and disadvantages of solid-state detectors and crystal spectrometers for electron beam instrumentation is given by Chandler (1977) and a comprehensive comparison of the two types of detector in both XRF and electron beam instruments is given by Potts (1987).

7.3 X-RAY FLUORESCENCE SPECTROMETRY

7.3.1 General

Because of the large number of elements (at least 80) and wide range of concentration that can be determined, XRF spectrometry is one of the most widely used routine instrumental methods of analysing geological materials. Both the major elements (Na, Mg, Al, Si, P, K, Ca, Ti, Mn and Fe) and a number of trace elements can be analysed rapidly and routinely in bulk specimens. Most XRF analyses are carried out using wavelength dispersive systems, because the technique has developed from this basis, but energy dispersive systems are becoming more commonly used as instrumentation improves with modern technological developments. Potts, Webb and Watson (1985) made a comparison of the advantages, application and performance of EDXRF (Ag tube) and WDXRF (Cr tube) for the analysis of silicate rocks for both major and trace elements, and concluded that although the limits of detection by EDXRF for the lighter elements are inferior to those by WDXRF, analytical precision at the levels normally encountered in silicate rocks is comparable. For the analysis of light elements from which structural formulae of clays may be calculated, WDXRF is still to be preferred because of its superior energy resolution for these elements.

Very high precision can be obtained from modern equipment and it is pertinent to note that the accuracy of XRF analysis depends on the reliability of the matrix correction program and the procedures used for calibration. The technique is a comparative one in which calibrations are generally carried out relative to recommended compositions of silicate rock reference materials. The accuracy obtained, therefore, depends critically on the materials used for calibration and the degree of confidence with which their compositions are known.

Because of the high precision obtainable with stable modern X-ray spectrometers, much attention has focused on sample preparation and the method of presentation of the sample to the instrument in order to minimize sample inhomogeneity effects.

7.3.2 Sample preparation

Accurate quantitative analysis by XRF spectrometry requires the use of uniform homogeneous samples, i.e. the specimen should behave towards X-rays as if it was a uniform glass. When different phases are present in a sample, as in most natural clays, fine grinding may produce the required result, but when a wavelength is > 3 Å, it can be very difficult to achieve sufficiently fine particle sizes by grinding. Long grinding times also increase the degree of contamination introduced into the sample from the grinding mill. For a sample to be uniform to Si $K\alpha$ or Al $K\alpha$ radiation, submicron particles are required, and these cannot be obtained by grinding.

Early analyses by XRF were performed on pressed powder pellets, sometimes with the use of a binding agent. This type of sample, however, is subject to particle-size, mineralogical and very marked absorption-enhancement effects. Indeed, the absorption-enhancement effects, sometimes called matrix or interelement effects, are one of the most serious problems to XRF analysis in that the intensity of a particular fluorescence line is not directly proportional to the concentration of the relevant element, but is affected by the amounts of all other elements in the sample. Many mathematical models have been designed to correct for these effects.

Although pressed powder pellets are still used for the determination of small amounts of heavy elements (atomic number > 26), where maximum sensitivity is required and critical depth of penetration is not a problem, fusion methods are now commonly used for accurate analyses of silicate materials such as clays; the sample is rendered uniform by dissolving it into a borate glass, eliminating particle-size and mineralogical effects. Interelement effects are still present in the glass, but they are markedly reduced by the dilution with flux, and can be corrected accurately.

The use of fusion techniques in XRF analysis was first proposed by Claisse (1956, 1957), who used sodium borate (borax) as a flux. After more than 35 years of application to geological materials, this basic technique has not changed except that as Na is now one of the elements analysed, samples are usually fused with fluxes consisting of lithium tetraborate and/or lithium metaborate, following development by Andermann and Allen (1961). Further suppression of matrix effects is obtained by adding a heavy absorber to the flux, and Rose, Adler and Flanagan (1962) used lanthanum oxide as the heavy absorber because lanthanum, which is not normally analysed with the major elements, causes the least spectrum line overlaps, and promotes glass formation in cast discs. Heavy absorber fluxes incorporating lanthanum oxide (15–20%) are now the standard sample preparation technique for major element analysis by WDXRF.

There are numerous variations in the details of fusion methods used in XRF analysis, with different laboratories developing different sets of conditions specific to particular applications; many papers have been published with details of particular 'recipes' and Potts (1987) has summarized the advantages and disadvantages of various methods, including those using low-dilution fusion mixtures. One of the problems in making stable glass discs involves the removal or prevention of excessive stress in the disc, and a detailed study of annealing conditions in glass-disc making has recently been carried out by Alvarez (1990).

Because sulphur is lost by volatilization during fusion, it is common to use powder techniques for its analysis by XRF. However, this is not very satisfactory because particle-size effects are maximized for such a light element, and Norrish and Thompson (1990) have described a fusion method for the analysis of sulphide samples which involves a short pre-oxidation treatment using sodium nitrate followed by fusion with an appropriate heat source.

7.3.2.1 Detailed fusion procedure

A widely used procedure for XRF analysis of fused glass discs is that described by Norrish and Hutton (1969). This involves fusion with a lithium borate flux incorporating lanthanum oxide as a heavy absorber and has been found to be suitable for a very wide range of geological samples. For laboratories interested in analysing geological samples of diverse origin and wide range of chemical composition, this method is strongly recommended. Discs with less than approximately 20% silica tend to be unstable and break easily, but this problem is easily overcome by adding high-purity silica to the sample and correcting the analysis for the known amount of added silica. This is necessary for the analyses of bauxites, for example. The Norrish and Hutton method also lists matrix correction coefficients for the common major and minor elements and application of these allows satisfactory analyses to be obtained for samples of extreme compositional variability.

To make a fused glass disc, 0.392 g finely ground sample pre-ignited to 1000°C is placed in a 95% Pt-5% Au crucible with 2.124 g flux (Spectroflux 105 as supplied by Johnson Matthey Chemicals Ltd). The mixture is heated in a muffle furnace at 1000°C until all the sample has dissolved (approximately 10 minutes). The homogeneous melt is poured on to a graphite platen on a hotplate at 220°C and an aluminium plunger (preheated to 200°C) immediately brought down to quench the melt and mould it into a glass disc. This procedure is shown diagrammatically in Figure 7.3. The glass disc is then placed between two asbestos-type mats on another hotplate at 200°C and left for at least 10 minutes to anneal. For speed and accuracy it is advisable to have the aluminium plunger permanently mounted on the hotplate in an arrangement such as that described by Harvey *et al.* (1973).

The original Norrish and Hutton method incorporated the use of a small amount of sodium nitrate added to the flux to ensure oxidizing conditions, but in the authors' experience this is not necessary. Sample

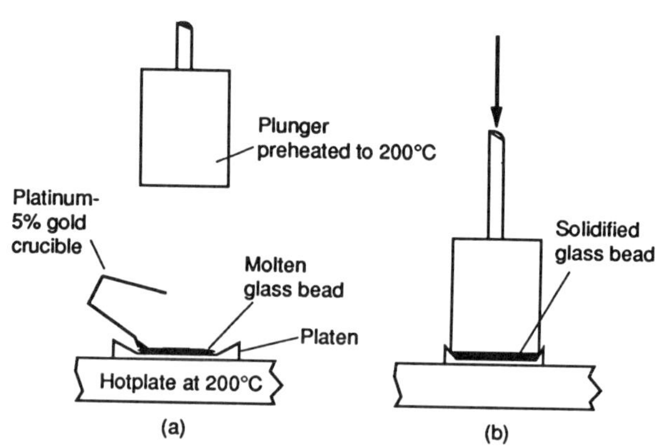

Figure 7.3 Apparatus for making glass fusion discs: (a) the melt is poured from a crucible on to the casting platen; (b) a preheated plunger is immediately lowered on to the platen to squash and quench the melt as a flat glass disc.

weights other than 0.392 may be used provided a correction is applied to the analysis, and discs have been successfully made with as little as 0.05 g.

Many trace elements can be determined in the Norrish and Hutton type discs despite dilution of the sample by a factor of about 6 : 1 by flux. If low levels or traces are to be determined, a modification of the method described by Hutton and Elliott (1980) using a dilution of 2.2 : 1 is more useful, and the same matrix correction coefficients can be applied.

Calibration for major and minor elements is carried out using a series of international rock standards, also pre-ignited to 1000°C.

7.3.3 Instrumentation

A schematic diagram of a conventional WDXRF spectrometer is shown in Figure 7.4. For a particular group of elements to be analysed, certain components can be varied to optimize analysing conditions. In most spectrometers there is usually a choice of two primary collimators, fine or coarse, to maintain adequate spectrum resolution, the coarse collimator being used for long wavelengths (light elements Na Kα to S Kα) and the fine collimator for shorter wavelengths from heavier elements. A variety of X-ray tubes are available including those with Si, Cr, Mo, Rh, Ag, Au and W anodes. In general terms, the lower atomic number anodes are best suited for exciting the light elements, and the higher atomic number anodes are used for the heavier elements. The Rh anode tube is commonly used for the analysis of a wide range of elements, and for those important in silicate rock analysis, the Rh tube lines cause few spectrum overlap interferences.

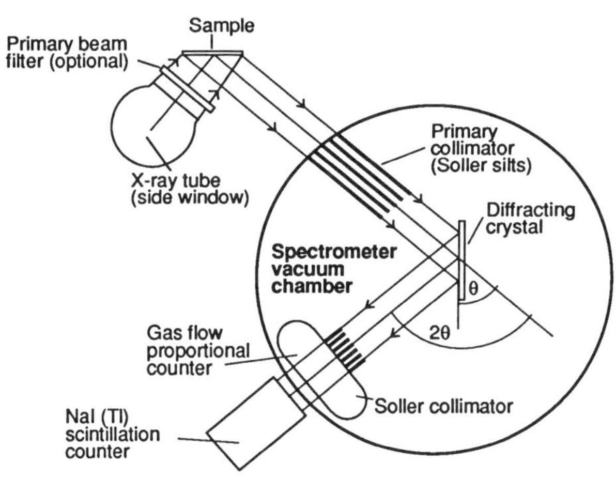

Figure 7.4 Schematic diagram of a wavelength dispersive X-ray fluorescence analyser showing the configuration of the X- ray tube, sample and spectrometer.

Table 7.1 Operating parameters and detection limits for wavelengths dispersive XRF (from Potts, 1987)

Element	X-ray line	Energy (keV)	Wavelength (Å)	Diffracting crystal[a]	Collimator	Tube kV	X-ray counter[b]	Count times[c]: Cr tube[d] (s)	Count times[c]: Mo tube[e] (s)	Count times[c]: W tube[e] (s)	Detection limits[f]: 1 : 5 glass disc (ppm)	Detection limits[f]: Powder pellet: Routine analysis (ppm)	Detection limits[f]: Powder pellet: 400 s count time per element (ppm)	
Na	Kα	1.041	11.910	TAP		50	Ar1	1892			2400	1600		Na_2O
Mg	Kα	1.253	9.890	TAP		50	Ar1	588			1200	1000		MgO
Al	Kα	1.486	8.340	TAP/PET		50	Ar1	211			600	492		Al_2O_3
Si	Kα	1.739	7.126	InSb/PET		50	Ar1/Ar2	200			1000	340		SiO_2
P	Kα	2.013	6.158	Ge/PET		50	Arl/Ar2	170			140	270		P_2O_5
S	Kα	2.307	5.373	Ge/PET		50	Arl/Ar2	30			160	20		S
K	Kα	3.312	3.742	PET		50	Ar2/Kr	12			140	8		K_2O
Ca	Kα	3.690	3.359	LiF(200)		50	Ar2/Kr	10			140	12		CaO
Ti	Kα	4.508	2.750	LiF(200)		50	Ar2/Kr	7			240	6		TiO_2
Ba	$L\beta_1$	4.827	2.568	LiF(220)	C	50	Ar2/Kr	140*1		230		24	7.4	Ba
Ba	$L\beta_2$	5.156	2.404	LiF(220)	C	50	Ar2/Kr						16.2	Ba
Cr	Kα	5.411	2.291	LiF(200)	C	50	Ar2/Kr			43		10	1.8	Cr
Mn	Kα	5.894	2.103	LiF(200)	C	50	Ar2/Kr Sc	171*2		138	240	13	2.3	MnO
Fe	Kα	6.398	1.937	LiF(200)	C	50	Ar2/Kr Sc	26		380	260	14	2.3	Fe_2O_3
Co	Kα	6.924	1.790	LiF(200)	C	60	Ar2/Kr Sc			724		6	1.2	Co
Ni	Kα	7.471	1.659	LiF(200)	C	60	Ar2/Kr Sc	37*1	23	73		6	1	Ni
Cu	Kα	8.040	1.542	LiF(200)	C	60	Ar2/Kr Sc	48*1	16			6	1	Cu

Zn	$K\alpha$	8.630	1.436	LiF(200)	C	60	Ar2/Xe	Sc	105*1	30		9	1	Zn
Rb	$K\alpha_1$	13.393	0.926	LiF(220)	C	100	Xe	Sc		22		3	0.8	Rb
Sr	$K\alpha_1$	14.163	0.875	LiF(200)	C	100	Xe	Sc		16		3	0.6	Sr
Zr	$K\alpha_1$	15.772	0.786	LiF(200)	C	100		Sc		20	133	6	1.6	Zr

[a] The most commonly used crystal is listed first. For elements above Ca, the selection of LiF(200) or LiF(220) follows the recommendations of Norrish and Chappell (1977). TAP, thallium acid phthalate; PET, pentaerythritol; InSb, indium antimonide; Ge, germanium; LiF, lithium fluoride.

[b] Ar1, argon flow counter with an ultra-thin window (1 μm polypropylene or 0.2 μm collodion) must be used. Ar2, argon flow counter with standard window (3 or 6 μm mylar). Kr, krypton sealed counter. Xe, xenon sealed counter. Se, sodium iodide scintillation counter.

[c] *1, per 1000 ppm element for Ba, Ni, Cu and Zn.

[d] Count times listed are those required to accumulate 100 000 counts for 1% element oxide calculated from the sensitivity data of Lee and McConchie (1982) for a 1 : 2 rock to flux glass disc excited at 50 kV, 35 mA. Mn*2 was excited with an aluminium primary beam filter, Ba was determined from the $L\alpha_1$ line.

[e] Count times listed are those required to accumulate 100 000 counts for 10 000 ppm of the element calculated from sensitivity data lised by (i) Lee and McConchie (1982) for 1 : 2 low dilution glass discs excited at 60 kV, 40 mA using a molybdenum tube, and (ii) Leake *et al.* (1969) for powder excited at between 40 and 100 kV (2 kW generator) using a tungsten tube. Some additional data for the Mo tube has been taken from Leake *et al.* (1969).

[f] Detection limits have been recalculated as the six-sigma limit of determination for routine operating conditions using 'typical' count times. The glass disc data are based on parameters specified by Norrish and Hutton (1969) and the powder pellet data (routine analysis) on data from Robinson and Bennett (1981). However, account has been taken of additional data presented by Leake *et al.* (1969), Nisbet, Dietric and Essenwein (1979), Hutton and Elliott (1980). Lee and McConchie (1982), and Vie le Sage *et al.* (1979). Data of Norrish and Chappell (1977), representative of 400 s total count time per samples are listed in the last column. Samples were excited with gold or molybdenum tubes as appropriate.

X-rays emitted from the sample are split into a spectrum by the diffracting crystal which, from Bragg's equation $n\lambda = 2d \sin \theta$, must have a $2d$ spacing of the same order of magnitude as the wavelength of the incident X-rays. To analyse a range of elements with a range of wavelengths with good angular dispersion (i.e. degree of separation between adjacent emission lines), it is necessary to use several crystals as mentioned above, each having high reflectivity and adequate angular dispersion for a few emission lines. A large number of crystals are available but only a few are used in routine work, e.g. for the ten major and minor elements commonly determined in a silicate analysis, five crystals are used: LiF(200) for Fe to K Kα lines, except for Mn Kα for which LiF(220) is used to reduce line overlap, Ge for S Kα, PE (pentaerythritol) for Si and Al Kα, and PX1, for Mg and Na Kα lines. For the lowest atomic number elements, such as Na and Mg, multilayered crystals or pseudocrystals, such as PX1, have recently been developed to produce higher reflected intensities than conventional single crystals.

Two types of detector are used to measure the intensity of the diffracted X-rays: the gas proportional detector and the scintillation counter, and these are commonly mounted in tandem along the axis of the diffracted beam (Figure 7.4). The gas proportional counter is used to detect X-rays of energy up to 6–8 keV (Fe Kα) and the scintillation detector for higher energy X-rays. A comprehensive description of instrumentation for wavelength dispersive XRF is given by Potts (1987, pp. 253–270).

Many modern spectrometers are fully automated, with a computer program designed to set the instrumental parameters for the elements to be determined, to measure the intensity of the emission lines on a peak minus background basis (counting for predetermined times at the peak position and at an offset position for background levels), to convert net intensities into concentration of the elements using a predetermined calibration, and finally to correct for interelement effects using matrix correction coefficients. An analysis of the ten major and minor elements in a silicate material measuring each element sequentially takes about 8 minutes.

7.3.4 Detection limits

In order to provide some indication of the levels of concentration down to which it is possible to determine elements under routine operating conditions, a selection of elements likely to be analysed in clay samples has been made from the table published by Potts (1987). These are presented in Table 7.1 and show levels of major and minor elements detectable in fused glass discs made by the Norrish and Hutton (1969) method, and some trace elements in pressed powder pellets. As detection limits are dependent to some extent on operating parameters, these are listed also. The use of new diffracting crystals may have led to improvements in limits of detection for some elements, e.g. Na and Mg, due to PX1 having replaced TAP (thallium acid phthalate) as the diffracting crystal because of its superior reflectivity.

7.4 MICROANALYSIS WITH SEM

7.4.1 Introduction

The usefulness of SEM in clay studies is much enhanced by the provision of an add-on energy dispersive X-ray analysis system. Not only can a rapid qualitative analysis of a morphological feature of interest be obtained but with appropriate specimen conditions quantitative chemical analysis is possible and elemental distribution maps may be obtained. The following sections describe some of the factors that need to be appreciated in the understanding of electron probe microanalysis and some of the ways the technique may be exploited.

7.4.2 Electron-beam–specimen interactions

The electrons from a beam striking a solid target will penetrate several micrometres into the sample, and before coming to rest they will influence a volume, as shown in Figure 7.5. It is apparent that the volume from which X-rays are produced is much greater than would be expected from the diameter of the electron beam striking the surface. The absorption and scattering of the primary electrons within this irradiated volume will determine the spatial resolution achieved. The relative importance of the various processes taking place has been discussed by

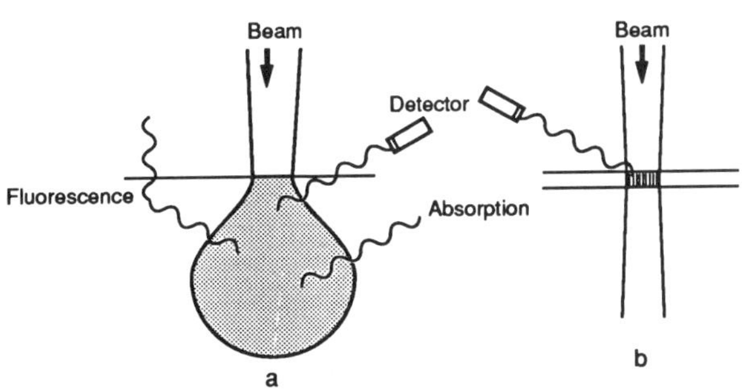

Figure 7.5 Diagrammatic representation of the process occurring in the volume of interaction between primary electron beam and specimen: (a) bulk specimen, (b) thin specimen.

Beaman and Isasi (1972). The **diameter** of the volume of interaction is a function of the primary beam accelerating voltage, whereas the **shape** is a function of the atomic number of the specimen material. Thus, the easiest way of improving analytical resolution is to decrease the accelerating voltage as much as possible, but this will be accompanied by a corresponding reduction in peak intensity. A useful guide to calculating spatial resolution is found in Beaman and Isasi (1972). For example, at an accelerating voltage of 25 kV, the penetration and X-ray resolution for Fe in a typical clay-mineral matrix would be 2–3 μm.

The effect of lateral beam spreading is quite small in a thin specimen and this is treated more fully in section 7.5. For bulk specimens within the volume of interaction between primary electron beam and specimen (Figure 7.5) a number of processes occur which affect the production and collection of X-rays. These effects will only be described briefly here. For a more detailed mathematical treatment the reader is referred to Beaman and Isasi (1972) or Long (1977).

7.4.3 X-ray absorption

In a bulk specimen, the emerging X-rays will be partially absorbed by the specimen before reaching the detector. High-energy X-rays will be absorbed less than low-energy X-rays, and the further the X-rays have to travel in the specimen the more they will be absorbed. This length of travel will depend on the angular relationships between the primary beam, specimen and detector, and a high take-off angle will therefore result in less absorption than a low take-off angle. There will, of course, be an added absorption factor if there is a window between the emergent X-rays and the detector.

7.4.4 X-ray fluorescence

Provided that their energy is high enough, X-rays passing through a specimen will excite X-rays characteristic of the other elements present. This secondary fluorescence will increase the total X-ray output.

7.4.5 Atomic number and back-scatter

Low atomic number elements have a greater number of orbital elements per unit mass, and as a result, absorb a disproportionately large amount of the incident electron energy. The generated intensity of an element in a light matrix is, therefore, less than the same amount of the element in a heavy matrix. On the other hand, electrons are back-scattered more in a heavy matrix, and consequently, less of their energy is available for producing characteristic radiation. These two effects produce a non-linear relationship between concentration and X-ray intensity (Sweatman and Long, 1969).

The four interactions mentioned in the last three sections form the correction procedure for the X-ray emission from a thick specimen, and collectively are known as the *Z* (atomic number), *A* (X-ray absorption) and *F* (X-ray fluorescence) correction.

One other effect must be considered, namely the 'white radiation' or bremsstrahlung produced by the deceleration of the primary electron beam in the specimen. Eventually, the electrons will have no energy, and so there will be a band of background radiation showing all energies from zero to the energy of the primary beam. The characteristic lines will be superimposed on this continuum, which is the major source of background in an electron beam excited X-ray spectrum. Its presence is the main reason for EDS systems having relatively high minimum detectability limits. In the WDS arrangement the analysing crystal placed in front of the X-ray detector screens out all but the desired wavelength. Thus, high peak to background ratios are obtained leading to much lower minimum detectability limits.

7.4.6 Qualitative analysis

In many applications all that is required is a rapid qualitative analysis of a particle or feature as observed in the SEM image. All modern systems have automatic peak identification routines of varying degrees of sophistication but they may not be complete and should not be relied on absolutely. The count rate should be kept below 5000 counts/s while accumulating spectra, to reduce the possibility of spectral artefacts. Identification should start with the largest peaks, and all possible KLM combinations should be checked. This is particularly important with elements such as Ba, Ti and Ce, where the sequence and relative intensity of minor peaks is diagnostic.

7.4.7 X-ray mapping

7.4.7.1 Analogue maps

If the X-ray spectrometer is tuned to the energy of a chosen element while the electron beam is scanning the specimen, and the resulting signal is used to modulate the microscope cathode-ray-tube display, then a map of that element's distribution can be obtained. The technique is best illustrated with an example. The scanning electron micrograph in Figure 7.6(a) is of a freshly cleaved biotite flake isolated from the C horizon of a soil profile. An image of the same field, produced by using X-rays characteristic of potassium to modulate the microscope display, shows clearly that the central part of the flake is markedly potassium-depleted (Figure 7.6(b)). As a general rule the exposure time required for recording such images can be based on the time taken for accumulation of

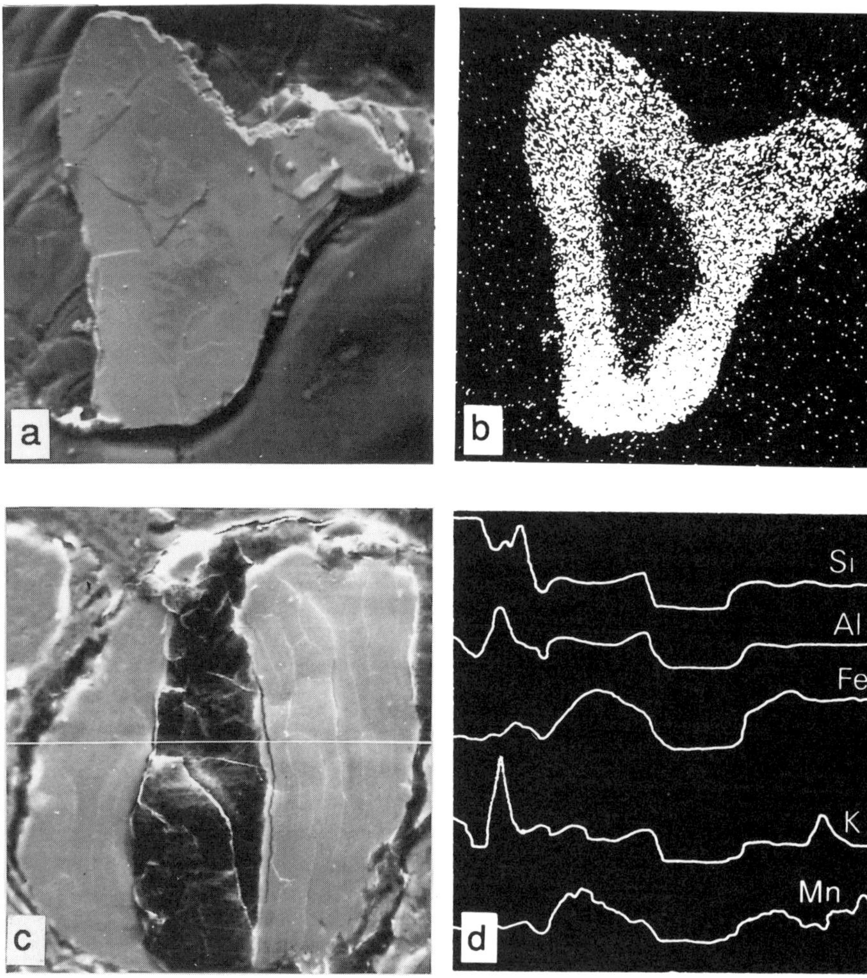

Figure 7.6 (a) Scanning electron micrograph of a partially weathered biotite particle, which has been freshly cleaved, mounted on a stub with double-sided tape and carbon coated; (b) K X-ray distribution image of the same field of view as (a); (c) resin-impregnated soil section, polished and carbon coated, featuring a partially infilled pore; (d) Si, Al, Fe, K and Mg concentration profiles pertaining to the line of analysis along the middle of (c).

250 000 X-ray counts from the area being scanned. If distribution images for several elements are required they have to be recorded sequentially.

7.4.7.2 Digital maps

With the incorporation of digital electronics into the design of modern SEMs, digital X-ray mapping is now the accepted method of displaying elemental distributions. Each pixel contains the number of X-ray counts instead of a brightness value, as in a secondary or back-scattered electron image. One advantage over the analogue system is that a number of elements may be mapped simultaneously, with the number of grey levels available in the system being divided between the number of elements to be mapped. On more sophisticated systems, however, up to 11 000 colours are available for each element. It is interesting that analogue maps commonly have a digital display, i.e. dot-on or dot-off, whereas digital maps can produce a continuous display, with areas of high concentration portrayed in brighter colours. Digital maps are an excellent way of displaying the spatial distribution of elements, but it is the availability of imaging software often included in modern X-ray microanalysis systems that provides the means of gaining new information about the sample. A much fuller account of X-ray mapping and computer-aided imaging is given by Friel and Barbi (1991).

7.4.8 Line profiles

Another way in which chemical information can be displayed is illustrated in Figure 7.6(c) and (d). The scanning electron micrograph shows a soil section where a pore has been partially filled with fine clay. The concentration profiles for Si, Al, Fe, K and Mg (Figure 7.6(d)) refer to the line across the centre of Figure 7.6(c). The profiles were obtained by driving the beam slowly along this line and modulating it (in the y-direction) by the output from a ratemeter tuned to the appropriate X-ray signal for each of the elements of interest.

A line profile has an advantage over a distribution image in that it can be superimposed on a secondary electron image, which allows an easy correlation between morphology and chemical analysis. It is more sensitive to small changes in concentration and it is also readily quantified.

7.4.9 Quantitative analysis

7.4.9.1 ZAF method

To a first approximation, the concentration of an element $A(C_A)$ is given by:

$$C_A = \frac{I_A}{I_S} C_S,$$

where I_A is the X-ray intensity or peak integral of the element A in the sample, I_S is the X-ray intensity of element A in the standard and C_S is the concentration of element A in the standard. If the standard is a pure element, then $C_S = 1$ and:

$$C_A = \text{constant} \times I_A.$$

When several elements are present, the ZAF corrections already mentioned must be taken into account. Thus, the corrected concentration of element A in a compound specimen is given by:

$$C_{A\,(\text{corr.})} = \frac{\text{constant} \times I_A}{Z \times A \times F},$$

where Z, A and F are the atomic number, absorption and fluorescence correction factors. These correction factors are applied to an initial concentration, given by the ratio of X-ray intensity of the element in the specimen to that in the standard, and a new corrected concentration results from this. The correction process is repeated using these new values, until the results of successive calculations converge. Where the standard is not a single element, matrix effects within the standard must also be taken into account. This kind of complex iterative calculation is obviously a task for a computer, either on-line, or using one of several readily available commercial routines. Myklebust, Fiori and Heinrich (1979) have described a data-evaluation procedure for carrying out quantitative electron-probe microanalysis with a lithium-drifted silicon detector. Their implementation of the *ZAF* method as FRAME-C has now become the standard in most small computers.

An important parameter in the foregoing mathematical treatment of X-ray spectra is the take-off angle, that is, the angle between the incident electron beam and the emergent X-rays. Best results are obtained from highly polished surfaces; rough fracture surfaces with poorly defined specimen geometry present special problems. Statham (1979) has used peak-to-background ratios in a correction procedure that minimizes the effect of specimen geometry, which could be a considerable improvement on conventional '*ZAF*' procedures. The peak-to-background approach is also insensitive to spectrometer parameters and beam current fluctuations.

7.4.9.2 φ–ρ–z Method

Whereas the method of *ZAF* is a mature, well-established method of quantifying electron-probe microanalysis the φ–ρ–*z* curve is an ionization distribution in which ionizations are plotted against mass–depth. These curves have been measured experimentally by various researchers, and Packwood and Brown (1981) suggested they could be used to calculate the *Z* and *A* factors for quantitative analysis. To do this, the shape of the curve must be known accurately, but this is where there is disagreement among the researchers developing the method. A more detailed discussion of the φ–ρ–*z* method is given by Friel and Barbi (1991).

The current opinion is that the choice of matrix correction model, i.e. *ZAF* or φ–ρ–*z* makes little difference in the analysis of heavier elements. Often the differences between matrix corrections are less than the uncertainties associated with counting statistics. However, it does appear that the φ–ρ–*z* method yields improved results for light elements, and it may be more accurate for non-conducting samples such as oxides and silicates.

7.4.9.3 Precision, accuracy and minimum detectability

The most important questions about electron-probe microanalysis are concerned with (1) the precision of the analysis, (2) the minimum concentration that can be detected and (3) the minimum area on the sample that can be analysed. Reasonable answers to these questions under average conditions are: (1) 2% relative, (2) 0.1% with EDS and 0.01% with WDS, and (3) *c.* 2 μm but real numbers can be readily calculated for any set of conditions. A thorough description and derivation of the relevant equations is given by Goldstein *et al.* (1981) and summarized by Friel and Barbi (1991). A suitable equation for estimating the minimum detectability limit is given by

$$C_{\mathrm{dl}} > \frac{3.29 \times ZAF \times (N_b)^{\frac{1}{2}}}{N_{p+b}},$$

where C_{dl} is the minimum detectability limit as a fraction, N is the number of counts in the background (N_b) or peak + background (N_{p+b}) measured on a pure element and *ZAF* is the total matrix correction factor. For example if $p = 102\,000$ counts, $b = 2000$ counts and $ZAF = 1$, then $C_{\mathrm{dl}} > 0.0015$ or 0.15 wt%.

The relevant equation for relative precision is

$$\frac{AC}{C} > \frac{2.33 \times 2 \times (N_{p+b})^{\frac{1}{2}}}{N_{p-b}},$$

where C is the concentration as a fraction and N_{p+b} and N_{p-b} as previously defined. Using the same numbers as for the detectability limit equation

$$\frac{AC}{C} = 0.015 \text{ or } 1.5\% \text{ relative.}$$

This is the precision, sometimes known as the analytical sensitivity, that may be expected from counting statistics where $N_p >> N_b$. A variety of other factors, however, can cause the analysis to be less accurate. These include uncertainties or inaccuracies in geometrical configuration, homogeneity of standards, beam stability, matrix correction factors and compatibility of phase size with volume of X-ray analysis.

7.4.10 Single-particle analysis

Microprobe analysis of clay minerals and related materials will often involve single particles on an inert substrate. Methods of data reduction for analysis of particles of variable geometry have been devised by Armstrong and Buseck (1975) and Aden (1981). If the clay mineral being analysed is a thin plate lying normal to the electron beam, the X-ray absorption path is readily calculated, provided that the particle thickness can be estimated by SEM. For more complicated particle shapes, thickness and density must be taken into account, and an estimate must be made of the idealized geometric model (Armstrong, 1978). Where two or more structurally similar but chemically distinct species are being studied, provided that the particle diameters are $< 5\ \mu m$, the composition ratios may be calculated without the use of matrix corrections.

7.4.11 Analysis using back-scattered electrons (BSE)

As previously stated, elements of atomic number < 11 cannot be detected by an ordinary X-ray energy dispersive detector with a beryllium window. This restriction can be partly overcome by exploiting the fact that the BSE signal is proportional to the average atomic number of the specimen. By standardizing the important operating conditions, and calibrating the back-scattered signal variation with atomic number, a quantitative measure of the apparent atomic number of any region of a specimen can be obtained (Ball and McArtney, 1981; Hall and Lloyd, 1981; Pye and Krinsley, 1983). The average atomic number will of course be influenced by the light elements, and if the heavier elements can be identified then a library of compounds can be matched with the BSE signal to find the closest fit to the average atomic number. Such an atomic number calibration will be valid only if the microscope conditions (for example kilovoltage) and the beam current are consistent, and it is necessary to have the microscope column controlled by software so that the correct conditions can be reinstated on the microscope and recalibrated for any change in beam current.

7.4.12 Specimen preparation

Specimen preparation methods for microanalysis are similar to those described for SEM imaging, although carbon-based conducting pastes should be used. Small samples of clay can be analysed after pressing a few milligrams of the clay into a plug and fixing it to a specimen stub with double-sided tape. The sample can then be recovered for further use. As a general rule, specimens should be flat, but this may not be possible and rough surfaces may have to be reoriented to leave an unobscured path to the detector. Morphological examination and microprobe analysis of individual clay flakes are greatly simplified if the clay can be deposited on to a carbon planchet; Berkheiser and Monsees (1982) have described a surfactant technique enabling clay particles to be so dispersed for X-ray microanalysis. If a specimen must be recovered uncontaminated, or if a quick qualitative analysis is all that is required, microanalysis may be carried out on an uncoated specimen. Generally, however, clay materials require a conductive coating, and carbon is commonly used because no interfering X-ray lines arise. Coatings of *c*. 20 nm are adequate. For quantitative analysis, reproducible film thickness is desirable, although serious errors attributable to film-thickness differences will arise only for light elements and low excitation voltages (Kerrick, Eminpizer and Villaume, 1973; Springer, 1974). An initial carbon coating can be useful for SEM microanalysis, followed by 20 nm of gold if good secondary electron images are required. Gold-coated specimens can be used quite successfully for qualitative microanalysis, provided that there are not peaks of interest likely to be masked by the characteristic gold peaks.

7.5 ANALYTICAL ELECTRON MICROSCOPY

7.5.1 Introduction

The addition of an X-ray analyser to TEM or scanning transmission electron spectroscopy (STEM) enables the clay mineralogist to obtain chemical analyses of extremely small particles. The technique is known as analytical electron microscopy (AEM) to distinguish it from the microprobe. The first AEM, EMMA4 (Cooke and Duncumb, 1968), had wavelength dispersive detectors, but now only energy dispersive detectors (EDX) are used. The thickness of transmission specimens is normally such that the beam spreading which occurs in the microprobe or SEM is effectively absent, so that the area analysed is defined by the beam diameter. This is illustrated in Figure 7.5. Goldstein *et al.* (1977), Newbury and Myklebust (1980) and, more recently, Michael *et al.* (1990) have shown that beam spreading in thin foils is a few tens of angstrom, which is small in relation to the sort of beam diameter required to give reasonable count rates. The short path-length of the emergent X-rays means that in many cases the effects of absorption and fluorescence may be neglected. The analysis of individual clay particles or of different areas of larger particles is thus possible. Although the accuracy and precision of the methods are often poor, the results are frequently far more meaningful than those obtained with bulk methods of chemical analysis. Functions such as element mapping at high resolution and automated particle detection and analysis are also available in STEM. Normal transmission specimen preparation methods are used, although dispersion aids must be chosen sensibly in order to avoid chemical contamination, and specimens should not be shadowed.

7.5.2 Instrumentation

The only difference between the EDX system in TEM or STEM and in the microprobe lies in the geometric configuration of the detector tip, which must fit in the limited space available in the TEM specimen chamber and also accommodate a shutter to protect the detector crystal from excessive radiation, as high X-ray and back-scattered electron fluxes can occur under certain circumstances; these are deleterious to the detector, particularly at high primary beam voltages. The high levels of back-scattered electrons introduce noise into the system so that dead times are high, with a consequent reduction in usable count-rate. Some manufacturers offer hardened detectors which do not require shutters. Windowless and ultra-thin window detectors are available as for the microprobe. Although the microscope itself need not have any extra lenses in order to function adequately as an AEM, it is advantageous to have either a condenser or objective mini-lens (as in STEM) in order to produce fine probes. Precautions must also be taken in order to reduce the generation of unwanted X-rays.

Two main configurations are common in microscopes with asymmetric objective lenses. The high take-off angle detector (Figure 7.7(a)) is the most useful for mineralogical analysis. The specimen is untilted, and the path length of the emergent X-rays through the specimen is minimized. The peak-to-background ratio is improved at high take-off angles (Zaluzec, 1979). Take-off angles of 60–70° are available. A take-off angle of 0° allows the detector to approach the specimen more closely (Figure 7.7(b)), increasing the sensitivity as the solid angle of X-ray accept-

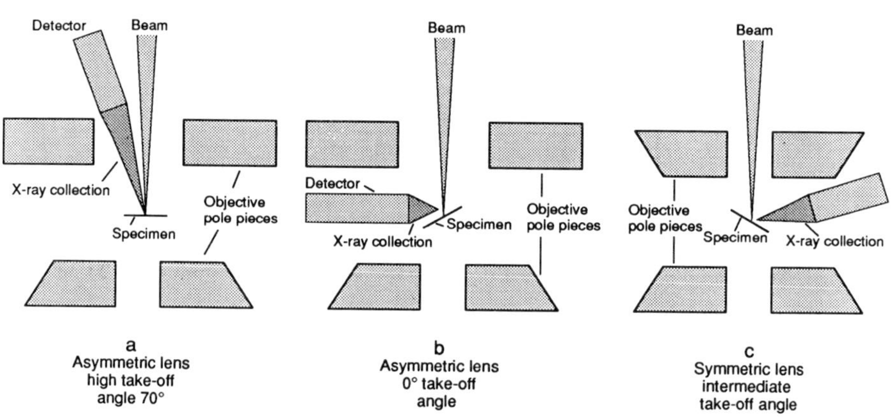

Figure 7.7 Different lens–detector configurations for AEM.

ance is greater, but the specimen must be tilted towards the detector, increasing the path length of the emergent X-rays in the specimen. This configuration is favoured by biologists, who must work with extremely low masses of material; it may also be of use where very thin particles are to be studied. In microscopes with symmetrical objectives, the take-off angle is intermediate between the two extremes, and the specimen need not be tilted so much (Figure 7.7(c)).

7.5.3 Quantitative analysis

In the microprobe, where all of the incident beam undergoes some sort of interaction with the specimen, the signal intensity of any element is a function of the concentration of that element. This is not the case in AEM, where most of the beam passes through the specimen without any interaction, and the signal intensity depends on the specimen thickness as well as on composition. The specimen mass can be estimated from the intensity of the white radiation, allowing the determination of elemental concentrations (e.g. Hall, 1971, 1991); this method is widely used by biologists to determine relatively low concentrations of heavy elements in fairly uniform matrices of light elements. This method is not well suited to chemical or mineralogical specimens, where the most commonly adopted procedure is to consider elemental ratios.

The characteristic X-ray intensity of any two elements is a function of their weight concentration ratio, provided the specimen is thin enough for absorption and fluorescence to be negligible (Cliff and Lorimer, 1975). For elements A and B the ratio is given by:

$$\frac{C_{\mathrm{A}}}{C_{\mathrm{B}}} = k_{\mathrm{A,\,B}} \times \frac{I_{\mathrm{A}}}{I_{\mathrm{B}}},$$

where C is the weight concentration, k is a constant and I is the characteristic X-ray intensity. The constant k may be calculated from first principles (e.g. Goldstein *et al.*, 1977; Szymanski and Lynch, 1986; Gauvin and L'Espérance, 1991).

In order to determine k, well characterized phases with two or more of the elements in question are required, one of them being common (the ratio standard element) to all the phases, if possible. In general it is advisable, but not always necessary (or indeed possible), for the standard phases to resemble the unknowns. Where suitable compounds are not available, crystallized droplets prepared by atomizing a solution containing the desired mixture of elements may be used (Graham and Steeds, 1984). In mineralogy it is convenient to use Si as the ratio standard element (Cliff and Lorimer, 1975; Cliff, Maher and Joy, 1984). In this case, a series of k factors, $k_{x,\mathrm{Si}}$, is accumulated. Wood, Williams and Goldstein (1984) recommend the use of Fe as the ratio standard element, as the efficiency of detection for Si is less than 1. Where these elements are not suitable for a particular study a more appropriate ratio standard may be chosen, using whichever combination of K, L or M lines gives the best results (Schreiber and Wims, 1981). For a standard containing the elements A and B, $k_{A,B}$ is given by:

$$k_{A,B} = \frac{C_A \times I_B}{C_B \times I_A}.$$

If $k_{A,B}$ and $k_{B,C}$ are known, then $k_{A,C}$ can be determined by simple arithmetic, but with increased statistical error.

Figure 7.8 shows the variation of $k_{x,Si}$ with atomic number (Z) for a typical Be-windowed detector. At low Z, absorption by the window, the gold coating on the detector crystal and the Si dead-layer reduce the sensitivity. The sensitivity is then relatively uniform until it again decreases at higher Z as the X-rays become sufficiently energetic for a significant fraction to pass right through the detector.

Most analyser programs allow the storage of the accumulated *k* factors in a data base so that quantitation is automatic.

7.5.4 Effect of specimen thickness

The maximum thickness at which specimens meet the thin-film criterion depends on the differences in mass absorption of the material for the particular X-rays being measured. Thus, difficulties in the analysis of phases containing both light and heavy elements may be avoided if the

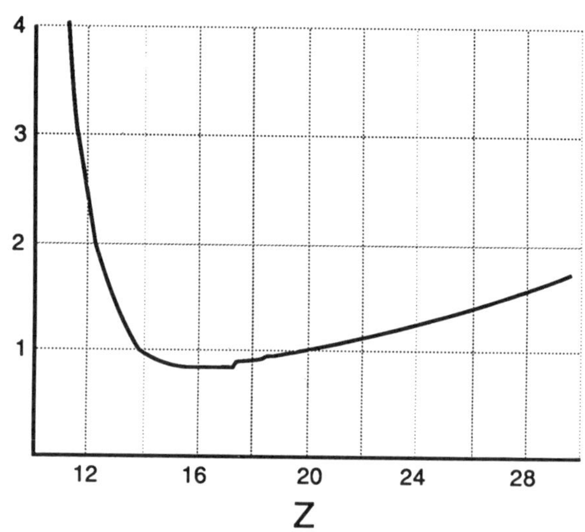

Figure 7.8 Schematic graph of $K_{x/Si}$ versus Z for a Be-windowed detector. There is considerable variability in the exact shape of the curve between different detectors, and with time.

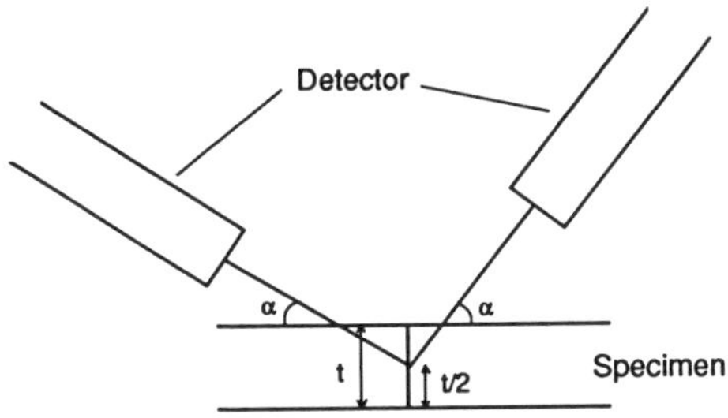

Figure 7.9 Effect of take-off angle α on mean path through the specimen.

M line of the heavy element has a similar energy to that of the K line of the light one. Tixier and Philibert (1969), Goldstein *et al.* (1977) and Williams and Goldstein (1991) have discussed the conditions under which the thin-film approximation breaks down. The absorption correction factor A for elements a and b in a given specimen is strongly dependent on the differences in mass absorption coefficients and is given by:

$$A = \frac{\mu/\rho^{a}}{\mu/\rho^{b}} \times \frac{1 - \exp(-\mu/\rho^{b} \operatorname{cosec} \alpha \times \rho t)}{1 - \exp(-\mu/\rho^{a} \operatorname{cosec} \alpha \times \rho t)}$$

where $\frac{\mu}{\rho}$ is the mass absorption coefficient, ρ the specimen density and t the thickness. The variable α is the take-off angle between the specimen and the detector, shown in Figure 7.9 (Goldstein *et al.*, 1977). In practice, the correction is automatic, provided specimen thickness and density are known, and allowance can be made for tilted samples. A mean thickness, $\frac{t}{2}$, as described by Champness, Cliff and Lorimer (1982), is used in most programs.

According to Williams and Goldstein (1991), if A lies between 0.97 and 1.03, corrections need not be made, as the errors due to counting statistics are about 3%. These authors give limiting thicknesses of 25 nm for O Kα in MgO and 9 nm for Al Kα in NiAl. The limit for Al Kα in kaolinite and Mg Kα in chlorite, calculated using the Link Analytical RTS2 program, is about 400 nm, which is outside the range of thickness normally encountered in transmission electron microscopy.

Fluorescence effects have been observed in metal alloys (e.g. Nockolds *et al.*, 1980) and a method for calculating them proposed. In clay minerals, however, where there is a high concentration of light elements, fluorescence is unlikely to be troublesome (Champness, Cliff and Lorimer, 1982; Williams and Goldstein, 1991).

7.5.5 Thickness measurement

Where it is considered necessary to measure the thickness of the specimen in order to apply corrections, a number of methods may be used.

If the clay particle has well-defined top and bottom edges oriented suitably to the tilt axis, the thickness may be determined simply by tilting the specimen through a known angle. Similarly, if edges are not visible, the separation between top and bottom contamination spots may be measured (Lorimer, Cliff and Clark, 1976). This method tends to overestimate the thickness (Love, Cox and Scott, 1977), and in any case it is preferable to avoid contamination as much as possible. With well-crystallized materials convergent beam diffraction gives very accurate thicknesses (Kelly *et al.*, 1975).

Thickness may also be determined from the X-ray spectra themselves. Morris, Ball and Statham (1980) used the variation in K to L intensity ratio to measure thickness, but both lines for the same element are not always present in a spectrum. These authors have also used the variation in intensity with tilt angle. The intensity of the continuum may also be used (Love, Cox and Scott, 1977). These methods require standards of known thickness which have compositions similar to those of the unknowns, and are less accurate than those described in the previous paragraph (Love, Cox and Scott, 1977).

7.5.6 Spurious peaks

When EDX analysis in TEM was first developed, a great deal of work was done to eliminate X-rays in the microscope column, generated principally by condenser apertures and the specimen chamber. These X-rays may be detected themselves, or they may fluoresce unwanted parts of the specimen. This work has been reviewed by Zaluzec (1979). Most of these problems have been eliminated, or at least reduced to tolerable levels, by the microscope manufacturers, who can supply hard X-ray kits if these have not already been installed. These consist of thick condenser apertures to self-absorb the hard X-rays excited by the intense primary beam, and graphite apertures and inserts in the objective lens. The objective aperture must be withdrawn during analysis.

Nevertheless, some spurious peaks are still generated in addition to the sum and escape peaks. Wilson and Lambrianidis (1990) have reported a broad peak overlapping Nb in a Jeol 2000FX microscope. This was found to be generated by Compton scattering of Mo radiation from the condenser aperture by the graphite objective pole piece insert (part of the hard X-ray kit). The specimen itself emits X-rays and scatters electrons, both of which can excite unwanted signals from other parts of the specimen, the holder and the grid. Parts of the specimen not in the field of view may also lie in the path to the detector; this is more likely with low take-off angles. In wedge-shaped specimens, such as are prepared by Ar milling, contributions to the X-ray spectrum arise from thick areas close to the thin area under analysis. A method for quantifying this has been described by Glas (1989). The influence of the holder may be reduced by using one made from Be or graphite, or by coating it with carbon paint (e.g. Leit-C). When Cu grids are unsuitable, Be or nylon grids may be used. Beryllium is toxic and expensive, however. The nylon grids give a Ti signal and are unstable in the beam, and there is a more insidious danger. The grids are woven from a thick thread, which can block the path to the detector unless the specimen is in the centre of a square. In some situations, differential absorption of lower energy X-rays can lead to erroneous results. Figure 7.10 illustrates the main sources of undesirable X-rays in a well-designed system.

The carbon support film itself may give a weak Si signal, especially if silicone oil is used in the coating plant used to prepare it. With specimens of reasonable thickness this is either lost in the background or swamped by the Si in the specimen, but it can be troublesome with very thin, Si-free minerals. In such situations, a holey support film may be more appropriate.

Crystalline specimens can give rise to coherent bremsstrahlung peaks (Reese, Spence and Yamamoto, 1984). This manifests itself as small, more-or-less Gaussian peaks which could be mistaken for the characteristic lines of trace elements. Vecchio and Williams (1987) have established that the peak positions are sensitive to crystal orientation, accelerating voltage and detector take-off angle. Thus the intensity of

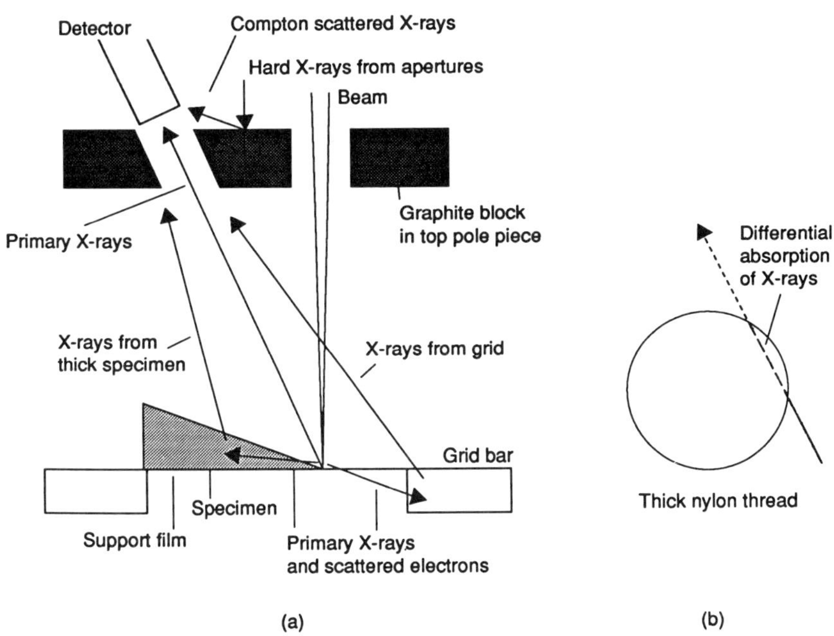

Figure 7.10 (a) Spurious X-ray generation in AEM, (b) differential absorption of X-rays by nylon grids.

coherent bremsstrahlung peaks can be minimized by using a high take-off angle and avoiding strongly diffracting orientations. Current analyser software packages have routines for calculating the peak positions. In hydrated minerals, which lose much of their crystallinity (see below), the effect is of little consequence.

7.5.7 Undesirable beam effects

There are a number of beam-induced phenomena in addition to those described above which may give erroneous peak intensities.

Although it is theoretically possible to obtain X-ray analyses from very small areas, the intensity of the radiation can cause severe damage, leading to compositional change as well as loss of crystallinity. The mechanisms of beam heating and radiolysis are reviewed by Hobbs (1979). Beam damage is lower at higher accelerating voltages, but hydrated materials remain problematic, many intumescing in the beam. Aluminium-loss has been reported from kaolinite (Chandler, 1973). Calcium-loss in gyrolite was found to be reduced after prolonged evacuation, presumably causing partial dehydration (Lachowski, Murray and Taylor, 1979). The role of water vapour in inducing compositional change is discussed by Hren (1979). An additional thin layer of carbon on top of montmorillonite was found to be beneficial in conducting away heat and electric charge (Chandler, 1973). Sodium and potassium rapidly migrate out of the irradiated zone of a specimen (Chandler, 1973, Hodson and Marshall, 1971). Van der Pluijm, Lee and Peacor (1988) have measured K-loss in clay minerals and have found that, in addition to the effect of beam intensity, loss is higher for crystals $< 100\,nm$ thick, but lower when (001) is parallel to the crystal surface. Halides, carbonates and sulphides are particularly susceptible to radiolysis (Hobbs, 1979). Some of these effects are illustrated for dickite in Figure 7.11. Two analyses were made on the crystal. In the first, the beam was spread over the circled area; little damage is apparent and the composition (excluding water and with O determined by stoichiometry) was found to be $Al_{1.93}Si_{2.05}O_7$. For the second, a fine probe was used, with the current adjusted to give a similar count rate to that of the first analysis. A hole has been burnt in the crystal. (The hole is elongated due to specimen or beam drift). There is also some intumescence in the region. The apparent composition thus determined was $Al_{1.33}Si_{2.50}O_7$, which is clearly nonsense. Thus it is often necessary to sacrifice analytical resolution by using a well-spread beam, or, in STEM, by rastering the beam over an area.

The use of contamination spots to measure thickness has already been mentioned. These are formed by the beam-induced polymerization of hydrocarbon molecules on the surface of the specimen and take the form of spikes of carbon on upper and lower surfaces, as shown in Figure 7.12. The mechanism is discussed by Hren (1979). As the carbon build-up

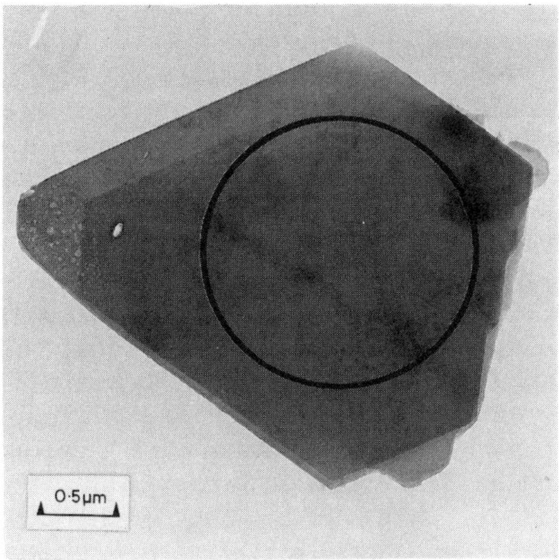

Figure 7.11 Dickite crystal showing effects of large and fine electron probes. Some drift, probably due to heating, has elongated the hole made by the fine probe, and minor damage is also apparent around it.

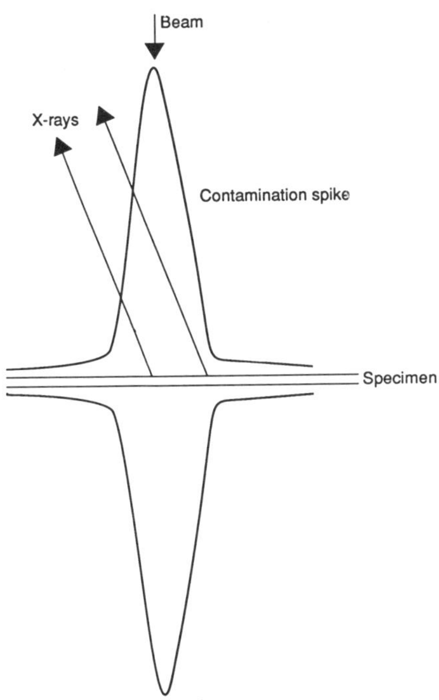

Figure 7.12 Absorption of X-rays in a carbon contamination spike.

increases during an analysis, differential absorption of low-energy X-rays may occur, giving erroneous intensities (Champness, Cliff and Lorimer, 1982). The effect is particularly troublesome in older microscopes and with fine probes. In modern machines the pumping system uses oil with very low vapour pressures, or may be dry in the case of turbomolecular or ion pumps, leading to very low contamination rates, provided the specimen is clean. Finger grease is another principal source of hydrocarbon vapour, so sensible handling of the specimen and holder is required. Liquid-nitrogen-cooled anticontaminators, which condense and thus trap the hydrocarbons, are standard fittings for all TEM.

Electron channelling in crystalline specimens can occur if a major zone is lying perpendicular to the beam, resulting in strong Bragg reflections. These can be recognized by the presence of extinction contours in the area of interest. (Faint extinction contours are visible as a star-shaped feature in Figure 7.11.) The electron flux through the specimen is no longer uniform so that differential X-ray generation occurs (Duncumb, 1962; Cherns, Howie and Jacobs, 1973). It is therefore better to avoid such orientations, although the effect is often small. The ALCHEMI technique, described by Spence and Taftø (1983), makes use of this effect to estimate site occupancies by comparing relative intensities at different orientations of the specimen. An improved method of data acquisition and analysis, which enhances the scope and accuracy of ALCHEMI, has been described by Turner *et al.* (1991). As with coherent bremsstrahlung, the loss of crystallinity by hydrated materials prevents the effect from occurring.

7.5.8 Precision and detection limits

In the unlikely event that all the above-mentioned sources of error have been eliminated, the precision of the analysis depends on the counting statistics of the X-ray signal. If the peaks are considered to be Gaussian, the standard deviation, σ, is given by $n^{1/2}$, where n is the number of counts in a given peak. Thus for a peak with 10 000 counts, σ is 100, or 1%. According to Goldstein and Colby (1975), however, if detector drift and electronic noise are included, σ may be as large as $2n^{1/2}$. The effect means that there is an error of at least 2% in k factors, which must be added to the error of an analysis of an unknown. It is clear that it is important to obtain the best possible counting statistics in the determination of the k factors. In the analysis of real specimens it is unlikely that such high numbers of counts are obtained for the minor constituents, with obvious consequences. A commonly used way to improve the statistics of analysis is to perform many analyses on different particles of the same phase. In STEM this procedure can be automated, but there is a greater risk of beam-induced compositional change when fine probes are used.

Warren and Ransom (1992) have reassessed analytical data for several clay minerals in the light of the statistical error expected and have shown that much of the reported compositional variation in clay minerals falls within the expected error envelopes of pure phases. They warn that interpretation of compositional variation is not as clear-cut as has been supposed hitherto.

Chandler (1977) has discussed the minimum detectable concentration of a given element. For a peak to be distinguished it must have a certain minimum intensity above the background. As the error in measuring the height of the peak (p), above the background (b), is given by $(p+b)^{1/2}$, the peak is significant when

$$p - b > n(p + b)^{\frac{1}{2}}.$$

The value of n is generally 1 or 2, depending on the choice of confidence limits. It can be shown that the minimum detectable concentration, C_m, of a particular element is given by

$$C_m = \frac{nC_x(2b_x)^{\frac{1}{2}}}{p_x - b_x},$$

where p_x and b_x are the peak and background from a sample with a concentration C_x of that element. Williams and Goldstein (1991) describe a slightly different derivation of the formula which uses 3 for the value of n.

The value of C_m is highly dependent on the background, which is itself dependent on a number of factors, such as the take-off angle of the detector and the design of the specimen holder as well as the energy of the characteristic peak. In general, C_m is of the order of about 0.1%, provided there is no peak overlap. Increasing counting times, beam current (not always wise with clay minerals) and accelerating voltage all help to increase $p - b$. With long counting times, specimen and beam drift are often troublesome (and not infrequently with short ones). In STEM the specimen can be tracked automatically if the analyser is reasonably sophisticated, but in TEM, tracking must be done manually. Fortunately, it is often the case that an element detected as a trace in a bulk analysis is concentrated in one particular phase, and AEM is ideally suited for its detection.

7.5.9 Examples

Although many of the foregoing paragraphs may give a somewhat negative impression of AEM, it is nevertheless an extremely powerful tool, principally because of its ability to analyse very small volumes of material. At a trivial level it offers rapid identification of individual phases. Figure 7.13 shows the X-ray fingerprints of two different chlorites, illite and kaolinite; each spectrum was obtained with 60 s live counting time from the areas circled on the accompanying micrographs. There is evidence of beam damage in the kaolinite.

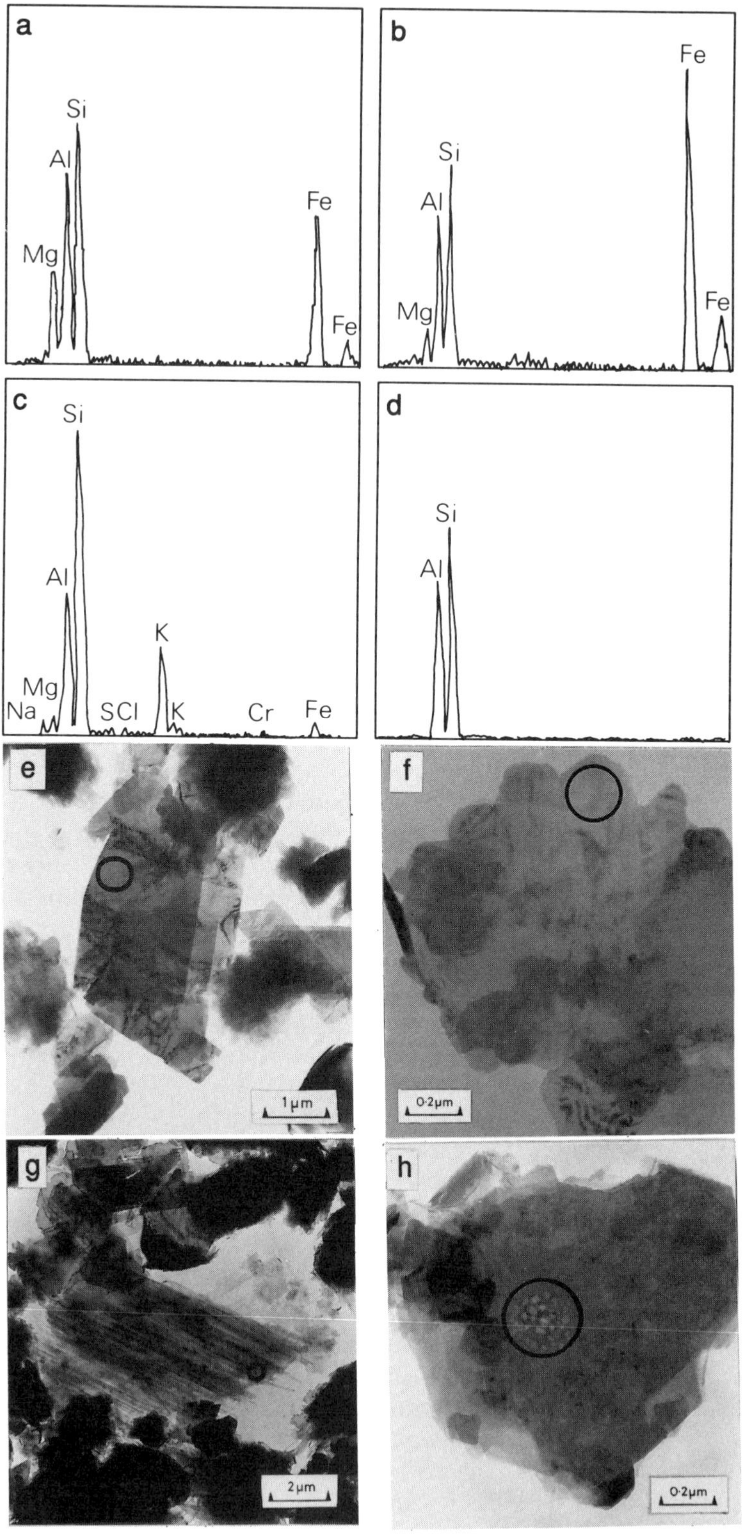

a
Si
Al
Mg
Fe
Fe
b
Fe
Si
Al
Mg
Fe
c
Si
Al
K
Mg
Na
SCl
K
Cr
Fe
d
Si
Al
e
1 μm
f
0·2μm
g
2μm
h
0·2μm

7.5.9.1 *Kaolinite*

Szymanski, Loeber and Durand (1990) made a statistical study of reference clays, using both experimental and theoretical *k* factors, and obtained a reasonable agreement in the range $Z = 12–30$. In spite of the fact that holes were drilled in their specimens by the beam, they obtained the following composition for kaolinite (with detected elements normalized):

	%	σ
Si	50.04	2.18
Al	48.83	2.08
Fe	0.82	0.69
K	0.31	0.43

Note the poor precision of the Fe and K analyses, which is inevitable with these levels.

Gilkes and Singh (1991) have reported the presence of 2.1–2.8% Fe_2O_3 substituting for Al_2O_3 in kaolinite crystals smaller than 100 nm.

7.5.9.2 *Glauconite*

Ireland, Curtis and Whiteman (1983) studied argon-milled specimens containing glauconite and illite. Some glauconite crystals were found to be very low in Fe, with a concomitant increase in Al concentration. They were closely associated with diagenetic pyrite, implying that exchange had accompanied recrystallization. This was on a scale such that the effect was not detectable by other means.

7.5.9.3 *Illite–Smectite*

Combined AEM and HREM studies of illite–smectite intergrowths have revealed a number of interesting features. Ahn and Peacor (1986) have

Figure 7.13 Energy dispersive X-ray spectra recorded at 200 kV accelerating voltage of (a) and (b) two different chlorites, (c) illite, (d) kaolinite with micrographs, (e) chlorite[spectrum (a)], (f) chlorite[spectrum (b)], (g) illite[spectrum (c)] and (h) kaolinite[spectrum (d)].

described a specimen of smectite containing well-defined packets of illite, recognizable by their 1 nm spacing. The boundaries between the two phases were discontinuous, and AEM of areas of each phase in excess of 30 nm diameter revealed the compositions to be those of pure end-member phases, although XRD data implied the presence of a mixed-layer phase. In another study, by Jiang *et al.* (1990), a mixed-layer illite–smectite was found as a replacement product of illite. By optimizing the conditions of the HREM they were able to distinguish the smectite layers from those of the illite layers (collapse of the smectite layer in the vacuum of the microscope makes this difficult). Analytical electron microscopy showed that the apparent illite had the composition $(K_{1.21}Na_{0.12})(Al_{3.36}Fe_{0.31}Mg_{0.33})(Si_{6.28}Al_{1.72})O_{20}(OH)_4$, whereas the mixed phase was $(K_{0.46}Na_{0.43})(Al_{3.75}Fe_{0.06}Mg_{0.19})(Si_{6.26}Al_{1.74})O_{20}(OH)_4$. In this example the beam was rastered over the specimen during analysis and identification of the phases was checked by consideration of the $(Ca + Na):K$ and $Al:Si$ ratios, generally applicable to this kind of clay material, in order to minimize the effects of element mobility.

7.6 CONCLUSIONS

This chapter has discussed three different ways in which the emission of X-rays can be exploited instrumentally for the study of clays. X-ray fluorescence is an established tool for chemical analysis. Although applicable to a wide range of materials it is well suited for the bulk chemical analysis of clays. Results of very high precision can be obtained routinely. Another feature is low detection limits, varying from about 2000 ppm for Na to 20 ppm or better for elements of atomic number $\geqslant 16$. However, XRF has no spatial resolution and this is the special advantage of electron-probe microanalysis, the principal role of which, in clay-mineral studies, is as an add-on EDS accessory to the scanning electron microscope whereby chemical information on features as small as 1 μm can be obtained. Good quantitative results are possible from highly polished specimens but it should be remembered that for clay minerals in a heterogeneous environment the analysed volume may not be exclusively occupied by one species. Only analytical transmission electron microscopy has the spatial resolution necessary to make quantitative chemical analysis of individual clay minerals particles possible. This is not easy but with due consideration of the factors described in the preceding section meaningful results should be attainable.

ACKNOWLEDGEMENT

Acknowledgement is given to the Scottish Office Agriculture and Fisheries Department for providing funding for this work.

REFERENCES

Aden, G. D. (1981) Quantitative energy dispersive analysis of small particles. PhD thesis, Arizona State University, 459 pp.

Ahn, J. H. and Peacor, D. R. (1986) Transmission and analytical electron microscopy study of the smectite to illite transition. *Clays and Clay Minerals*, **34**, 165–179.

Alvarez, M. (1990) Glass disk fusion method for the X-ray fluorescence analysis of rocks and silicates. *X-ray Spectrometry*, **19**, 203–206.

Andermann, G. and Allen, J. D. (1961) X-ray emission analysis of finished cements. *Analytical Chemistry*, **33**, 1695–1699.

Armstrong, J. T. (1978) Methods of quantitative analysis of individual micro-particles with electron beam instruments, in *Scanning Electron Microscopy*, Vol. 1 (ed. O. Johari). AMF O'Hare, SEM Inc., Chicago, pp. 455–468.

Armstrong, J. T. and Buseck, P. R. (1975) Quantitative chemical analysis of individual microparticles using the electron microprobe: theoretical. *Analytical Chemistry*, **47**, 2178–2192.

Ball, M. D. and McArtney, D. G. (1981) The measurement of atomic number and composition in an SEM using backscattered detectors. *Journal of Microscopy*, **124**, 57–68.

Beaman, D. R. and Isasi, J. A. (1972) Electron beam microanalysis, in *STM Special Publication, 506*. American Society for Testing and Materials, Philadelphia, Pa, 80 pp.

Berkheiser, V. E. and Monsees, M. B. (1982) Dispersion of clays on graphite support for X-ray microprobe analysis. *Soil Science Society of America, Journal*, **46**(3), 663–666.

Castaing, R. and Guinier, A. (1950) Sur l'exploration et l'analyse elementaire d'un échantillon par une sonde electronique, in *Proceedings, First International Conference on Electron Microscopy*. Martinus Nijhoff, Delft, pp. 60–63.

Champness, P. E., Cliff, G. and Lorimer, G. W. (1982) Quantitative analytical electron microscopy of metals and minerals. *Ultramicroscopy*, **8**, 121–132.

Chandler, J. A. (1973) Recent developments in analytical electron microscopy. *Journal of Microscopy*, **98**, 359–378.

Chandler, J. A. (1977) X-ray microanalysis in the electron microscope, in *Practical Methods in Electron Microscopy*, Vol. 5(2), (ed. A. M. Glauert). Northern-Holland, Amsterdam, 230 pp.

Cherns, D., Howie, A. and Jacobs, M. H. (1973) Characteristic X-ray production in thin crystals. *Zeitschrift fuer Naturforschung*, **A28**, 565–571.

Claisse, F. (1956) *Accurate X-ray Fluorescence Analysis Without Internal Standard*. Preliminary Report No. 327, Department of Mines, Quebec.

Claisse, F. (1957) Letter to the editor. *Norelco Reporter*, **4**, 95–96.

Cliff, G. and Lorimer, G. W. (1975) The quantitative analysis of thin specimens. *Journal of Microscopy*, **103**, 203–207.

Cliff, G., Maher, D. M. and Joy, D. C. (1984) Determination of UTW $k_{x.Si}$ factors for seven elements from oxygen to iron. *Journal of Microscopy*, **136**, 219–225.

Cooke, C. J. and Duncumb, P. (1968) Performance analysis of combined electron microscope and electron-probe microanalysis, EMMA, in *Proceedings, 5th International Conference on X-ray Optics and Micro-analysis, Tubingen* (eds G. Mollenstedt and K. H. Gaukler). Springer-Verlag, Berlin, pp. 245–247.

Cosslett, V. E. and Duncumb, P. (1956) Microanalysis by a flying spot X-ray method. *Nature*, **177**, 1172–1173.

Duncumb, P. (1962) Enhanced X-ray emission from extinction contours in a single crystal gold film. *Philosophical Magazine*, **7**, 2102–2105.

Friel, J. J. and Barbi, N. C. (1991) *X-ray Microanalysis and Computer-aided Imaging*. Princeton Gamma-Tech, Princeton, NJ 08540.

Gauvin, R. and L'Espérance. G. (1991) Determination of the C_{n1} parameter in the Bethe formula for the ionization cross-section by use of Cliff–Lorimer k_{AB} factors obtained in a TEM at different accelerating voltages. *Journal of Microscopy*, **163**, 295–306.

Gilkes, R. J. and Singh, B. (1991) An analytical transmission electron microscopic study of soil kaolinite. *Proceedings, 7th Euroclay Conference, Dresden*, (eds M. Störr, K.-H. Henning and P. Adalphi), Ernst-Moritz-Arndt Universität, Greifswald, pp. 411–416.

Glas, F. (1989) Quantitative study of spurious X-ray production in thin film microanalysis. *Journal of Microscopy*, **154**, 19–29.

Goldstein, J. I. and Colby, J. W. (1975) Special techniques in the X-ray analysis of samples, in *Practical Scanning Electron Microscopy* (eds J. I. Goldstein and H. Yakowitz). Plenum Press, New York, pp. 435–490.

Goldstein, J. I., Costley, J. C., Lorimer, G. W. and Reed, S. J. B. (1977) Quantitative X-ray in the electron microscope. *Scanning Electron Microscopy*, **1**, 315–324.

Goldstein, J. I., Newbury, D. E., Echlin, P., *et al.* (1981) *Scanning Electron Microscopy and X-ray Microanalysis*. Plenum, New York.

Graham, R. J. and Steeds, J. W. (1984) Determination of Cliff–Lorimer factors by analysis of crystallized droplets. *Journal of Microscopy*, **133**, 275–280.

Hall, M. G. and Lloyd, G. E. (1981) The SEM examination of geological samples with a semi-conductor back-scattered electron detector. *American Mineralogist*, **66**, 362–368.

Hall, T. A. (1971) The microprobe assay of chemical elements, in *Physical Techniques in Biological Research*, Vol. 1A (ed. G. Osten). Academic Press, New York, 393 pp.

Hall, T. A. (1991) Suggestions for the quantitative X-ray microanalysis of thin sections of frozen-dried and embedded biological tissues. *Journal of Microscopy*, **164**, 67–79.

Harvey, P. K., Taylor, D. M., Hendry, R. D. and Bancroft, F. (1973) An accurate fusion method for the analysis of rocks and chemically related materials by X-ray fluorescence spectrometry. *X-ray Spectrometry*, **2**, 33–44.

Hobbs, L. W. (1979) Radiation effects in analysis of inorganic specimens by TEM, in *Introduction to Analytical Electron Microscopy* (eds J. J. Hren, J. I. Goldstein and D. C. Joy). Plenum, New York, pp. 437–480.

Hodson, S. and Marshall, J. (1971) Migration of potassium out of electron microscope specimens. *Journal of Microscopy*, **93**, 49–53.

Hren, J. J. (1979) Barriers to AEM: contamination and etching, in *Introduction to Analytical Electron Microscopy* (eds J. J. Hren, J. I, Goldstein and D. C. Joy). Plenum, New York, pp. 481–505.

Hutton, J. T. and Elliott, S. M. (1980) An accurate XRF method for the analysis of geochemical exploration samples for major and trace elements using one glass disc. *Chemical Geology*, **29**, 1–11.

Ireland, B. J., Curtis, C. D. and Whiteman, J. A. (1983) Compositional variation within some glauconites and illites and implications for their stability and origins. *Sedimentology*, **30**, 769–786.

Jiang, W. T., Peacor, D. R., Merriman, R. J. and Roberts, B. (1990) Transmission and analytical electron microscopic study of mixed layer illite/smectite formed as an apparent replacement product for diagenetic illites. *Clays and Clay Minerals*, **38**, 449–468.

Kelly, P. M., Jostons, A., Blake, R. G. and Napier, J. G. (1975) The determination of foil thickness by scanning transmission electron microscopy. *Physica Status Solidi A*, **31**, 771–780.

Kerrick, D. M., Eminpizer, L. B. and Villaume, J. F. (1973). The role of carbon film thickness in electron microprobe analysis. *American Mineralogist*, **58**, 920–925.

Lachowski, E. E., Murray, L. M. and Taylor, H. F. W. (1979) Truscottite: composition and ionic substitutions. *Mineralogical Magazine*, **43**, 333–336.

Leake, B. E., Hendry, G. E., Kemp, A. *et al.* (1969) The chemical analysis of rock powders by automatic X-ray fluorescence. *Chemical Geology*, **5**, 7–86.

Lee, R. F. and McConchie, D. M. (1982) Comprehensive major and trace element analysis of geological material by X-ray fluorescence using low dilution fusions. *X-ray Spectrometry*, **11**, 55–63.

Long, J. V. P. (1977) Electron probe microanalysis, in *Physical Methods in Determinative Mineralogy* (ed. J. Zussman). Academic Press, London, pp. 273–341.

Lorimer, G. W., Cliff, G. and Clark, J. N. (1976) Determination of the thickness and spatial resolution for the quantitative analysis of thin foils, in *Developments in Electron Microscopy and Analysis* (ed. J. A. Venables). Academic Press, London, pp. 153–156.

Love, G., Cox, M. G. C. and Scott, V. D. (1977) Foil thickness measurement in transmission electron microscopy, in *Developments in Electron Microscopy and Analysis 1977* (ed. T. Mulvey). Institute of Physics, London, pp. 347–350.

Michael, J. R., Williams, D. B., Klein, C. F. and Ayer, O. R. (1990) The measurement and calculation of the X-ray spatial resolution obtained in the analytical electron microscope. *Journal of Microscopy*, **160**, 41–53.

Morris, P. L., Ball, M. D. and Statham, P. J. (1980). *Electron Microscopy and Analysis. Conference Series of the Institute of Physics*, **52**, 413–416.

Moseley, H. G. J. (1913) The high-frequency spectra of the elements. *Philosophical Magazine*, **26**, 1024.

Myklebust, R. L., Fiori, C. E. and Heinrich, K. F. L. (1979) *Frame C: a Compact Procedure for Quantitative Energy-dispersive Electron Probe X-ray Analysis.* National Bureau of Standards Technical Note 1106, Government Printing Office, Washington, US, 106 pp.

Newbury, D. E. and Myklebust, R. E. (1980) Calculation of electron beam spreading in composite thin foil targets, in *Microbeam Analysis 1980* (ed. D. B. Wittry), San Francisco Press, San Francisco, pp. 173–175.

Nisbet, E. G., Dietric, V. J. and Essenwein, A. (1979) Routine trace element determination in silicate minerals and rocks by X-ray fluorescence. *Fortschritte der Mineralogie*, **57**, 264–279.

Nockolds, C., Nasir, M. J., Cliff, G. and Lorimer, G. W. (1980) X-ray fluorescence correction in thin foil analysis and direct methods for foil thickness measurement, in *Developments in Electron Microscopy and Analysis 1977* (ed. T. Mulvey). Institute of Physics, London, pp. 417–420.

Norrish, K. and Chappell, B. W. (1977) X-ray fluorescence spectrometry, in *Physical Methods in Determinative Mineralogy*, 2nd edn (ed. J. Zussmann). Academic Press, London, pp. 201–272.

Norrish, K. and Hutton, J. T. (1969) An accurate X-ray spectrographic method for the analysis of a wide range of geological samples. *Geochimica et Cosmochimica Acta*, **33**, 431–453.

Norrish, K. and Thompson, G. M. (1990) XRS analysis of sulphides by fusion methods. *X-ray Spectrometry*, **19**, 67–71.

Oatley, C. W., Nixon, W. C. and Pease, R. F. W. (1965) Scanning electron microscopy. *Advances in Electronics and Electron Physics*, **21**, 181–247.

Packwood, R. H. and Brown, J. D. (1981) A Gaussian expression to describe $\varphi(\rho z)$ curves for quantitative electron probe microanalysis. *X-ray Spectrometry*, **10**, 138–146.

Potts, P. J. (1987) X-ray fluorescence analysis: principles and practice of wavelength dispersive spectrometry, in *A Handbook of Silicate Rock Analysis*. Blackie, Glasgow, pp. 226–285.

Potts, P. J., Webb, P. C. and Watson, J. S. (1985) Energy dispersive X-ray fluorescence analysis of silicate rocks: comparison with wavelength-dispersive performance. *Analyst (London)*, **110**, 507–513.

Pye, K. and Krinsley, D. H. (1983) Mudrocks examined by backscattered electron microscopy. *Nature*, **301**, 412–413.

Reese, G. M., Spence, J. C. H. and Yamamoto, N. (1984) Coherent bremsstrahlung from kilovolt electrons in zone axis orientations. *Philosophical Magazine*, **A49**, 697–716.

Robinson, D. and Bennett, M. C. (1981) XRF determinations of 19 trace elements in international geochemical reference samples. *Geostandards Newsletter*, **5**, 175–181.

Rose, H. J., Adler, I. and Flanagan, F. J. (1962) Use of La_2O_3 as a heavy absorber in X-ray fluorescence analysis of silicate rocks. *US Geological Survey, Professional Paper*, **450-B**, 106–121.

Schreiber, T. P. and Wims, A. M. (1981) A quantitative X-ray microanalysis thin film method using K-, L- and M-lines. *Ultramicroscopy*, **6**, 323–334.

Spence, J. H. C. and Taftø, J. (1983) ALCHEMI: a new technique for locating atoms in small crystals. *Journal of Microscopy*, **130**, 147–154.

Springer, G. (1974) The role of carbon film thickness in electron microprobe analysis: a comment. *American Mineralogist*, **59**, 1121–1122.

Statham, P. J. (1979) Measurement and use of peak-to-background ratios in X-ray analysis. *Mikrochimica Acta (Wien), Supplement*, **8**, 229–242.

Sweatman, T. R. and Long, J. V. P. (1969) Quantitative electron-probe microanalysis of rock-forming minerals. *Journal of Petrology*, **10**(2), 332–379.

Szymanski, R. and Lynch, J. (1986) Quantitative X-ray microanalysis of divided soils in the STEM, in *Proceedings, 11th International Congress on X-ray Optics and Microanalysis* (eds J. D. Brown and R. H. Pacwood). University of Western Ontario, London, Ontario, pp. 412–415.

Szymanski, R., Loeber, L. and Durand, D. (1990) Local composition of reference clays as studied statistically by analytical scanning transmission electron microscopy, in Proceedings, 9th International Clay Conference, Strasbourg 1989 (eds V. C. Farmer and V. Tardy). *Sciences Geologiques Memoire*, **89**, 149–158.

Tixier, R. and Philibert, J. (1969) Analyses quantatives d'échantillons minces, in *Proceedings, 5th International Congress on X-ray Optics and Microanalysis* (eds G. Mollenstedt and K. H. Gaukler). Springer-Verlag, Berlin, pp. 180–186.

Turner, P. S., White, T. J., O'Connor, A. J. and Rossouw, C. J. (1991) Advances in ALCHEMI analysis. *Journal of Microscopy*, **162**, 369–378.

Van der Pluijm, B. A., Lee, J. H. and Peacor, D. R. (1988) Analytical electron microscopy and the problem of potassium diffusion. *Clays and Clay Minerals*, **36**, 498–504.

Vecchio, K. S. and Williams, D. B. (1987) Experimental conditions affecting coherent bremsstrahlung in X-ray microanalysis. *Journal of Microscopy*, **147**, 15–35.

Vie le Sage, R., Quisefit, J. P., Dejean de La Batie, R. and Faucherre, J. (1979) Utilisation du rayonnement primaire diffuse par l'échantillon pour une détermination rapide et précise des elements traces dans les roches. *X-ray Spectrometry*, **8**, 121–128.

Warren, E. A. and Ransom, B. (1992) The influence of analytical error upon the interpretation of chemical variations in clay minerals. *Clay Minerals*, **27**, 193–209.

Williams, D. B. and Goldstein J. I. (1991) Quantitative X-ray microanalysis in the analytical electron microscope, in *Electron Probe Quantitation* (eds K. F. J. Heinrich and D. E. Newbury). Plenum, New York, pp. 371–398.

Wilson, A. R. and Lambrianidis, L. T. (1990) Compton scattering X-ray artefacts. *Journal of Microscopy*, **160**, 1–7.

Wood, J. E., Williams, D. B. and Goldstein, J. I. (1984) Experimental and theoretical determination of k_{A-Fe} factors for quantitative X-ray analysis in the analytical electron microscope. *Journal of Microscopy*, **133**, 225–274.

Zaluzec, N. J. (1979) Quantitative X-ray microanalysis: instrumental considerations and applications to materials science, in *Introduction to Analytical Electron Microscopy* (eds J. J. Hren, J. I. Goldstein and D. C. Joy), Plenum, New York, pp. 121–167.

CHAPTER 8

Chemical analysis

D. C. Bain and B. F. L. Smith

8.1 INTRODUCTION

Of the determinative methods used to study clay minerals, chemical analysis is the oldest and most established. Before the development of the earliest instrumental techniques, such as X-ray diffraction and thermal methods, the identification of clay-mineral phases was accomplished by chemical analysis, supplemented where possible by optical data and physical characteristics such as specific gravity, colour, hardness, etc. However, fine-grained materials cannot be characterized reliably using these methods alone, and misidentifications were common. With the routine application of infrared spectroscopy and electron microscopy, in addition to X-ray diffraction and thermal analysis, instrumental techniques are now unquestionably superior to chemical analysis for clay-mineral identification and a clay is often described and identified without recourse to chemical analysis. Nevertheless, for complete characterization, a chemical analysis is still essential, as this is the only way that a structural formula can be calculated and the distribution of cations in the structure defined directly. A full characterization should include the determination of all the major and minor elements (with the proportion of the iron present in the ferrous and ferric forms) and also the determination of the cation-exchange capacity and, if possible, the anion-exchange capacity.

Classification of clay minerals has always depended heavily on chemical data, and many clay-mineral groups, such as smectites, illites and chlorites, are subdivided according to composition. The characterization of a clay mineral as 'dioctahedral' or 'trioctahedral' can be determined from a *b*-dimension measurement using X-ray or electron-diffraction techniques, but these terms also have a chemical connotation, indicating the valency of the cations occupying the octahedral sheets. The type and

Clay Mineralogy: Spectroscopic and Chemical Determinative Methods.
Edited by M. J. Wilson. Published in 1994 by Chapman & Hall, London.
ISBN 0 412 53380 4.

proportions of cations, however, are best determined by chemical analysis. For example, a trioctahedral mineral contains divalent octahedral cations, which could be Fe^{2+}, Mg^{2+}, or both, in various proportions. Sometimes there is a complete gradation between two species of clays in the same group and these are best distinguished by their chemistry, e.g. montmorillonite and beidellite.

Thus, despite the many and sophisticated instrumental techniques available for studying clays, a good and complete chemical analysis is still necessary for the comprehensive study and characterization of any clay mineral. This chapter outlines the approach required to obtain such a chemical analysis, from the preparatory stages of sample handling, grinding and cation saturation, to the methods of analysis. Analysis of clay is, in effect, a silicate analysis, but there are special problems associated with clays that are not encountered with some other types of silicate materials, and particular attention is paid to overcoming these problems. The section on methods of analysis is not comprehensive, but draws attention to techniques of particular interest in the study of clays and lists references with more details of other relevant methods. The analysis of ferrous iron and determination of exchange capacity are dealt with in more practical detail because these aspects of silicate analysis are often neglected. The final section details the procedure for calculating the structural formula from the chemical analysis of a clay mineral, indicating how impurities can affect the results obtained.

8.2 SAMPLE PREPARATION

No matter how much time is spent on a particular analytical method, or how accurate and precise the method may be, the quality of the final analysis is primarily dependent on the quality of the sample preparation; i.e. the analysis is only as good as the sample it represents. This fact cannot be overemphasized, yet it is too often ignored. First of all, the sample should be properly selected in the field, to ensure that it is representative of the total material. The analyst and the individual requiring the analysis should ensure that the sample is handled and prepared so that it is not altered by oxidation or contamination with the elements to be determined. Often an analyst is presented with a sample and a certain type of analysis is requested without prior consultation. To obtain the maximum benefit from a chemical analysis, the person requiring the analysis should consult the analyst at the earliest possible stage, so as to obtain guidance on the methods and equipment for separation, handling and grinding.

The many problems involved in sampling rocks and minerals, and the subsequent treatment and preparation of these samples for chemical analysis are described in detail by Johnson and Maxwell (1981).

Fortunately for the clay mineralogist, many of these problems are minimized if truly clay-sized material (i.e. < 2 μm) is under investigation. If the clay fraction has been separated by sedimentation or centrifugation, and has been disaggregated by gentle grinding (by hand in an agate mortar after drying) then the problems of grain size, homogeneity, representative subsampling, contamination and alteration by prolonged grinding are avoided. However, as this book is concerned with clay minerals in a broader sense, i.e. layer-silicate minerals, some of which may occur in particles considerably larger than clay size (e.g. mica, chlorite, vermiculite and serpentine) some details are presented relating to grinding and representative sampling.

8.2.1 Grinding

The use of a grinding mill to reduce the particle size of the sample inevitably causes contamination. The aim is to minimize this contamination, or to ensure that any elements introduced in the grinding process are not important in the analysis. One factor that affects the level of contamination is the length of the grinding time. With its high efficiency and consequent short grinding times (less than 10 minutes) the swing-mill or disc-mill is commonly used, with a lining material suitable for the analysis. A choice can be made between tungsten carbide, nickel–chromium alloy or agate. However, appreciable oxidation of chlorite and biotite can occur when using this method (Reay, 1981), and in heterogeneous materials some of the more resistant components may not be crushed to the same size as the less resistant components. Consequently, this method is not suitable for small quantities of material.

Particle shape can affect efficiency of grinding, and in a mixture, minerals of certain shapes will be ground more effectively than others. Platy minerals are notoriously difficult to grind and this can cause difficulties in the analysis. For example, in the analysis of pressed powder pellets of a rock containing biotite, using X-ray fluorescence spectrometry, Volborth (1964) found that the increase in the biotite surface relative to the other non-platy minerals gave serious inaccuracies in the analysis.

In the authors' experience, one of the most effective mills for grinding platy particles is the *McCrone Micronizing Mill*. The grinding vessel is filled with an ordered array of some 48 identical, cylindrical grinding elements (corundum or agate), which on vibration rotate and grind the powder circulating between the grinding surfaces. To minimize oxidation of ferrous iron and sulphur, alcohol or acetone can be added to produce a slurry. Platy minerals can be reduced to less than 10 μm diameter in about 15–30 minutes by this method.

Health risks can result from inhalation of small fibres, and therefore it is advisable to wet grind any sample known to contain fibrous minerals.

Materials high in clay content can be difficult to grind in the dry state due to caking, but this can be alleviated by pre-drying at 140°C.

Whatever method is chosen, excessive grinding should always be avoided as it may cause changes in the mineral structure – important if the sample is to be examined by XRD, IR and EM methods – and contamination increases with grinding time. If it is necessary to sieve the sample at any stage, brass and stainless-steel sieves should be avoided, particularly if trace elements are to be determined. Nylon mesh sieves are recommended.

8.2.2 Representative sampling

If the sample of interest is a mineral separate, often the main problem is how to obtain a sufficient quantity of material for analysis. This does, however, eliminate the difficulty in acquiring a representative portion, as the total available material is analysed. Such a situation can arise with the clay fraction of a coarse-textured soil, where the clay fractions may comprise only a small percentage of the total soil.

If, however, the sample of interest is available in bulk, then it must be reduced to a smaller representative amount that is statistically equivalent to the whole sample. The size of a representative sample is influenced by the size of the particles in the material. Jackson (1962) has shown that for samples containing particles up to 100 µm in diameter the optimum sample weight is about 5 mg, so that precautions are not necessary to ensure a representative sample for clay-size material (< 2 µm). If, however, the sample contains very coarse particles, the sample must be split (after grinding if necessary) in an unbiased manner, using a sample splitter or riffler. This should be used, repeatedly if necessary, on the subdivided portions, until the desired quantity is obtained. Allen and Khan (1970) critically evaluated several sampling techniques, and concluded that the spinning riffler is by far the best technique for subsampling powders.

8.2.3 Cation saturation

Most clays, particularly those containing smectites and vermiculites, have exchangeable cations which will vary according to the environment in which the clay was formed or located. Sometimes all the exchange sites are occupied by the same cation, but this is often not the case, and varying proportions of Na^+, K^+, Mg^{2+} and Ca^{2+} commonly occur as exchangeable cations. If a chemical analysis is required for calculation of the structural formula of a particular clay mineral, it is advisable to saturate the clay with an index cation not present in the sample. Most clays contain some non-clay impurities or small amounts of another clay mineral, and these other phases may contain some of the elements that are possible exchangeable cations. For example, a smectite-rich clay may contain impurities such as carbonates (calcite and dolomite) or feldspars, and as a result, allocation of elements such as Na, K, Mg and Ca to the smectite formula becomes difficult, if not impossible. An additional complication is that Mg^{2+} occurs in octahedral sites as well as in exchangeable sites.

Thus, saturating the clay sample with an index cation simplifies calculation of the structural formula, and, by choosing an appropriate cation, correction of the final analysis for the presence of impurities is also simplified. Sometimes a less common element, such as Ba or Sr, should be chosen if it is known that the impurities contain all the common saturating cations (Na^+, K^+, Ca^{2+}, Mg^{2+}). In the analysis of bentonites in which interstratified illite–smectite was the dominant clay mineral, and contained K, Mg and possibly Na, with calcite as the suspected impurity, Nadeau and Bain (1986) saturated the clays with Sr^{2+}.

Saturation with the chosen cation is performed by treating the clay with a 1 M salt solution of a cation that will not enter the clay-mineral structure. After centrifuging and decanting off the excess salt solution, the treatment is repeated, and the clay is shaken overnight in fresh salt solution, to ensure that all naturally occurring exchangeable cations have been replaced with the chosen cation. Finally, the sample is washed free of excess salt; however, the sample should not be washed unnecessarily, because desaturation can occur, along with alteration by hydrolysis (Chaussidon, 1982). If chloride solutions are used, removal of chloride may be conveniently monitored by testing with silver nitrate. Washing with alcohol is advisable in the latter stages to remove the excess exchange solution, and if the clay is very finely dispersed (particularly if Na^+ or Li^+ is the exchangeable cation), dialysis is strongly recommended to prevent loss of the very fine particulate material.

8.3 METHODS OF MAJOR ELEMENTAL ANALYSIS

Major elemental analysis methods have changed dramatically over the last 50 years. Initially, the 'Classical Scheme' of analysis was adopted, as devised by Lundell and Hoffman (1938), Groves (1951) and Hillebrand *et al.* (1953). These methods consisted of a decomposition step, followed by gravimetric determination of the constituent elements, and although recognized by many analysts to be accurate, the procedure is tedious, very time-consuming and requires quite a large sample (*c.* 1 g), which in many cases is not available. The development of colorimetric methods has led to their introduction into more rapid analytical schemes, whereby much smaller sample weights (*c.* 0.1 g) can be utilized (Shapiro and Brannock, 1956).

More recently, techniques such as flame photometry, atomic emission and absorption, inductively coupled plasma and X-ray fluorescence spectroscopy have resulted in greatly reduced analysis times and sample weights, sometimes at the expense of a small reduction in accuracy, which could be minimized by replication.

The elements normally analysed and, by convention, reported in the form of percentage oxides, are: SiO_2, Al_2O_3, Fe_2O_3, FeO, CaO, MgO, K_2O, Na_2O, TiO_2, MnO, H_2O^+, H_2O^-, P_2O_5 and CO_2. The summation should result in a total in the range 99.75–100.50% for 'classical analyses' and 99.0–101.0% for 'rapid' techniques for an acceptable analysis. Chalmers and Page (1957) recommended that reports of elemental analyses should include an estimate of the precision and accuracy of the results. The latter should always be monitored by carrying out a complete analysis of a mineral or rock of similar known composition.

A quick qualitative analysis is a very useful precursor to any quantitative determination. This was previously achieved by standard qualitative analysis, but microprobe analysis or X-ray fluorescence now gives not only a rapid indication of the range of elements present, but also a measure of their relative amounts.

Two forms of analytical techniques are available for the measurement of the elemental composition of clay minerals:

- non-destructive – where no further sample preparation is necessary and the sample can be reused for other analyses;
- destructive – where the sample must undergo decomposition into a suitable form for analysis.

8.3.1 Sample decomposition

Few of the wide range of decomposition methods available can cause complete dissolution of all types of clay minerals. Dolezal, Povondra and Sulick (1968) comprehensively reviewed these methods, and their book (translated into English) is invaluable for any analyst. They described three main decomposition methods: acid, fusion and sintering, but only the acid and fusion techniques are in current use. A more recent technique is microwave acid decomposition.

8.3.1.1 Acid decomposition

Samples are dissolved in hydrochloric, hydrofluoric, nitric, sulphuric, perchloric or phosphoric acid or combinations of these. The most widely used acid-dissolution technique is decomposition with hydrofluoric acid, either alone or in conjunction with nitric or sulphuric acid, in special Teflon decomposition vessels. These allow rapid decomposition of the sample without loss of silica by volatilization of SiF_4, essential if silicon is one of the required elements (Bernas, 1968; Lim and Jackson, 1982). Previously, a separate decomposition technique was necessary for the determination of silica (Shapiro and Brannock, 1956).

(i) Decomposition by hydrofluoric acid (Bernas, 1968)
Fifty (± 0.01) mg of a prepared, representative sample (< 200 mesh) are placed in a Teflon decomposition vessel with 0.5 ml aqua regia. The latter is particularly necessary if any organic material is present. Three millilitres of HF (48%) are added and the vessel is tightly sealed and placed in an oven at 110°C for 1 h. After cooling, the dissolution products are quantitatively transferred to a 50-ml plastic beaker containing 2.8 g of HBO_3. Ten millilitres distilled water are added and the mixture stirred magnetically to dissolve any insoluble fluorides. When clear, the solution is diluted to 100 ml and stored in a plastic bottle ready for major elemental analysis.

8.3.1.2 Fusion decomposition

Decomposition by fusion at high temperature, with alkali carbonates, hydroxides, borates or pyrosulphates, usually in the sample:flux ratio of 1 : 7, results in a melt which can then be dissolved in water or acid to give a solution containing the constituent elements.

(i) Fusion with lithium metaborate (Ingamells, 1966; Shapiro, 1967; Verbeek, Mitchell and Ure, 1982)
One hundred (± 0.01) milligrams of the prepared representative sample (< 200 mesh) are weighed into a small platinum crucible, 700 mg $LiBO_3$ added and thoroughly mixed, before heating gently over a Bunsen burner. The temperature is gradually increased to around 900°C, and the heating is continued for a further 10 minutes to ensure complete decomposition. The crucible and melt are allowed to cool and then placed in a 50 ml beaker. Thirty-five millilitres distilled water and 4 ml nitric acid (conc.) are added to the crucible, and the contents stirred magnetically, using a Teflon follower, for 1.5 h to ensure complete dissolution. The solution is then quantitatively transferred to a 100-ml standard flask, diluted to the mark, and is reserved for major elemental analysis.

8.3.1.3 Microwave acid decomposition

Recently, following the development of microwave ovens and their adaptation to the dissolution of material for instrumental analysis, this technique has been modified and applied to the acid dissolution of a wide range of materials (Lamonthe, Fries and Consul, 1986; Papp and Fischer, 1987; Berman, 1988; Kemp and Brown, 1990) prior to their elemental analysis.

Microwave heating offers several advantages to the chemical decomposition of geological material (Fischer, 1986). Instant uptake of radiation throughout the sample solution eliminates the heat conduction stage because energy is transferred to the sample by absorptive polarization rather than molecular collisions. Microwave heating can, therefore, be expected to produce more efficient sample dissolution by the effect of better contact between acids and sample. The main advantages of microwave digestion are faster (5 × 1) sample decomposition, better safety, smaller consumption of acids and the prevention of the loss of volatile components. When large numbers of samples are to be prepared the microwave system is limited to 6–12 samples at a time.

These advantages are further advanced by the use of pressurized containers, which result in higher sample–solution temperatures and pressures and minimal loss of volatile components.

8.3.2 Analytical determinations

A wide range of analytical techniques is currently available to quantify the elemental composition of clays. The methods are too numerous and diverse to be detailed in this handbook, but the main methods in use are listed in Table 8.1, with references to more detailed information.

Table 8.1 Spectrochemical instrumental methods

Technique	*References*
UV/visual	Shapiro and Brannock (1956), Langmyhr and Graff (1965), Shapiro (1967), Jeffery (1981)
Atomic absorption	Shapiro (1967), Bernas (1968)
Flame emission	Verbeek, Mitchell and Ure (1982), Jeffery (1981), Baker and Suhr (1982), Ure (1983)
Inductively coupled plasma	Thomson and Walsh (1983)
X-ray fluorescence	Norrish and Chappell (1977), Jones (1982)
Neutron activation analysis	De Soete, Gijbels and Hoste (1972), Helmke (1982)

The choice of method depends on a number of factors, such as availability of equipment, size of sample, adequate sensitivity and accuracy of the required analysis. The UV/visual methods are cheap and fairly simple to use, but suffer from interference effects. Conventional atomic-absorption and flame-emission spectrometry have the disadvantage of being single-element techniques but the advantages of easy sample presentation and low cost. Inductively coupled plasma (ICP) has several advantages when dealing with clay samples: it has a linear dynamic range over several orders of magnitude and can analyse small samples. However, the inherently high sensitivity of ICP is lost in the dilution process. All these techniques are solution-analysis methods, and consequently, are prone to losses and contamination during the process of clay dissolution. X-ray fluorescence (XRF) and neutron activation analysis (NAA) are multi-element techniques for solid samples. X-ray fluorescence can be non-destructive, but for major elemental analysis it is usually advisable to fuse the sample and dilute with flux to reduce interelement effects. It is very accurate, but cannot differentiate between ferrous and ferric iron. Neutron activation analysis can also be non-destructive, has high sensitivity and accuracy, but suffers the major drawback of requiring a nuclear reactor.

8.4 ANALYSIS FOR FERROUS IRON

Ferrous iron frequently forms the major part of the iron content of rocks and minerals. Measurement of its content with regard to the environment in which these materials have formed can give valuable information, and can, in some cases, differentiate between members of an isomorphous series, e.g. micas. It is very important to know the proportion of iron in the ferrous form when the structural formulae of clay minerals are calculated. If all the iron present is assumed to be in the ferric state, misleading estimations of the layer charge can occur. Atmospheric oxidation rapidly changes ferrous iron to the ferric state, a process which is well illustrated in the case of poorly drained gley soils, where the blue-grey colour, due to ferrous iron, changes with improved drainage conditions, e.g. along root channels, to rusty brown ferric stains.

8.4.1 Sample preparation

Great care must be taken in preparing small representative samples for the determination of ferrous iron. Prolonged grinding can result in the oxidation of ferrous to ferric iron, and to overcome this, Hillebrand (1919) recommended very gentle grinding to < 70 mesh. Kolthoff and Sandell (1956) suggested gentle grinding under alcohol, and this is currently the adopted procedure.

8.4.2 Decomposition

At this stage too, precautions must be taken to prevent oxidation of the ferrous iron components, by carrying out the decomposition under an inert atmosphere or liquid. In practice, this is accomplished by decomposing the material in a boiling mixture of hydrofluoric and sulphuric acids; the steam so generated prevents any aerial oxidation. The decomposition, however, must not be prolonged, because ferrous iron can be oxidized by concentrated sulphuric acid.

An alternative technique is to carry out the decomposition in the presence of an oxidant such as dichromate or vanadate (Wilson, 1960), which will immediately oxidize any ferrous iron to ferric. This can then be reversed stoichiometrically (see following equation) by altering the pH, and the regenerated ferrous iron can then be determined:

$$V^{5+} + Fe^{2+} \rightleftharpoons V^{4+} + Fe^{3+}.$$

8.4.3 Sources of interference

The presence of sulphides (pyrrhotite, pyrites), organic matter and oxidants (manganates, vanadates) can result in substantial errors in the determination of ferrous iron. Graphite itself does not interfere, but other forms of organic matter can cause poor results.

8.4.4 Procedures

The determination of ferrous iron in rocks and minerals is carried out on separate subsamples of the material used for total elemental analysis. The samples are decomposed by hydrofluoric or hydrofluoric/sulphuric acid mixtures and the ferrous iron is measured in the resulting solution. Titrimetric and spectrophotometric methods have been used to quantify the ferrous iron released. Titrations with oxidants, such as permanganate, dichromate or vanadate, or back titrations of excess oxidants have been utilized. A spectrophotometric method for the determination of ferrous and ferric iron in minerals as *ortho*-phenanthroline complexes has been extensively studied by Stucki (1981). He determined Fe^{2+} conventionally, using a fluorescent lamp for the total iron and the Fe^{3+} by difference. Hey (1941) suggested the use of iodine monochloride as an oxidant, in a medium consisting of hydrochloric and hydrofluoric acids, measuring the ferrous iron by titrating the liberated iodine with iodate, or by back titrating the excess iodate with thiosulphate (van Loon, 1965). This technique has been used by Nicholls (1960) to determine ferrous iron in samples containing organic matter.

Other instrumentally orientated techniques for measuring ferrous iron are: potentiometric titrations, ion chromatography, polarography and NMR or Mössbauer spectroscopy. The latter is used mainly for unstable compounds and, as the measurements are very time-consuming, it is used only for special cases (Goodman, 1980).

Ion chromatography, however, offers a method of determining both Fe^{2+} and Fe^{3+} simultaneously by means of a post-column-reaction method (Kanai, 1990). The results obtained were in good agreement with conventional methods and with geological reference rocks. This technique can also be used with rocks containing large amounts of organic carbon, which causes difficulties with conventional redox titrimetric methods.

8.4.5 Analytical methods

Two main methods in current use for the determination of ferrous iron in rocks and minerals are detailed.

8.4.5.1 *Shapiro and Brannock (1952) method*

The mineral sample (< 70 mesh) is decomposed in a platinum crucible using a mixture of hydrofluoric and sulphuric acids at boiling point. The oxidation of any ferrous iron present is minimized because air is excluded by the steam that is generated. When decomposition is complete, the crucible and contents are plunged into a boric acid solution and the ferrous iron liberated is then measured quantitatively by titration with an oxidant.

Reagents Sulphuric acid – 4.5 M.
Diphenylamine sulphonate indicator: dissolve 0.2 g sodium diphenylamine sulphonate in 1 l distilled water, then add 1 l 85% phosphoric acid (analytical reagent: AR) and mix well.
Standard dichromate solution (0.0167 M).

Decomposition and determination Place 0.500 g sample (< 70 mesh) in a platinum crucible (100 ml capacity), add 20 ml of 4.5 M H_2SO_4, cover, and carefully boil over a gas burner until the steam generated excludes atmospheric oxygen. Carefully add 10 ml hydrofluoric acid (AR) to the crucible, cover and continue the boiling for a further 15 minutes. Plunge the crucible and contents into 400 ml of 2.5% boric acid in a 800-ml beaker, add 10 ml diphenylamine sulphonate indicator, and titrate immediately with standard dichromate solution to a purple end-point. Determine a reagent blank.

$$1 \text{ ml of } 0.0167\,\text{M}\ K_2Cr_2O_7 \equiv 5.58 \text{ mg } Fe^{2+}$$

8.4.5.2 *Wilson (1960) method*

In this method, the decomposition of the sample takes place in the presence of an oxidant – vanadate at room temperature. Two methods of determining the amount of ferrous iron released will be described: titrimetric and spectrophotometric.

Decomposition Three to twenty milligrams of ground sample (< 70 mesh) are weighed into a 40-ml polystyrene vial, and 0.100 ml of 0.139 M ammonium vanadate solution are added using a microsyringe burette. If the FeO content is > 1 mg, more vanadate solution must be used. One millilitre of hydrofluoric acid is added and the vial stoppered and allowed to stand until complete decomposition is achieved; for most minerals overnight (16 h) is sufficient. The resulting solution is often cloudy, due to the presence of insoluble sulphates and fluorides.

(i) Titrimetric method
Reagents Ammonium vanadate 0.139 M – dissolve 1.6260 g ammonium vanadate (NH_4VO_3) in 20 ml of 5 M H_2SO_4, and dilute to a final volume of 100 ml.
Ammonium ferrous sulphate 0.07 M – dissolve 2.75 g $(NH_4)_2Fe(SO_4)_2 \cdot 6H_2O$ in 100 ml of 0.5 M H_2SO_4 (1 ml ≡ 10 mg FeO).
Barium diphenylamine sulphate indicator – 0.01% w/v aqueous.
Hydrofluoric acid – (AR) 40% w/v.
Sulphuric acid (5 M).

Determination Three millilitres of 5 M H_2SO_4 are added to the solution obtained after the decomposition procedure, and the mixture stirred magnetically to dissolve any previously insoluble products. Five drops of barium diphenylamine sulphonate and 3 ml distilled water are then added, and the mixture immediately titrated to a pale green end-point with 0.07 M ammonium ferrous sulphate, using a microburette for increased accuracy. Blank titrations should always be carried out on the reagents, and the difference between the blank titration and the sample titration is related to the ferrous iron content of the sample.

(ii) Spectrophotometric method

The decomposition of the sample is carried out in a similar manner to that used for the titrimetric determination. Ferrous iron released is immediately oxidized to the ferric state by the vanadate oxidizing reagent, the latter being reduced to the vanadous form. The reaction is reversible, stoichiometric and pH-dependent. Wilson (1960) used a solution of beryllium sulphate to complex any excess hydrofluoric acid. An increase in pH causes the vanadous form to be oxidized to vanadate, and the appropriate amount of ferric iron to be reduced to ferrous iron, which can then be determined by forming a coloured complex with dipyridyl.

Shapiro (1960) preferred to use a solution of boric acid, instead of the highly carcinogenic beryllium sulphate, to complex excess fluoride ions. Organic matter and sulphide minerals do not interfere in the determination of ferrous iron by this method.

Reagents 2,2′-Dipyridyl – 0.15% w/v aqueous.
Beryllium sulphate – 50% w/v aqueous (*NB carcinogenic*).
Ammonium acetate buffer – 50% w/v aqueous.
Standard iron solution (stock) – dissolve 0.5458 ferrous ammonium sulphate (AR) in 100 ml water and add 5 ml of 3.5 M sulphuric acid. Slowly add a dilute solution of potassium permanganate until the mixture is slightly pink, then dilute to 1 l. (1 ml ≡ 0.1 mg FeO.)
Standard iron solution (working) – dilute 100 ml stock solution to 1 l. (1 ml ≡ 10 μg FeO.)
Preparation of standard ferrous iron calibration curve – pipette a range (0–80 μg FeO) into a series of 100-ml standard flasks, add *c.* 100 mg ascorbic acid to each flask to reduce the iron to the ferrous state, and after 10 minutes develop the dipyridyl complex as in the spectrophotometric determination.

Determination When the decomposition is complete, add 5 ml of 50% beryllium sulphate solution *carefully* (by safety pipette) to the decomposed mixture in the polystyrene vial, and stir until any insoluble fluorides dissolve completely. Then add 5 ml 2, 2′-dipyridyl solution and 10 ml of 50% ammonium acetate buffer to bring the pH to *c.* 5. Quantitatively transfer the solutions to 100-ml standard flasks and make up to the mark with distilled water. Measure the absorptions of the red ferrous dipyridyl complex at 525 nm, and compare with the standard calibration curve. A standard rock (e.g. US Geological Survey: Flanagan, 1973) of similar composition – for comparison – and a reagent blank, should be carried through the full procedure.

8.5 EXCHANGE CAPACITY

Clay minerals can adsorb cations and anions and retain them in an exchangeable form. Exchangeable ions normally associated with clay minerals are: the cations Ca^{2+}, Mg^{2+}, H^+, K^+, NH_4^+, Na^+ and Al^{3+} and the anions SO_4^{2-}, Cl^-, PO_4^{3-} and NO_3^-. Exchange capacity can be defined as the total ionic charge of the adsorption complex active in the adsorption of ions. It is normally measured at pH 7 and reported as milli-equivalents per 100 g (meq/100 g).

The exchange reaction of minerals is of great importance in agriculture, plant-nutrient availability, soil structure, rock-weathering processes, ceramic and concrete work, and in the oil industry, where bentonites are widely used in drilling.

The exchange capacity of minerals can be considered from two main viewpoints: cation-exchange capacity (CEC) and anion-exchange capacity (AEC).

8.5.1 Cation-exchange capacity

Cation-exchange capacity can be defined as the sum of the exchangeable cations that a mineral can adsorb at a specific pH, i.e. a measurement of the negative charges carried by the mineral. These negative charges in clay minerals can be derived from:

- isomorphous substitution within the structure;
- broken bonds at edges and external surfaces;
- the dissociation of accessible hydroxyl groups.

Negative charges due to isomorphous substitutions within mineral structures are known as permanent charge, and are independent of conditions such as pH and ion activity, whereas those from the latter two sources are so dependent. For this reason, the type of CEC methods used should always be reported with the results. The high levels of CEC for clay minerals (Table 8.2) can, to some extent, be attributed to their large areas of external surface accessible to hydrated cations.

Table 8.2 Cation-exchange capacities (CEC) of clay minerals (Grim, 1968)

Mineral	*CEC (meq/100 g)*
Kaolinite	3–15
Halloysite	5–50
Illite	10–40
Chlorite	10–40
Allophane	25–50
Montmorillonite	60–150
Vermiculite	100–150

Kelley (1948) and Thomas (1977) have comprehensively reviewed the background of CEC measurements, Metson (1956) and Grim (1968) reviewed the methods for clay minerals, and Chapman (1965), Rhoades (1982) and Begheyn (1987) those for soils.

Rhoades (1982) suggested four types of method:

- summation method – measurement, and summation of, displaced exchangeable cations, i.e. exchangeable Na + Ca + Mg + K + exchangeable acidity;
- direct displacement of saturating index cation;
- displacement of saturating index cation after washing free from saturating salt;
- radioactive tracer measurement of index cations (^{45}Ca: Blume and Smith, 1954; ^{133}Ba: Bache, 1970).

For most minerals, the measurement of CEC is not dependent on the pH of saturation. Methods used for soils are mostly based on saturation, with ammonium as the index cation, but for some minerals low results are obtained (Chapman, 1965). For 1 : 1 minerals (kaolinite, halloysite) ammonium cannot completely replace aluminium and hydrogen, and for 2 : 1 minerals (vermiculite, micas, illite) the ammonium index ion becomes fixed and is not exchangeable. Sodium and barium have been suggested as more suitable saturating ions because they can be easily exchanged and determined. The use of divalent barium as a saturating cation is considered by many workers to be more effective than that of monovalent ions.

Recently, two methods have been proposed that are quicker and more reliable for a wider range of soil types than the conventional methods. These are both one-step extraction methods – (1) LiEDTA or LiBaEDTA (Begheyn, 1987) and (2) AgTu (silver thiourea) (Searle, 1986). These have been recommended by the International Soil Reference and Information Centre as possible improvements on the conventional CEC determinations (Pleijsier, 1986).

No general method exists that can be reliably used for all clay minerals, and the method chosen should therefore be appropriate to the type of mineral to be analysed. Details of three methods in current use are given.

8.5.1.1 Ammonium saturation (Chapman, 1965, modified)

Reagents Ammonium acetate (1 M) pH 7 – 58 ml acetic acid (conc.) and 55 ml ammonium hydroxide diluted to 1 l and the pH adjusted to 7 + 0.01.
Isopropyl alcohol – 99%.
Sodium chloride – aqueous w/v solution acidified to be 0.005 M in HCL.

Methods Weigh accurately 10–100 mg clay mineral into a 15-ml centrifuge tube. Add 10 ml of 1 M ammonium acetate solution (pH 7), shake and leave overnight.

Centrifuge and remove the clear supernatant liquid by suction. Add a further 10 ml ammonium acetate solution, and repeat the saturation process another four times. Wash free from excess ammonium salts by centrifuging with six 10-ml portions of isopropyl alcohol.

Shake the ammonium-saturated clay with 10 ml NaCl (acidified) solution, centrifuge and siphon off the supernatant liquid into a 100-ml standard flask. Repeat the process another six times and dilute to 100 ml with distilled water. The ammonium ion replaced can then be measured by micro-Kjeldahl distillation, spectrophotometrically by the isocyanurate method (Fraser and Russell, 1969), or by using a selective-ion electrode (Busenberg and Clemency, 1973).

8.5.1.2 Sodium saturation

Reagents Sodium acetate (1 M) – dissolve 136 g $NaOAc \cdot 3H_2O$ in 1 l distilled water, and adjust the pH to 8.2 with HOAc or NaOH. Other reagents as in ammonium saturation method.

Method The saturation technique is the same as described above, using sodium acetate as the saturating salt and ammonium acetate to displace the index sodium ion. This can then be determined easily by flame photometry or atomic-emission spectroscopy. The results are normally reported as meq/100 g oven-dry clay.

8.5.1.3 Barium saturation (Gillman, 1979)

Reagents Barium saturating solution (0.1 M) – dissolve 24.4 g $BaCl_2 \cdot 2H_2O$ in 1 l distilled water. Barium equilibrating solution (0.002 M) – dissolve 0.4889 g $BaCl_2 \cdot 2H_2O$ in 1 l distilled water and adjust the pH to 7 with $Ba(OH)_2$ or HCl. Magnesium displacing solution (0.005 M).

Method Place 10–100 mg (accurately weighed) clay mineral in a 15-ml (weighed) centrifuge tube, add 10 ml of 0.1 M barium chloride and shake for 2 h. Centrifuge, siphon off the supernatant liquid and equilibrate the clay with 3 × 10 ml portions of 0.002 M barium chloride, shaking for 1 h between centrifugations and discarding the supernatants. Then weigh the tube plus wet clay to measure the amount of entrained barium chloride solution. Add 10 ml of 0.005 M magnesium sulphate solution and shake the suspension gently for 1 h. Adjust the conductivity of the suspension to the conductivity of the 0.0015 M magnesium sulphate reference solution by adding 0.005 M magnesium sulphate or distilled water. Shake the suspension gently for a further 16 h (overnight) and adjust the conductivity again if necessary. Then note the weight of the centrifuge tube plus the suspension, to estimate the volume of magnesium sulphate or distilled water added. Centrifuge the suspension and determine the magnesium and chloride contents of the supernatant liquid by atomic-absorption spectrometry (Mg) and spectrophotometry or ion chromatography (Cl).

Calculation If V_1 = volume of entrained $BaCl_2$ (ml), V_2 = volume of added $MgSO_4$ (ml), V_3 = final supernatant volume, and C_1 = Mg concentration of supernatant in meq/ml.

Then

$$\text{(MgSO}_4\text{ added)} \qquad \text{CEC (meq/100 g)} = \frac{100\,(0.01\,V_2 - C_1 V_3)}{\text{wt of sample (g)}}$$

$$\text{(if water added)} \qquad \text{CEC (meq/100 g)} = \frac{100\,(0.1 - C_1 V_3)}{\text{wt of sample (g)}}$$

Measurement of the chloride content of the supernatant also allows the estimation of AEC. If C_2 = chloride concentration of 0.002 M $BaCl_2$ and C_3 = chloride concentration of supernatant in meq/ml then:

$$\text{AEC (meq/100 g)} = \frac{100\,(C_3 V_3 - C_2 V_1)}{\text{wt of sample (g)}}$$

For minerals with higher CEC values, lower sample weights and higher concentrations of magnesium sulphate solutions (e.g. 0.01 M) should be considered.

8.5.2 Anion-exchange capacity

Anion-exchange capacity can be defined as the sum of exchangeable anions that a clay mineral can adsorb, usually expressed as milli-equivalents per 100 g clay. It is considered to be a measure of the positive charges carried by a clay mineral. Grim (1968) suggested three types of anion exchange in clay minerals:

- replacement of hydroxyl groups – a surface reaction, e.g. adsorption of phosphate by kaolinite;
- anions of similar geometric size to silica tetrahedra (e.g. phosphate, arsenate, borate) may be adsorbed at the edges of clay minerals – a surface reaction;
- free exchangeable aluminium and iron may form insoluble salts with exchanging anions, making it uncertain whether the exchange is due to this effect or to the clay mineral.

In the replacement of surface hydroxyls some destruction of the structure may occur, resulting in inaccurate measurements. This can, however, be minimized by using low concentrations of saturating salts. Anion-exchange capacity has been shown to be proportional to the surface area and to vary with the degree of crystallinity of the clay mineral. The range of anion-exchange values obtained for the common clay minerals is illustrated in Table 8.3.

Parfitt (1978) has reviewed the mechanisms of anion exchange for a wide range of clay minerals. He concluded that:

- minerals with surface hydroxyls associated with aluminium and iron would adsorb anions;
- the adsorption of anions is dependent on pH, with maximum adsorption occurring at lower pH;

Table 8.3 Anion-exchange capacities (AEC) of clay minerals (Grim, 1968)

Mineral	*AEC (meq/100 g)*
Vermiculite	4
Illite	4–17
Chlorite	5–20
Kaolinite	7–20
Montmorillonite	20–30

- the order of adsorption of anions by minerals can generally be considered to be phosphate > arsenate > selenite > molybdate > sulphate = fluoride > chloride > nitrate;
- the presence of other anions can cause competition for adsorption sites, and exchangeable cations such as calcium and aluminium can, in some cases, result in the formation of insoluble by-products.

Investigations of the anion-exchange capacity of clay minerals have been concerned mainly with phosphate adsorption. The release of hydroxyl ions by fluoride ions has also been utilized by Weiss *et al.* (1956), and Wada and Okamura (1977) used ammonium chloride as the saturating salt to measure the CEC and AEC for clay minerals and soils. They showed that there are differences between soils and clays in the dependence of CEC and AEC on the pH and ion concentrations, these being most marked in the presence of allophane and imogolite.

Methods for the determination of the anion-exchange capacity of minerals usually involve saturation with an index anion and measurement of the amount adsorbed. Details of two methods are given:

8.5.2.1 Phosphate (Mehlich, 1948)

Reagents Triethanolamine solution – dilute 22.5 ml triethanolamine to 250 ml with distilled water and adjust the pH to 8.1 with HCl. Mix with 500 ml of 5% $BaCl_2{\cdot}2H_2O$ (w/v) and dilute to 1 l.
Calcium chloride solution – dissolve 50 g $CaCl_2{\cdot}2H_2O$ in 1 l distilled water and adjust the pH to 8.0 with $Ca(OH)_2$ solution.
Phosphate saturating solution – 0.001 M H_3PO_4.
Molybdate reagent (Murphy and Riley, 1962 – dissolve 12 g ammonium molybdate in 200 ml distilled water. Dissolve 0.2670 g antimony-potassium tartrate in 500 ml of 5 M H_2SO_4. Mix throughly, dilute to 1 l and store in a dark bottle.
Working reagent – dissolve 1.056 g ascorbic acid in 100 ml ammonium molybdate reagent and mix well (this is stable for 24 h).

Methods Accurately weigh 10–100 mg clay mineral into a 15-ml centrifuge tube. Add 10 ml triethanolamine solution and shake for 30 minutes. Centrifuge and wash five times with 95% ethanol, shake with 10 ml calcium chloride solution for 30 minutes, centrifuge and wash again five times with 25% ethanol. Add 10 ml phosphate saturating solution (0.001 M), shake for 30 minutes and allow to stand for 24 h. Shake again, centrifuge and reserve the supernatant liquid for the determination of adsorbed phosphate (i.e. concentration in saturating solution – concentration in supernatant). Extractable phosphate is determined on a separate sample with 0.03 M ammonium fluoride in 0.1 M HCl, using a 1 : 20 mineral–solution ratio.

Phosphate determination Pipette a suitable aliquot into a 100-ml standard flask, dilute to 80 ml with distilled water, add 8 ml working reagent and make up to 100 ml. Shake well and measure the absorption at 880 nm after 10 minutes, comparing results with those obtained for a series of phosphate standards in the range 0–400 µg P.

Then AEC in meq/100 g clay = (extractable P + P adsorbed) meq/100 g.

8.5.2.2 Chloride (Wada and Okamura, 1977)

A 0.3–0.5 g sample (accurately weighed in a weighed 15-ml centrifuge tube) is first saturated with Cl^- index anion, using 1 M NH_4Cl (10 ml) at pH 6, and is shaken for 30 minutes. The suspension is centrifuged, the supernatant discarded, and the residue equilibrated by washing 5 × 10 ml portions of 0.1 M NH_4Cl. The residual excess saturating Cl^- ion is estimated from the increase in weight (Cl_{res}). The Cl^- ion is then replaced by washing with (5 × 10 ml portions) 1 M KNO_3 and the Cl^- released (Cl_{rel}) is measured by selective-ion electrode potentiometry or by spectrophotometry.

AEC (meq/100 g) clay = ($Cl_{rel} - Cl_{res}$) in meq 100 g.

8.6 DETERMINATION OF A STRUCTURAL FORMULA

8.6.1 Correction for impurities

Before calculating the structural formula of a clay mineral, it is essential to ensure that the chemical analysis represents only the mineral of interest. Clay samples are rarely monomineralic, and several impurities are normally present, which may be crystalline, amorphous, or both as is particularly common in soil clays. Sometimes, authors simply note that certain impurities are present and make no allowance in the formula calculations, but this can only lead to a formula of dubious validity.

If all the impurities have been identified, corrections can be made to the analysis before calculating the structural formula of the studied mineral, provided that:

- the impurities are present only in relatively small amounts;
- the amounts of the impurities can be determined;
- the chemical composition of each impurity is known with some certainty.

If the impurities are present in substantial quantities, then the chemical composition and amount of each impurity must be known with a very high degree of accuracy, and this is not always possible. Large corrections are best avoided, as the resulting formula – and any interpretations based on it – are open to question. For example, Evans and Adams (1975) corrected for up to 40% of impurities (consisting of quartz, albite, orthoclase and rutile) in the analysis of mudstones, to obtain illite and chlorite structural formulae. These were queried by Bjørlykke and Jorgensen (1976), on the basis that feldspar contents had been underestimated.

8.6.1.1 Identification of impurities

Once the decision has been made that a structural formula is to be calculated for a clay mineral, it is necessary to detect and identify all the other phases present in the sample. All available techniques should be used, as no single technique is suitable for detection and identification of the wide range of possible impurities. X-ray diffraction is clearly suitable for most crystalline components, but detection limits are not as low as IR for kaolinite and quartz for example, or as low as DTA for gibbsite and goethite. Mössbauer spectrometry is particularly useful for identifying and quantifying the amounts of any iron phases present.

8.6.1.2 Removal of impurities

Ideally, it is desirable to eliminate impurities from the sample before the chemical analysis is performed. This is almost always preferable to applying corrections to the analysis, and should be pursued if at all possible. Several methods can be used to purify the phase to be analysed. For smectite-rich samples, impurities can often be eliminated by size fractionation, as smectite is usually concentrated in the very finest fraction. However, the smectite in the finest fractions does not necessarily have a chemical composition identical to that in the coarser fractions – a beidellite described by Nadeau *et al.* (1985) was found to be more montmorillonitic in the $< 1.4\,\mu m$ fraction than in the $> 3\,\mu m$ fraction. Nevertheless, size fractionation is particularly useful in removing the non-phyllosilicates such as quartz, feldspars, carbonates, zeolites and amphiboles.

Another physical method of separating phases associated with clays is magnetic separation, which can be used either to remove iron-containing impurities, or to concentrate an iron-containing phyllosilicate such as chlorite. Magnetic separation rarely produces a single pure phase, but has been successfully applied to clay suspensions to reduce the level of impurities (Schultz and Dixon, 1979; Russell, Birnie and Fraser, 1984).

Chemical methods can also be used to remove impurities, but these must be monitored carefully, as the reagent used may attack the mineral of interest. Carbonate impurities can be eliminated by acid dissolution, but this is not to be recommended if trioctahedral clays are present because they are very susceptible to acid attack. A greatly superior method of removing carbonate is to pass a stream of carbon dioxide through a suspension of the clay, thereby converting the carbonate to bicarbonate, which is soluble in water. This method is slow and probably unsuitable for routine use, but it ensures freedom from acid attack on clays. It was successfully used to remove calcite from a Jordanian clay, thereby allowing the structural formula to be calculated more accurately for the chromium-rich smectite–volkonskoite (Khoury *et al.*, 1984).

If X-ray-amorphous materials are present, chemical dissolution methods are necessary for their removal. The most common amorphous phases are iron oxides and hydroxides, which can be removed (along with crystalline forms such as goethite and hematite) by treatment with sodium dithionite or ammonium oxalate, and aluminosilicate phases, which can be dissolved with sodium carbonate solutions. The use of these selective chemical treatments for dissolving or characterizing poorly ordered phases is discussed in detail in Chapter 9. The effects of these treatments on the crystalline clays should always be monitored by XRD, IR or EM if possible, as changes can be induced in the clays, e.g. Mössbauer spectrometry has revealed that deferration treatment can alter the Fe^{2+}/Fe^{3+} ratio in vermiculite and smectite (Ericsson, Linares and Lotse, 1984).

8.6.1.3 Application of corrections for impurities

Once the chemical composition has been obtained, any impurities that have not been removed prior to analysis must be corrected for before the analysis is used to calculate a structural formula. To do this, the chemical composition and the quantity of each impurity must be known. For many minerals, the composition is close to the theoretical composition, e.g. for quartz, cristobalite, carbonates, kaolinite, gibbsite, goethite, hematite, lepidocrocite, anatase and ilmenite. Difficulties arise, however, with minerals in which isomorphous substitution is common. Any feldspar present is unlikely to be a pure orthoclase, albite or anorthite end-member, and XRD or IR should be used to identify the type of feldspar. Chlorite composition is also variable, but the relative intensities of the basal reflections on the XRD trace should give some indication of the chlorite species present. Smectites vary from highly aluminous types to iron-rich varieties, and an 060 measurement by XRD usually indicates whether the species is di- or trioctahedral. Illite is probably the most difficult mineral to correct for, as the K content varies widely. Use of the theoretical muscovite formula would allow for the minimum amount of illite present, but the K_2O content of illites can vary from 11.8% (as in muscovite) to as low as 4.8%.

The lower the quantity of any one impurity the less important it is to know its composition accurately. Provided that the amount of the impurity is less than 5% (for example), approximate compositions should not produce appreciable errors in the subsequent calculation.

Any Ti present in the analyses of clays is most likely to occur as anatase or possibly ilmenite. There have been some reports of Ti^{3+} occurring in the structure of kaolinite (Dolcater, Syers and Jackson, 1970; Rengasamy, 1976) and in octahedral sites in biotite and phlogopite (Evans and Raftery, 1980). In the absence of direct evidence of a particular Ti-mineral, the quantity of TiO_2 (normally < 2%) is simply subtracted. Titanium can occur in clays in an X-ray-amorphous form of anatase, in which case its presence is only revealed by chemical analysis (Bain, 1976).

Small amounts of P_2O_5, SO_3 and CO_2 are commonly deducted from the analysis, on the assumption that they are present in unidentified impurities.

An example of corrections applied to the analysis of a clay is presented in Table 8.4. The sample is a bentonite containing an interstratified illite–smectite, and the purpose of the analysis was to calculate a structural formula for this mineral. To obtain as accurate a formula as possible, the following steps were undertaken:

- a fine fraction (< 0.5 μm) was separated by centrifugation to concentrate the illite–smectite;
- as calcite and gypsum were possible impurities, the sample was saturated with Sr^{2+}, so that any Ca detected in the analysis would indicate the presence of these phases;
- an examination by Mössbauer spectroscopy indicated 1.63% goethite, and allowed the structural iron in the illite–smectite to be partitioned accurately between ferrous and ferric;
- an examination by IR spectroscopy revealed the presence of kaolinite and quartz, estimated at 1.5% and 0.5% respectively.

The percentages of Fe_2O_3, SiO_2 and Al_2O_3 corresponding to the amounts of goethite, kaolinite and quartz determined, were subtracted from the chemical analysis (performed on the sample previously ignited at 1000°C). The TiO_2 percentage (0.36%) was assumed to be due to anatase, and was subtracted along with the small amounts of MnO, CaO, P_2O_5 and SO_3, which were presumed to be present in impurities at levels below the detection limit. The sum of the impurities was 4.3%, which is fairly small, but still produces a significant difference in the calculated structural formula (see section 8.6.2.1). The normalized analysis is also listed, so that the illite–smectite composition can be compared with others.

8.6.2 Calculation of formulae

Ideally, the structural formula of a mineral should be determined by first calculating the unit-cell volume from the unit-cell dimensions, then

Table 8.4 Corrections for impurities in < 0.5 μm fraction of a bentonite, Sr^{2+}-saturated

Analysis		*Corrections*						*Corrected analysis*	*Normalized analysis*
		Goethite (1.63%)	*Kaolinite (1.5%)*	*Quartz (0.5%)*	*Anatase*	*Others*	*Sum*		
SiO_2	53.23	—	0.81	0.5	—	—	1.31	51.92	53.93
TiO_2	0.36	—	—	—	0.36	—	0.36	—	—
Al_2O_3	31.87	—	0.69	—	—	—	0.69	31.18	32.39
Fe_2O_3	3.97	1.63	—	—	—	—	1.63	2.34	2.43
FeO	0.20	—	—	—	—	—	—	0.20	0.21
MnO	0.07	—	—	—	—	0.07	0.07	—	—
MgO	1.81	—	—	—	—	—	—	1.81	1.88
CaO	0.01	—	—	—	—	0.01	0.01	—	—
Na_2O	ND	—	—	—	—	—	—	—	—
K_2O	5.92	—	—	—	—	—	—	5.92	6.15
P_2O_5	0.17	—	—	—	—	0.17	0.17	—	—
SO_3	0.06	—	—	—	—	0.06	0.06	—	—
SrO	2.90	—	—	—	—	—	—	2.90	3.01
Total	100.57						4.30	96.27	100.00

ND = not determined.

calculating the mass of the unit cell after determination of the density of the mineral, and finally apportioning the elements from the chemical analysis, according to the percentage of each present (Newman and Brown, 1987).

If the type of clay mineral has been established from X-ray diffraction, its structural formula is calculated by assuming the anion composition of the mineral, e.g. if the mineral is identified as a chlorite, its anion composition is $O_{20}(OH)_{16}$ with an anion framework charge of 56. Alternatively, if the mineral is a smectite, the calculation is based on a unit cell with $O_{20}(OH)_4$ and 44 negative charges.

The amount of water associated with clays varies according to humidity, and so it is best to express the analysis of a clay on an ignited basis, either by analysing the clay after ignition (at say 1000°C) or by recalculating the analysis to an ignited basis using the figure for weight loss at 1000°C.

The procedure for calculating the structural formula is as follows:

- divide the weight percentage of each oxide by its molecular weight – this yields 'moles' of oxide;
- multiply 'moles' for each oxide by the number of positive charges per oxide to produce the number of gram equivalents of the cation;
- divide the charge from the anion framework by the total of the gram equivalents of all the cations to produce a proportionality factor;
- multiply the gram equivalent of each cation by the proportionality factor, and divide by the valency of the cation to produce the number of cations in the unit cell.

An example of this procedure is illustrated in Table 8.5. The calculation is performed as for a smectite, i.e. based on a unit cell with 44 negative charges, and the corrected analysis normalized to 100% is used. However, because of the introduction of the proportionality factor, the corrected analysis could be used without normalization, producing the same structural formula.

8.6.2.1 Assignment of cations to structural sites

In order to obtain the structural formula, the numbers of cations have to be allocated to tetrahedral, octahedral and exchangeable positions in the silicate structure. Based on considerations of ionic radii, it is assumed that all cations having a coordination number of six are in octahedral sites, and those cations having coordination numbers greater than six are considered to be in sites between the aluminosilicate layers. The procedure is as follows.

1. All Si^{4+} atoms are assigned to tetrahedral sites, and the remainder of the eight sites are filled with Al^{3+} atoms, which can have coordination

Table 8.5 Calculation of the structural formula of illite–smecite, from its chemical analysis, after corrections for impurities

Oxide	*Weight(%)*		*Molecular weight*		*Gram equivalent of cation*		*Proportionality factor*		*Gram equivalent cation per unit cell*		*Cation valence*		*Cations per unit cell*
SiO_2	53.93	÷	60.06	×4 =	3.5917	×	7.4867	=	26.8899	÷	4	=	6.72
Al_2O_3	32.39	÷	101.94	×6 =	1.9064	×	7.4867	=	14.2726	÷	3	=	4.76
Fe_2O_3	2.43	÷	159.70	×6 =	0.0913	×	7.4867	=	0.6835	÷	3	=	0.23
FeO	0.21	÷	71.85	×2 =	0.0058	×	7.4867	=	0.0434	÷	2	=	0.02
MgO	1.88	÷	40.32	×2 =	0.0932	×	7.4867	=	0.6978	÷	2	=	0.35
K_2O	6.15	÷	94.19	×2 =	0.1306	×	7.4867	=	0.9778	÷	1	=	0.98
SrO	3.01	÷	103.62	×2 =	0.0581	×	7.4867	=	0.4350	÷	2	=	0.22
					5.8771								

numbers of four or six. For the illite–smectite in Table 8.5, there are 6.72 Si^{4+} atoms and 1.28 (i.e. 8 – 6.72) Al^{3+} in the tetrahedral sheet.

2. The remaining Al^{3+} atoms, and all Fe^{3+}, Fe^{2+} and Mg^{2+} are assigned to the octahedral sites. Thus, for the illite–smectite, there are 3.48 (4.76 – 1.28) Al^{3+} octahedral atoms, and the octahedral sheet is $(Al^{3+}_{3.48}Fe^{3+}_{0.23}Fe^{2+}_{0.02}Mg^{2+}_{0.35})$.
3. Any Ca, Na and K, and any other cations present, known to have coordination numbers greater than six are assigned to sites between the aluminosilicate layers. For the illite–smectite there are 0.98 K^{+} atoms and 0.22 Sr^{2+} atoms.

It will be clear that unequivocal knowledge of the nature of the exchangeable cation obtained by saturating with a suitable cation is a distinct advantage at the cation assignment stage. In the natural state, Na, K, Mg, Ca, and even Al, can occur in exchangeable positions, and because Mg and Al usually occur in octahedral sites, this makes the assignment impossible. Saturating with a cation that is too large to fit in an octahedral site enables a confident assignment to be made. For the illite–smectite, Sr^{2+} was the saturating cation, which enabled the octahedral assignment to be completed, and also revealed that K^{+} is present as an interlayer cation in 12 coordination sites. The final structural formula of the illite–smectite is:

$$K_{0.98}Sr_{0.22}(Si_{6.72}Al_{1.28})(Al_{3.48}Fe^{3+}_{0.23}Fe^{2+}_{0.02}Mg_{0.35})O_{20}(OH)_4.$$

The total number of octahedral cations is 4.08, thereby indicating that the mineral is dioctahedral.

To enable the structural formula to be calculated, several assumptions are necessary. One assumption is that all the ferric iron is situated in the octahedral sheet, and this will usually be valid, although there have been reports (based mainly on Mössbauer studies) that iron can occur in interlayer positions (Cardile and Johnston, 1985) and that Fe^{3+} can occur in tetrahedral sites (Güven, 1988; Stucki, 1988).

If the formula had been calculated from the chemical analysis uncorrected for impurities, the resultant formula would have been:

$$K_{0.95}Sr_{0.21}(Si_{6.67}Al_{1.33})(Al_{3.38}Fe^{3+}_{0.37}Fe^{2+}_{0.02}Mg_{0.34})O_{20}(OH)_4.$$

This is significantly different from the formula obtained from the corrected analysis, indicating different degrees of isomorphous substitution, and affecting the layer-charge calculations. This illustrates the importance of correction for impurities, even when their total is less than 5%.

8.6.2.2 Calculation of layer charge

Once the structural formula has been obtained, the charge distribution within the structure can easily be calculated. In an 'ideal' structure (such as pyrophyllite or talc) the tetrahedral sheet would have a positive charge of 32 (eight quadrivalent silicon cations) and the octahedral sheet a positive charge of 12 (either six divalent cations or four trivalent cations), the combined positive charge of 44 balancing the 44 negative charges on the anion framework. However, because of ionic substitutions this 'ideal' arrangement is rarely attained.

In the illite–smectite, the tetrahedral charge is $(6.72 \times 4) + (1.28 \times 3) = 30.72$, and so the tetrahedral sheet has a net negative charge of $(32 - 30.72) = 1.28$. The octahedral charge is $(3.48 \times 3) + (0.23 \times 3) + (0.02 \times 2) + (0.35 \times 2) = 11.87$ and the octahedral sheet has a net negative charge of $(12 - 11.87) = 0.13$. Consequently, the total charge deficit on the unit cell (tetrahedral + octahedral sheets) is $1.28 + 0.13 = 1.41$. This negative charge is balanced by the interlayer and exchangeable cations – in the illite–smectite this is $(0.98 \times 1) + (0.22 \times 2) = 1.42$. Considering the assumptions involved in the structural formula calculation, together with the unavoidable experimental errors in the analytical methods, a difference of 0.01 between layer charge and interlayer charge is not significant, and is less than is often encountered.

8.6.2.3 Calculation of cation-exchange capacity

As the exchange is known to be represented by the strontium content (3.01% SrO) in the illite–smectite, the cation-exchange capacity is:

$$(1000 \times 3.01)/(103.62 \div 2) = 58 \text{ meq Sr per 100 g ignited clay.}$$

8.6.2.4 Examples of calculated structural formulae

There are numerous examples in the literature of structural formulae of clay minerals calculated from chemical analyses. Useful compilations of such analyses and formulae for a wide range of different clay groups are given in Weaver and Pollard (1973) and Newman and Brown (1987); both include a discussion of the chemistry of each group, and Newman and Brown also have a section with details of the calculation of several structural formulae, including those for more unusual examples, such as Al-interlayered montmorillonite. A selection of papers with analyses and formulae of particular types of clay from various environments and origins is listed in Table 8.6; many of those included give details of impurities present in the sample and explain how the appropriate corrections were applied.

The importance of obtaining good, accurate structural formulae for a mineral before embarking on interpretations and speculations, cannot be overemphasized. This was highlighted by Schultz (1982), who critically examined several papers in which illite or highly illitic interstratified clay was deduced to have a much greater octahedral than tetrahedral charge. He doubted the existence of such phases, because of:

Table 8.6 Examples in the literature of clay minerals and their calculated structural formulae

Mineral type	*Reference*
Saponite	Kohyama, Shimoda and Sudo (1973)
Saponite	Cowking *et al.* (1983)
Saponite	Ianovici, Neacsu and Neacsu (1983)
Beidellite	Chen, Wan and Brindley (1976)
Ferruginous beidellite	Özkan and Ross (1979)
Ferriferous beidellite	Wilson and Mitchell (1979)
Smectites	Curtin and Smillie (1981)
Iron-rich smectites	Brigatti (1983)
Volkonskoites	Mackenzie (1984)
Chlorite–smectite and corrensite	Beaufort *et al.* (1991)
Chlorite–smectites	Beaufort (1987)
Corrensite	Morrison and Parry (1986)
Smectites, palygorskite and sepiolite	Paquet *et al.* (1987)
Illite–smectites	Bouchet *et al.* (1988)
Illite and illite–smectites	Środoń et al. (1986)
Vermiculite–smectite	Craw (1984)
Chlorites	Foster (1962)
Chlorite	Bain (1977)
Chlorite	Proust (1982)
Chlorites	Whittle (1986)
Chlorites	Hillier and Velde (1991)
Chlorites and illites	Jahren and Aagaard (1989)
Chlorite and illite	Bodine and Standaert (1977)
Chlorite and chlorite–vermiculite	Noack and Colin (1986)
Swelling chlorite	Macneill (1978)
Swelling chlorite	Bain and Russell (1981)
Swelling chlorite	Curtis *et al.* (1984)
Vermiculite	Ildefonse, Capin and Velde (1979)
Vermiculite	De la Calle and Suquet (1988)
Illites	Warren and Curtis (1989)
Intergradient mineral	Wada *et al.* (1991)

'inappropriate basic assumptions on calculation of structural formulas, failure to correct for impurities, and/or possible inaccuracies in chemical analyses'. More recently, Warren and Ransom (1992) computed the errors in chemical analyses of 2 : 1 clay minerals arising from (1) contamination by phases present in amounts not detectable by X-ray diffraction, and (2) imprecision in analyses by analytical transmission electron microscopy. They concluded that variations in the composition of clay minerals reported in the literature do not necessarily represent the actual chemical variation of the minerals, with consequent implications for interpretations based on accurate descriptions of mineral chemistry and compositional variation.

REFERENCES

Allen, J. and Khan, A. A. (1970) Critical evaluation of powder sampling procedures. *Chemical Engineer*, **May**, CE108–CE112.

Bache, B. W. (1970) Barium isotope method for measuring cation-exchange capacity of soils and clays. *Journal of the Science of Food and Agriculture*, **22**, 169–171.

Bain, D. C. (1976) A titanium-rich clay. *Journal of Soil Science*, **27**, 68–70.

Bain, D. C. (1977) The weathering of ferruginous chlorite in a podzol from Argyllshire, Scotland. *Geoderma*, **17**, 193–208.

Bain, D. C. and Russell, J. D. (1981) Swelling minerals in a basalt and its weathering product from Morvern, Scotland: II. Swelling chlorite. *Clay Minerals*, **16**, 203–212.

Baker, D. E. and Suhr, N. H. (1982) Atomic absorption and flame emission spectrometry. *Agronomy*, **9**, 13–27.

Beaufort, D. (1987) Interstratified chlorite/smectite ('metamorphic vermiculite') in the Upper Precambrian greywackes of Rouez, Sarthe, France, in *Proceedings, International Clay Conference, Denver* (eds L. G. Schultz, H. van Olphen and F. A. Mumpton). The Clay Minerals Society, Bloomington, IN, pp. 59–65.

Beaufort, D., Meunier, A., Thomassin, J. H. and Fouillac, A. M. (1991) Corrensite, chlorite/smectite mixed layered minerals and saponite in the metamorphic basement of the Basin of Paris (France), in *Proceedings 7th Euroclay Conference, Dresden* (eds M. Störr, K.-H. Henning and P. Adolphi). Ernst-Moritz-Arndt Universität, Greifswald, pp. 65–69.

Begheyn, L. T. (1987) A rapid method to determine cation exchange capacity and exchangeable bases in calcareous, gypsiferous, saline and sodic soils. *Communications in Soil Science and Plant Analysis*, **18**, 911–932.

Berman, S. S. (1988) Acid digestion of marine samples for trace element analysis using microwave heating. *Analyst*, **113**, 159–163.

Bernas, B. (1968) A new method for decomposition and comprehensive analysis of silicates by atomic spectrometry. *Analytical Chemistry*, **40**, 1682–1686.

Bjørlykke, K. and Jorgensen, P. (1976) Comments on the paper: Chlorite and illite in some Lower Palaeozoic mudstones of mid-Wales. *Clay Minerals*, **11**, 165–167.

Blume, J. M. and Smith, E. (1954) Equilibration of exchangeable calcium and cation exchange capacity by equilibration with ^{45}Ca. *Soil Science*, **77**, 9–17.

Bodine, M. W. and Standaert, R. R. (1977) Chlorite and illite compositions from Upper Silurian rock salts, Retsof, New York. *Clays and Clay Minerals*, **25**, 57–71.

Bouchet, A., Proust, D., Meunier, A. and Beaufort, D. (1988) High-charge to low-charge smectite reaction in hydrothermal alteration processes. *Clay Minerals*, **23**, 133–146.

Brigatti, M. F. (1983) Relationships between composition and structure in Fe-rich smectites. *Clay Minerals*, **18**, 177–186.

Busenberg, E. and Clemency, C. V. (1973) Determination of cation exchange capacity of clays and soils using an ammonia electrode. *Clays and Clay Minerals*, **21**, 213–217.

De la Calle, C. and Suquet, H. (1988) Vermiculite, in *Hydrous Phyllosilicates (Exclusive of Micas)* (ed. S. W. Bailey), *Reviews in Mineralogy*, Vol. 19. Mineralogical Society of America, Washington, DC, pp. 455–496.

Cardile, C. M. and Johnston, J. H. (1985) Structural studies of nontronites with different iron contents by ^{57}Fe Mössbauer spectroscopy. *Clays and Clay Minerals*, **33**, 295–300.

Chalmers, R. A. and Page, E. S. (1957) The reporting of chemical analyses of silicate rocks. *Geochimica et Cosmochimica Acta*, **11**, 247–251.

Chapman, H. D. (1965) Cation exchange capacity, in *Methods of Soil Analysis* (*Agronomy*, 9) (eds C. A. Black, D. D. Evans, J. L. White, L. E. Ensminger and F. E. Clark). American Society of Agronomy, Madison. WI, pp. 891–901.

Chaussidon, J. (1982) The exchange complex: the general laws of anion and cation exchange, in *Constituents and Properties of Soils* (eds M. Bonneau and B. Souchier). Academic Press, London, pp. 377–398.

Chen, P.-Y., Wan, H.-M. and Brindley, G. W. (1976) Beidellite clay from Chang-Yuan. Taiwan: geology and mineralogy. *Clay Minerals*, **11**, 221–233.

Cowking, A., Wilson, M. J., Tait, J. M. and Robertson, R. H. S. (1983) Structure and swelling of fibrous and granular saponitic clay from Orrock quarry, Fife, Scotland. *Clay Minerals*, **18**, 49–64.

Craw, D. (1984) Ferrous-iron-bearing vermiculite–smectite series formed during alteration of chlorite to kaolinite. Otago schist, New Zealand. *Clay Minerals*, **19**, 509–520.

Curtin, D. and Smillie, G. W. (1981) Composition and origin of smectite in soils derived from basalt in Northern Ireland. *Clays and Clay Minerals*, **29**, 277–284.

Curtis, C. D., Ireland, B. J., Whiteman, J. A. *et al.* (1984) Authigenic chlorites: problems with chemical analysis and structural formulae calculation. *Clay Minerals*, **19**, 471–481.

De Soete, D. G. R., Gijbels, R. and Hoste, J. (1972) *Neutron Activation Analysis*. Wiley-Interscience, New York.

Dolcater, D. L., Syers, J. K. and Jackson, M. L. (1970) Titanium as free oxide and substituted forms in kaolinites and other soil minerals. *Clays and Clay Minerals*, **18**, 71–79.

Dolezal, J., Povondra, P. and Sulick, Z. (1968) *Decomposition Techniques in Inorganic Analysis*. Iliffe Books, London.

Ericsson, T., Linares, J. and Lotse, E. (1984) A Mössbauer study of the effect of dithionite/citrate/bicarbonate treatment on a vermiculite, a smectite and a soil. *Clay Minerals*, **19**, 85–91.

Evans, L. J. and Adams, W. A. (1975) Chlorite and illite in some Lower Palaeozoic mudstones of mid-Wales. *Clay Minerals*, **10**, 387–397.

Evans, S. and Raftery, E. (1980) X-ray photoelectron studies of titanium in biotite and phlogopite. *Clay Minerals*, **15**, 209–218.

Fischer, L. B. (1986) Microwave dissolution of geologic material: application to isotope dilution analysis. *Analytical Chemistry*, **58**, 261–263.

Flanagan, F. J. (1973) 1972 values for international geochemical reference samples. *Geochimica et Cosmochimica Acta*, **37**, 1189–1200.

Foster, M. D. (1962) Interpretation of the composition and a classification of the chlorites. *US Geological Survey, Professional Paper*, **414A**, 1–33.

Fraser, A. R. and Russell, J. D. (1969) A spectrophotometric method for determination of cation exchange capacity of clay minerals. *Clay Minerals*, **8**, 229–230.

Gillman, G. P. (1979) A proposed method for the measurement of exchange properties in highly weathered soils. *Australian Journal of Soil Research*, **17**, 129–139.

Goodman, B. A. (1980) Mössbauer spectroscopy, in *Advanced Chemical Methods for Soil and Clay Research* (eds J. W. Stucki and W. L. Banwart). D. Reidel, Dordrecht.

Grim, R. E. (1968) *Clay Mineralogy*. McGraw-Hill, New York, pp. 185–233.

Groves, A. W. (1951) *Silicate Analysis*. Allen and Unwin, London.

Güven, N. (1988) Smectites, in *Hydrous Phyllosilicates (Exclusive of Micas)* (ed. S. W. Bailey), *Reviews in Mineralogy*, Vol. 19. Mineralogical Society of America, Washington, DC, pp. 497–559.

Helmke, P. A. (1982) Neutron activation analysis. *Agronomy*, **9**, 67–84.

Hey, M. H. (1941) The determination of ferrous iron in resistant silicates. *Mineralogical Magazine*, **26**, 116–118.

Hillebrand, W. F. (1919) The analysis of silicate and carbonate rocks. *US Geological Survey Bulletin*, **700**.

Hillebrand, W. F., Lundell, G. E. F., Bright, H. A. and Hoffman, J. I. (1953) *Applied Inorganic Analysis*. John Wiley, New York.

Hillier, S. and Velde, B. (1991) Octahedral occupancy and the chemical composition of diagenetic (low-temperature) chlorites. *Clay Minerals*, **26**, 149–168.

Ianovici, V., Neacsu, G. and Neacsu, V. (1983) Saponite from Moldova Noua, Romania. *Clay Minerals*, **18**, 205–207.

Ildefonse, P., Capin, E. and Velde, B. (1979) A soil vermiculite formed from a meta-gabbro, Loire-Atlantique, France. *Clay Minerals*, **14**, 201–210.

Ingamells, C. O. (1966) Absorptiometric methods in rapid silicate analysis. *Analytical Chemistry*, **38**, 1228–1234.

Jackson, M. L. (1962) *Soil Chemical Analysis*. Taylor, Garnett, Evans and Co., Watford, p. 33.

Jahren, J. S. and Aagaard, P. (1989) Compositional variations in diagenetic chlorites and illites, and relationships with formation-water chemistry. *Clay Minerals*, **24**, 157–170.

Jeffery, P. G. (1981) *Chemical Methods of Rock Analysis*. Pergamon Press, Oxford.

Johnson, W. M. and Maxwell, J. A. (1981) *Rock and Mineral Analysis*. John Wiley, New York.

Jones, A. A. (1982) X-ray fluorescence spectrometry. *Agronomy*, **9**, 85–121.

Kanai, Y. (1990) Simultaneous determination of iron(II) and iron(III) oxides in geological materials by ion chromatography. *Analyst*, **115**, 809–812.

Kelley, W. P. (1948) *Cation Exchange in Soils*. Rheinhold, New York.

Kemp, A. J. and Brown C. J. (1990) Microwave digestion of carbonate rock samples for chemical analysis. *Analyst*, **115**, 1197–1199.

Khoury, H. N., Mackenzie, R. C., Russell, J. D. and Tait, J. M. (1984) An iron-free volkonskoite. *Clay Minerals*, **19**, 43–57.

Kohyama, N., Shimoda, S. and Sudo, T. (1973) Iron-rich saponite (ferrous and ferric forms). *Clays and Clay Minerals*, **21**, 229–237.

Kolthoff, I. M. and Sandell, E. B. (1956) *Text book of Quantitative Inorganic Analysis*. Macmillan, New York.

Lamonthe, P. J., Fries, T. L. and Consul, J. J. (1986) Evaluation of a microwave oven system for the dissolution of geologic samples. *Analytical Chemistry*, **58**, 1881–1886.

Langmyhr, F. G. and Graff, P. R. (1965) A contribution to analytical chemistry of silicate rocks; a scheme of analysis for eleven main constituents based on decomposition by hydrofluoric acid. *Norges Geologiske Undersolkelse*, **230**, 1–128.

Lim, C. H. and Jackson, M. L. (1982) Dissolution for total elemental analysis. *Agronomy*, **9**, 1–11.

Lundell, G. E. F. and Hoffman, J. I. (1938) *Outlines of Methods of Chemical Analysis*. Wiley, New York.

Mackenzie, R. C. (1984) Constitution of and relationships among volkonskoites. *Clay Minerals*, **19**, 669–671.

Macneill, S. (1978) A chemical investigation of a chlorite intergrade mineral in the Keuper Marl. *Clay Minerals*, **13**, 357–365.

Mehlich, A. (1948) Determination of cation- and anion-exchange properties of soils. *Soil Science*, **66**, 429–445.

Metson, A. J. (1956) Methods of chemical analysis for soil survey samples. *New Zealand Department of Scientific and Industrial Research*, **12**, 83–90.

Morrison, S. J. and Parry, W. T. (1986) Dioctahedral corrensite from Permian Red Beds, Lisbon Valley, Utah. *Clays and Clay Minerals*, **34**, 613–624.

Murphy, J. and Riley, J. P. (1962) A modified single solution method for the determination of phosphate in natural waters. *Analytica Chimica Acta*, **27**, 31–36.

Nadeau, P. H. and Bain, D. C. (1986) Composition of some smectites and diagenetic illitic clays and implications for their origin. *Clays and Clay Minerals*, **34**, 455–464.

Nadeau, P. H., Farmer, V. C., McHardy, W. J. and Bain, D. C. (1985) Compositional variations of the Unterrupsroth beidellite. *American Mineralogist*, **70**, 1004–1010.

Newman, A. C. D. and Brown, G. (1987) The chemical constitution of clays, in *Chemistry of Clays and Clay Minerals* (ed. A. C. D. Newman). Mineralogical Society, London, pp. 1–128.

Nicholls, G. D. (1960) Techniques in sedimentary geochemistry. Determination of ferrous iron contents of carbonaceous shales. *Journal of Sedimentary Petrology*, **30**, 603–612.

Noack, Y. and Colin, F. (1986) Chlorites and chloritic mixed-layer minerals in profiles on ultrabasic rocks from Moyango (Ivory Coast) and Angiquinho (Brazil). *Clay Minerals*, **21**, 171–182.

Norrish, K. and Chappell, B. W. (1977) X-ray fluorescence spectrometry, in *Physical Methods in Determinative Mineralogy* (ed. J. Zussman). Academic Press, London, pp. 201–272.

Özkan, A. I. and Ross, G. J. (1979) Ferruginous beidellites in Turkish soils. *Soil Science Society of America, Proceedings*, **43**, 1242–1248.

Papp, C. S. E. and Fischer, L. B. (1987) Application of microwave digestion to the analysis of peat. *Analyst*, **112**, 337–338.

Paquet, H., Duplay, J., Valleron-Blanc, M. M. and Millot, G. (1987) Octahedral compositions of individual particles in smectite–palygorskite and smectite–sepiolite assemblages, in *Proceedings, International Clay Conference, Denver*, pp. 73–77.

Parfitt, R. L. (1978) Anion adsorption by soils and soil material. *Advances in Agronomy*, **30**, 1–50.

Pleijsier, L. K. (1986) *Proceedings of an International Workshop on the Laboratory Method and Data Exchange Programme*. International Soil Reference and Information Centre, (ISRIC), Wageningen, The Netherlands.

Proust, D. (1982) Supergene alteration of metamorphic chlorite in an amphibolite from the Massif Central, France. *Clay Minerals*, **17**, 159–173.

Reay, A. (1981) The effect of disc mill grinding on some rock-forming minerals. *Mineralogical Magazine*, **44**, 179–182.

Rengasamy, P. (1976) Substitution of iron and titanium in kaolinites. *Clays and Clay Minerals*, **24**, 265–266.

Rhoades, J. D. (1982) Cation exchange capacity, in *Methods of Soil Analysis* (*Agronomy*, **9**) (eds A. L. Page, R. H. Miller and D. R. Keeney), American Society of Agronomy, Madison, WI, pp. 149–157.

Russell, J. D., Birnie, A. and Fraser, A. R. (1984) High-gradient magnetic separation (HGMS) in soil clay mineral studies. *Clay Minerals*, **19**, 771–778.

Schultz, D. G. and Dixon, J. B. (1979) High-gradient magnetic separation of iron oxides and other magnetic minerals from soil clays. *Soil Science Society of America, Proceedings*, **43**, 793–799.

Schultz, L. G. (1982) Chemical characteristics and origin of Ordovician K-bentonites along the Cincinnati Arch: a discussion. *Clays and Clay Minerals*, **30**, 445–446.

Searle, P. L. (1986) The measurement of soil cation exchange properties using the single extraction silver thiourea method: an evaluation using a range of NZ soils. *Australian Journal of Soil Research*, **24**, 193–200.

Shapiro, L. (1960) A spectrophotometric method for determination of FeO in rocks. *US Geological Survey Research*, **400B**, B496–B497.

Shapiro, L. (1967) Rapid analysis of rocks and minerals by a single-solution method. *US Geological Survey Professional Paper*, **575B**, B187–B191.

Shapiro, L. and Brannock, W. W. (1952) Rapid analysis of silicate rocks. *US Geological Survey Circular*, **165**, 17 pp.

Shapiro, L. and Brannock, W. W. (1956) Rapid analysis of silicate rocks. *US Geological Survey Bulletin*, **1036-C**, 56 pp.

Srodon, J., Morgan, D. J., Eslinger, E. V., Eberl, D. D. and Karlinger, M. R. (1986) Chemistry of illite/smectite and end-member illite. *Clays and Clay Minerals*, **34**, 368–378.

Stucki, J. W. (1981) The quantitative assay of minerals for Fe^{2+} and Fe^{3+} using 1,10-phenanthroline: II, a photochemical method. *Soil Science Society of America, Journal*, **45**, 638–641.

Stucki, J. W. (1988) Structural iron in smectites, in *Iron in Soils and Clay Minerals* (eds J. W. Stucki, B. A. Goodman and U. Schwertmann). NATO ASI Series, Reidel, Dordrecht, pp. 625–675.

Thomas, G. W. (1977) Historical developments in soil chemistry: ion exchange. *Soil Science Society of America, Journal*, **41**, 230–238.

Thomson, M. and Walsh, J. N. (1983) *Handbook of Inductively Coupled Plasma Spectrometry*. Blackie, Glasgow and London.

Ure, A. M. (1983) Atomic absorption and flame emission spectrometry, in *Soil Analysis* (ed. K. A. Smith). Marcel Dekker, New York.

Van Loon, J. C. (1965) Titrimetric determination of the iron (II) oxide content of silicates using potassium iodate. *Talanta*, **12**, 599–603.

Verbeek, A. A., Mitchell, M. C. and Ure, A. M. (1982) The analysis of small samples of rock and soil by atomic absorption and emission spectrometry after a lithium metaborate fusion/nitric acid dissolution procedure. *Analytica Chimica Acta*, **135**, 215–228.

Volborth, A. (1964) Biotite mica effect in X-ray spectrographic analysis of pressed rock powders. *American Mineralogist*, **49**, 634–643.

Wada. K. and Okamura, Y. (1977) Measurements of exchange capacities and hydrolysis as a means of characterizing cation and anion retention by soils, in Proceedings, International Seminar on Soil Environment and Fertility Management in Intensive Agriculture, Tokyo, *Society of Scientific Soil Manure, Japan, Tokyo*, pp. 811–815.

Wada, K., Kakuto, Y., Wilson, M. A. and Hanna, J. V. (1991) The chemical composition and structure of a 14 Å intergradient mineral in a Korean Ultisol. *Clay Minerals*, **26**, 449–461.

Warren, E. A. and Curtis, C. D. (1989) The chemical composition of authigenic illite within two sandstone reservoirs as analysed by TEM. *Clay Minerals*, **24**, 137–156.

Warren, E. A. and Ransom, B. (1992) The influence of analytical error upon the interpretation of chemical variations in clay minerals. *Clay Minerals*, **27**, 193–209.

Weaver, C. E. and Pollard, L. D. (1973) *The Chemistry of Clay Minerals*. Elsevier, Amsterdam.

Weiss, A., Mehler, A., Koch, G. and Hoffman, U. (1956) Anion exchange capacity of clay minerals. *Zeitschrift fuer Anorganische und Allgemeine Chemie*, **284**, 247–271.
Whittle, C. K. (1986) Comparison of sedimentary chlorite compositions by X-ray diffraction and analytical TEM. *Clay Minerals*, **21**, 937–947.
Wilson, A. D. (1960) The micro-determination of ferrous iron in silicate minerals by a volumetric and a colorimetric method. *Analyst*, **85**, 823–827.
Wilson, M. J. and Mitchell, B. D. (1979) Comparative study of a vertisol and an entisol from the Blue Nile plains of Sudan. *Egyptian Journal of Soil Science*, **19**, 207–220.

CHAPTER 9

Characterization of poorly ordered minerals by selective chemical methods

B. F. L. Smith

9.1 INTRODUCTION

At the beginning of this century, ideas on the composition of clay-size material were greatly influenced by van Bemmelen (1910) who suggested that this material existed as a completely amorphous colloidal complex. However, many mineralogists thought that clays could be a mixture, containing varying proportions of crystalline material (Mackenzie, 1963). The mixed-oxide hypothesis of van Bemmelen (1910), associating the colloidal state with non-crystallinity persisted until Hendricks and Fry (1930), using X-ray methods showed the presence of large amounts of crystalline material in soil clays. This resulted in the virtual rejection of ideas advocating an amorphous nature for clays. Also, researches were directed almost exclusively towards elucidation of the physico-chemical reactions within the crystal structure of clay minerals. Despite this emphasis on crystallinity, Mattson, in a series of papers on the colloidal chemistry of soil, continued to recognize the importance of inorganic, poorly ordered components in soils – a balanced view which has now become generally accepted (Rich and Thomas, 1960; Mitchell, Farmer and McHardy, 1964). Soil clay can, therefore, best be regarded as consisting of both crystalline and non-crystalline minerals. Non-crystalline material is referred to by a number of names, such as X-ray-amorphous oxides and hydrous oxides, allophane, poorly ordered inorganic material and inorganic gel material. In 1972, Cradwick *et al.*, showed that imogolite, a poorly ordered, gel-like, hydrous aluminium

Clay Mineralogy: Spectroscopic and Chemical Determinative Methods.
Edited by M. J. Wilson. Published in 1994 by Chapman & Hall, London.
ISBN 0 412 53380 4.

silicate, first identified by Yoshinaga and Aomine (1962), actually had a well-defined crystalline structure, which they characterized by electron diffraction and trimethylsilylation. Recently, Parfitt (1990) redefined allophane as a group of clay-size minerals containing silica, alumina and water in chemical combination and with short-range order, whereas imogolite, a poorly ordered gel-like, hydrous aluminium silicate with a well-defined crystalline structure, comprises bundles of fine tubes with long-range order in one dimension (Farmer and Russell, 1990). Studies such as these have led to a reappraisal of amorphous material in soils.

9.2 PRINCIPLES

A wide range of methods exist for the identification of well-defined crystalline clay minerals in soils. However, until recently, definitive information on the nature of poorly ordered components in soils has been rather sparse. This very reactive material, often occurring in small amounts, is difficult to characterize. For some time, it was thought to be mainly confined to soils in regions of recent volcanic activity (Fieldes, 1955; Yoshinaga, 1966) but now it has been shown to be an important constituent of temperate soils (Mitchell and Farmer, 1962; McKeague and Day, 1966; Tait, Yoshinaga and Mitchell, 1978). In soils, it often occurs in association with the surface of primary soil particles (Follett *et al.*, 1965; Jones and Uehara, 1973; Farmer, Russell and Smith, 1983) and can, therefore, have an influence incommensurate with the amount present (Kirkman, Mitchell and Mackenzie, 1966).

Poorly ordered material in soils consists mainly of oxides and/or hydroxides of silicon, aluminium and iron (Guillet and Souchier, 1982), either discrete or in various combinations. The occurrence of non-crystalline forms of silica, usually termed opaline silica, is frequently associated with immature soils of volcanic origin, although instances have been reported in Spanish soils, on granite (Hoyos and Pino, 1958) and in organic surface horizons of Scottish soils (Mitchell and Farmer, 1962). This type of material is derived by weathering of primary silicate minerals or by recycling vegetation, where it occurs in humus-rich horizons in the form of phytoliths.

Poorly ordered oxides and hydroxides of aluminium in soils are not as widespread as those of silicon, being noted to a small extent in the early weathering state of New Zealand soils (Yoshinaga, Tait and Soong, 1973) and in B horizons of Scottish soils (Tait, Yoshinaga and Mitchell, 1978). Non-crystalline iron oxides are generally stabilized by organic and silicon complexes, the rate of crystallization to goethite and hematite being dependent on environmental factors (Schwertmann, 1978). Combinations of the three components are found in many soils. Amorphous aluminosilicates include gel-like compounds, containing mainly aluminium, silicon, oxygen and water, but usually also some iron, magnesium, hydroxyl, and occasionally phosphate. Allophane is a general term for amorphous aluminosilicate gels, where the Al_2O_3: SiO_2 mol ratio falls in the range 0.5–2.0 (Ross and Kerr, 1934). Farmer, Russell and Berrow (1980), Parfitt (1990) and Farmer *et al.* (1991) have shown that the composition of allophanic material in soils varies according to the pH of formation and concentration of silicic acid in solution. Earlier, Yoshinaga and Aomine (1962) separated two mineral colloids from a soil clay fraction. One colloid remained dispersed in both acid and alkali, while the other flocculated in alkali. The former was amorphous to X-rays, and was considered to be allophane, and the latter, which possessed a low degree of crystallinity, was termed imogolite.

Conventional methods can also provide some indication of non-crystalline inorganic material in soils. It is not always clear, however, whether there is a sharp division between the highly ordered crystalline core and the disordered surface material, or if there is a transition zone of increasing order inwards, as envisaged by Mattson in the 1930s. The only methods of obtaining even a semiquantitative estimation of highly disordered (X-ray amorphous) inorganic material in soils are those based on selective chemical dissolution.

9.3 SELECTIVE CHEMICAL METHODS

Poorly ordered material in soils is normally characterized by ascertaining its relative resistance to dissolution in various chemical reagents. Aluminosilicate gels form soluble silicates and aluminates in alkaline solutions. Both sodium hydroxide and sodium carbonate have been used for many years to dissolve siliceous and aluminous cements, which frequently bind clay particles into aggregates. Treatment with these reagents not only improves clay dispersion but also 'cleans up' the crystalline components, thereby assisting in their identification by instrumental techniques, particularly XRD. These selective chemical methods depend on the high specific surface area and lower thermodynamic stability of poorly ordered material in soils, which results in a higher dissolution rate than for crystalline minerals. A high ratio of dissolution reagent to sample weight ensures much better dissolution, and avoids saturation of the solution with silicon or aluminium.

A range of specified selective chemical methods will be given for estimating amounts of poorly ordered material in soils. Also, detailed analytical information will be provided on the methods in current use. Quantitative measurement of extracted aluminium, iron and silicon can be ascertained by either spectrophotometric or atomic-absorption methods. The latter, however, are not sensitive enough for the low amounts of silicon frequently obtained when extracting poorly ordered material from soils. Details of atomic-absorption methods are not given here, but these can be successfully used for most of the procedures, provided that the solutions are diluted to reduce interference from very high levels of salts in some extractants.

9.3.1 Alkali dissolution

Boiling with 2% Na_2CO_3 has been used (Jackson, Pennington and Mackie, 1950) to remove the finely divided siliceous material that cements clay particles. Hashimoto and Jackson (1960) showed that digestion with Na_2CO_3 solution may not completely remove poorly ordered aluminosilicates, perhaps because of the slow rate of reaction and saturation of the extracting solution. The effect of hot and cold 0.5 M Na_2CO_3 solutions on pure clay minerals, accessory minerals, gels and soil clays (Follett *et al.*, 1965) showed:

- little reaction with the crystalline minerals;
- appreciable solubility of silica gel and allophane;
- increased dispersion.

Considerable amounts of SiO_2 and Al_2O_3 were extracted from the soil clays examined. The effects of hot and cold 0.5 M Na_2CO_3 treatment were compared with that of 0.5 M NaOH (Hashimoto and Jackson, 1960) and it was concluded that the amounts of Si and Al extracted vary with the reagent in the sequence 0.5 M NaOH > hot 0.5 M Na_2CO_3 > cold 0.5 M Na_2CO_3. This result is in agreement with the calculated reactivity of the reagents.

Treatment with hot 0.5 M Na_2CO_3 not only removes 'less readily soluble' aluminosilica gel material, but also some crystalline gibbsite (Jorgensen *et al.*, 1970) and can result in the formation of zeolite-like aluminosilicate products (Farmer, Smith and Tait, 1977). Hashimoto and Jackson (1960) recommended 2.5 minutes of boiling with 0.5 M NaOH for the removal of allophane, free silicon and aluminium from soils. The reaction time was curtailed because prolonged treatment has been shown to attack crystalline components. Hashimoto and Jackson also pointed out that dehydroxylated kaolinite and halloysite were dissolved by this technique, whereas chlorite and montmorillonite, heated to 500°C, were virtually unaffected. A NaOH solution buffered at pH 9.7 has been used by Bruckert and Souchier (1975) to extract the more mobile organic complexes of iron and aluminium.

The two commonly used methods of alkali dissolution are now described.

9.3.1.1 2.5 minutes boiling with 0.5 M NaOH (Hashimoto and Jackson, 1960)

Reagents Sodium hydroxide (0.5 M) – dissolve 20 g NaOH (AR) pellets in 1 l distilled water and store in a polyethylene bottle.

Dissolution procedure Place 50-mg sample (< 200 mesh) in a nickel crucible, add 50 ml of 0.5 M NaOH and boil the suspension for 2.5 minutes. After cooling, the latter is quantitatively transferred to a 100-ml polypropylene centrifuge tube and centrifuged at 2200 *g* for 10 minutes. The supernatant solution is then siphoned off into a 200-ml standard flask, the residue is washed with distilled water, recentrifuged, and the wash water added to the solution in the 200-ml flask, diluted to the mark with distilled water and retained for the determination of aluminium and silicon.

(i) Aluminium determination – Aluminon (Robertson, 1950)

Reagents Ammonium acetate buffer, pH 4.5 – dilute 230 ml of 100% glacial acetic acid (AR) and 216 ml of 35% ammonium hydroxide (AR) to 1 l with distilled water.
Dissolve 10 g Acacia in 100 ml distilled water.
Dissolve 107 g ammonium chloride (AR) in 150 ml concentrated hydrochloric acid.
Mix the above solutions, dilute to 2 l and adjust the pH to 4.5 ± 0.01.
Aluminon (0.1%) – dissolve 1 g Aurine tricarboxylic acid in 150 ml distilled water to which five drops of ammonia have been added. Boil the solution to expel excess ammonia, cool and dilute to 1 l with distilled water.
Aluminium standard solution (stock) – dissolve 1.7584 g potassium aluminium sulphate $AlK(SO_4)_2 12H_2O$ in distilled water containing 50 ml of 5 M hydrochloric acid, and dilute to 1 l with distilled water.
Aluminium working standard – dilute 50 ml of stock Al standard to 1 l with distilled water, giving a solution of concentration 10 μg/ml Al.
Thioglycollic acid 80%.

Procedure Pipette a suitable aliquot (0–30 μg Al) into a 100-ml conical flask, adjust the pH to within the range 4–5, then dilute with distilled water to 50 ml total volume.

Add 10 ml of 1 M NH_4OAc buffer, then one drop of thioglycollic acid to reduce any ferric iron present to the ferrous form – which does not form a complex with Aluminon. Then add 2 ml Aluminon reagent and place in a boiling water bath for exactly 30 minutes. Cool and dilute to 100 ml with distilled water and measure the absorption on a spectrophotometer at 525 nm using distilled water in the reference cell. The absorptions are then compared with a standard aluminium curve prepared in the same manner for each set of determinations, in the range 0–30 μg Al. A blank determination should always be carried out.

(ii) Silicon determination – molybdate reaction (Webber and Wilson, 1969)

Ammonium molybdate at *c.* pH 1.2 reacts with silica and any phosphate present to produce heteropoly acids. Tartaric acid is added to destroy the phosphomolybdic complex, thus allowing silicon to be determined as the silicomolybdic complex in the presence of phosphate. The sensitivity of this method can be greatly increased by measuring this complex as the reduced molybdenum-blue form, which can be achieved by the addition of an amino acid.

Reagents Ammonium molybdate solution:
solution A – dissolve 89 g ammonium molybdate $(NH_4)_6Mo_7O_{24}{\cdot}4H_2O$(AR) in *c.* 800 ml distilled water;

solution B – add 63 ml of 98% sulphuric acid (d_{20} 1.84) (AR) cautiously with stirring to *c.* 100 ml distilled water and cool to room temperature.

Add solution B to A, cool to room temperature and dilute to 1 l with distilled water and store in a polyethylene bottle. This solution is stable for at least 6 months.

Tartaric acid – dissolve 280 g tartaric acid $C_4H_6O_6$ (AR) in distilled water to make up to 1 l and store in a polyethylene bottle. This solution is stable for at least 6 months.

Reducing solution – dissolve 2.8 g of sodium sulphite ($Na_2SO_3 \cdot 7H_2O$) and 0.6 g 1-amino-2-naphthol-4-sulphonic acid in *c.* 100 ml distilled water. Add 30 g of sodium metabisulphite ($Na_2S_2O_5$), dilute to 400 ml with distilled water and store in a polyethylene bottle. This solution is best kept in the dark and should be prepared fresh every month.

Silicon stock solution – weigh 1 g pure, dried silica into a platinum crucible, add 5 g anhydrous sodium carbonate (AR) and heat slowly over a Meker burner, gradually increasing the temperature until red heat, and then continue heating for a further 10 minutes, until the melt becomes clear. Allow the melt to cool before dissolving it in distilled water and diluting to 1 l. The solution, stored in a polyethylene bottle, is stable for at least 2 years. Silica concentration ≡ 1000 μg/ml SiO_2.

Silicon working standard – dilute 10 ml of the stock solution to 1 l with distilled water. Store in a polyethylene bottle. The solution is stable for at least 3 months. Silica concentration ≡ 10 μg/ml SiO_2.

Silicon flasks – for accurate and reproducible silicon determinations a set of 100-ml standard flasks should be specially reserved. These flasks should initially be cleaned carefully with chromic acid (*hazardous oxidizing reagent*) by soaking overnight, then washed with 1 : 1 hydrochloric acid and washed with distilled water, and should be kept full of distilled water after each determination to prevent the formation of siliceous coatings on the flasks.

Procedure Pipette a suitable aliquot (5–80 μg/SiO_2) into the 'silicon' flasks, adjust the pH to 3–4, and dilute to *c.* 80 ml with distilled water. Add 2.5 ml ammonium molybdate reagent and mix immediately. After 10 minutes add 5 ml tartaric acid, and after a further 5 minutes add 2.5 ml reducing solution. Dilute to 100 ml with distilled water. Stand the solution for 1 h and measure the absorption at 820 nm in a 4-cm cell using distilled water as a reference. A wavelength of 650 nm may be used, with a 40% reduction in sensitivity over that obtained at 820 nm.

The results obtained are compared with a standard graph, prepared for a range of silicon solutions (0–80 μg SiO_2). A blank determination should be carried out on the reagents.

9.3.1.2 0.5 M sodium carbonate extraction (cold) – Follett et al. (1965)

Reagents Sodium carbonate (0.5 M) – dissolve 106 g Na_2CO_3 anhydrous (AR) in 2 l of distilled water and store in a polyethylene bottle.

Dissolution procedure Weigh 100 mg sample (< 200 mesh) into a 100-ml polypropylene centrifuge tube, add 80 ml of 0.5 M Na_2CO_3 solution, stopper well and place in an end-over-end shaker overnight (16 h). Centrifuge for 10 minutes at 2200 *g*, and siphon off the supernatant solution into a 200-ml standard flask. Make up to the mark with distilled water and reserve the solution for the determination of aluminium and silicon. For soils, a 100-ml flask can be used, as the amounts of aluminium and silicon extracted are much smaller than from clay samples.

Aluminium and silicon contents are then determined as in the procedures already discussed.

9.3.2 Sodium dithionite

Free iron oxides and crystalline iron oxides are frequently removed from soil to improve mineral identification and quantification. Most methods used to remove these oxides involve the reduction and mobilization of iron. Deb (1950) was among the first to use sodium dithionite to measure the free iron oxide content of soils. This method was modified and improved by Mitchell and Mackenzie (1954) to measure 'free' iron, by Mehra and Jackson (1960) to determine the release of iron and aluminium, and by Mitchell, Smith and de Endredy (1971) to determine three major components of inorganic gel material in soils and soil clays, namely, silicon, aluminium and ferric iron. More recently (Parfitt and Childs, 1988), in a review of dissolution methods, showed that goethite, lepidocrocite, hematite and ferrihydrite are dissolved by dithionite thus allowing their Fe content to be estimated. The Fe attributable to goethite and hematite can be estimated by subtracting the oxalate extractable Fe (Fe_{ox}) from the dithionite extractable values Fe_d, i.e. goethite = $Fe_d - Fe_{ox}$. Lepidocrocite, if present in microcrystalline form, is partially soluble in the oxalate reagent.

Mehra and Jackson (1960) buffered the dithionite extraction of soils at pH 7 with 1 M sodium bicarbonate, and used sodium citrate to chelate the iron and aluminium released. However, the presence of citrate and dithionite in the extract causes difficulty in the determination of its silicon and aluminium content. Weaver, Syers and Jackson (1968) improved the determination of silicon by using a much higher concentration of ammonium molybdate reagent. Mitchell, Smith and de Endredy (1971) modified the above procedure to ensure a more reliable determination of silica and alumina in the dithionite extract of soils. They did not use sodium citrate, but instead followed the buffered dithionite treatment with a dilute acid wash to dissolve the precipitated iron sulphide. Then, aliquots from the extracts were treated with alkaline hypobromite, to destroy the dithionite and allow easy and rapid spectrophotometric determination of silicon and aluminium.

9.3.2.1 Dithionite – citrate – bicarbonate method – (Mehra and Jackson, 1960)

Reagents Sodium dithionite ($Na_2S_2O_4$).
Sodium citrate (0.3 M) – 88 g $Na_3C_6H_5O_7 \cdot 2H_2O$ (AR) 1 l.
Sodium bicarbonate (1 M) – 84 g $NaHCO_3$(AR) 1 l.
Sodium chloride (AR) saturated solution.
Acetone (AR).

Dissolution procedure Place a suitable weight of sample (100 mg clay or 400 mg soil) in a 100-ml polypropylene centrifuge tube. Add 40 ml of 0.3 M sodium citrate and 5 ml of 1 M sodium bicarbonate and heat the mixture on a water-bath at 80°C for 30 minutes.

Add 0.5 g of sodium dithionite to the hot suspension, stirring frequently (or agitating by means of an ultrasonic probe) for 10 minutes. The mixture is then allowed to cool, before adding 10 ml saturated sodium chloride and 10 ml acetone (to encourage flocculation). After cooling to room temperature, the mixture is centrifuged at 2200 *g* for 10 minutes, and then the supernatant is siphoned off into a 200-ml volumetric flask, the residue washed with 50 ml distilled water, recentrifuged and the wash liquid added to the previous supernatant. The solution is then made up to the mark with distilled water and reserved for iron, aluminium and silicon determinations.

(i) Iron determination – 1, 10-phenanthroline (de Endredy, 1963)

Reagents 1,10-phenanthroline $C_{12}H_8N_2 \cdot H_2O$ (0.25%) solution in water.
Sodium citrate $Na_3C_6H_5O_7 \cdot 2H_2O$ (AR) (20%) solution in water.
Ascorbic acid (L).
Hydrochloric acid (2 M) – dilute 172 ml concentrated 35% HCl (AR) to 1 l with distilled water.
Iron standard (stock) – dissolve 0.7022 g ferrous ammonium sulphate $(NH_4)_2SO_4FeSO_4 \cdot 6H_2O$ in 100 ml distilled water, containing 5 ml of 6 M sulphuric acid, add a dilute solution of potassium permanganate dropwise until the mixture is just pink in colour, then dilute to 1 l with distilled water (1 ml = 100 μg Fe).
Iron standard (working) – dilute 100 ml of iron stock solution to 1 l (1 ml = 10 μg Fe).

Procedure Pipette a suitable aliquot (0–60 μg Fe) into a 50-ml volumetric flask, and add 2 ml of 2 M hydrochloric acid. Dilute the solution to about 25 ml with distilled water, add 100 ± 10 mg of ascorbic acid, mix thoroughly and allow the mixture to stand for 10 minutes. Then add 2 ml of 0.25% 1,10-phenanthroline solution and 5 ml of 20% sodium citrate buffer, and after 1 h measure the colour of the ferrous iron–phenanthroline complex at 525 nm. Compare this with a standard iron calibration curve prepared in a similar manner in the range 0–80 μg Fe for the 4-cm cell, and 50–400 μg Fe when a 1-cm cell is used. NB. A reagent-blank determination should be carried through the complete extraction process.

(ii) Aluminium and silicon determinations

If aluminium and silicon must be measured in the sodium dithionite extracts, then pretreatment must be carried out to overcome possible interference from citrate and dithionite. Weaver, Syers and Jackson (1968) proposed a method for the determination of silica in citrate–bicarbonate–dithionite extracts of soils by destroying the dithionite, and oxidizing the extract by bubbling air through it. They also eliminated the interference of citrate by utilizing an excess of molybdate in the determination of silica. Mitchell, Smith and de Endredy (1971) used a different approach – citrate was not utilized in the extracting solution, and the dithionite was destroyed by alkaline hypobromite oxidation; any excess bromine was removed by acidification and heat treatment. An added bonus to this approach is that any organic colour developed in the extract is also destroyed, thus eliminating a possible source of interference in spectrophotometric determinations of aluminium (Mitchell and Smith, 1974).

Aluminium and silicon can then be determined, after pH adjustment of the treated solutions, by the procedures detailed in this section.

9.3.3 Ammonium oxalate (Smith, 1984)

Tamm (1922) was the first to use oxalate at pH 3.3 to extract amorphous oxides and hydrous oxides from soils, showing that it has only a minimal effect on crystalline oxides and clay minerals. The method depends essentially on the pH and complexing ability of oxalate to extract colloid complexes of aluminium, iron, silicon and manganese. Schwertmann (1964) reassessed this reagent quantitatively, finding that when used in the dark it dissolved only amorphous iron oxides. Magnetite and microcrystalline lepidocrocite suffered some dissolution, but hematite and goethite were inert. Oxalate-extractable iron and aluminium are used as a criterion in the classification of soils (McKeague and Day, 1966) and as a means of monitoring the mobility of iron and aluminium within the soil profiles (Schwertmann, 1973). Parfitt (1989) reassessed the optimum conditions for acid oxalate extraction of poorly ordered material in soils. He concluded that for most soils a 4-h extraction with 0.15 M acid oxalate pH 3 (soil : solution ratio of 1 : 100) in the dark resulted in maximum extractable Al, Fe and Si. Soils rich in extractables (> 5% Al + Fe) require a 0.2 M oxalate and a 1 : 200 ratio. The ferrihydrite, a hydrous oxide of iron, content of the soil can be ascertained by multiplying the oxalate extractable Fe, Fe_{ox} by 1.7 (Childs, Matsue and Yoshinaga, 1991), provided no other oxalate-soluble iron species are present. Following the widespread recognition of allophanic material in soils (Childs, Parfitt and Lee, 1983) the value of quantifying oxalate extractable silicon has become more important. When pure imogolite or proto-imogolite allophane (Al/Si : 2/1) is treated with oxalate, a value of 13.3% is obtained for extractable silicon (Si_{ox}). Estimation of allophane using oxalate and pyrophosphate extractants depends on their similar chemical behaviour. An approximate estimation of allophane can be obtained (Parfitt and Henmi, 1982) – when its presence is indicated by the ratio of inorganic-extractable aluminium ($Al_{ox} - Al_{py}$) to oxalate-extractable silicon (Si_{ox}) falling within the range 1.5–2.6 – by multiplying the value Si_{ox} by 7.5, i.e.

$$\text{percentage allophane} = \frac{100 \times \%Si_{ox}}{13.3}$$

$$= 7.5 \times \%Si_{ox}$$

where Al_{py} = pyrophosphate-extractable Al, Al_{ox} = oxalate-extractable Al, and Si_{ox} = oxalate-extractable Si. Care must be taken with the estimation of allophane in soils when large amounts of ferrihydrite, which contains Si (or possibly structural Al) are present (Parfitt and Childs, 1988). In these cases it is inappropriate to assign Si_{ox} or Al_{ox} entirely to allophane.

Table 9.1 Al : Si ratio $\left(\frac{Al_{ox} - Al_{py}}{Si_{ox}}\right)$ and the factors for allophane estimation from Si_{ox}

Al : Si (atomic ratio)	*Factor × Si_{ox}*
1.0	5
1.5	6
2.0	7
2.5	10
3.0	12
3.5	16

Recently Parfitt (1990), in a review of allophane in New Zealand, suggested a range of multiplication factors to estimate the allophane content of soils from oxalate extractable silicon (Table 9.1).

9.3.3.1 *Ammonium oxalate extraction*

Reagents Ammonium oxalate (0.2 M) pH 3 – mix 700 ml ammonium oxalate solution (28.3 g/litre) with 535 ml of 0.2 M oxalic acid (25.2 g/litre) and adjust to pH 3, storing the solution in a polyethylene bottle.

Procedure Place 200 mg soil (< 100 mesh) in a 100-ml polypropylene centrifuge tube, add 80 ml of 0.2 M ammonium oxalate at pH 3 and shake the mixture for 4 h in the dark. The supernatant solution, obtained after centrifuging at 2000 *g* for 10 minutes, is siphoned off into a 200-ml volumetric flask and diluted to the mark with distilled water. The solution is then reserved for the determination of iron, aluminium and silicon.

Pretreatment This is only necessary when aluminium has to be measured. Iron and silicon can be determined in the untreated extract using the methods detailed previously. For the silicon determination, excess ammonium molybdate reagent is utilized (Smith, 1984) for measuring the reduced silicon molybdate complex by a manual spectrophotometric method. This technique has now been automated (Wang and Schuppli, 1986) using a continuous flow procedure. Flow injection analysis would also be a possible alternative method of silicon determination (Thomson, Johnson and Petty, 1983). These techniques have been shown to be more accurate and reliable than atomic absorption for the detection of silicon in oxalate extracts at 0.1 μg/g levels.

(i) Aluminium

Treat a suitable aliquot (0–30 μg Al) with 2 ml 69% nitric acid, 1 drop 98% sulphuric acid and 20 ml distilled water, and then evaporate slowly on a hot-plate (*c.* 200°C) to destroy the oxalate. Then add 2 ml of 5 M hydrochloric acid to the residue, evaporate to dryness and redissolve the residue in 40 ml warm distilled water. Adjust the pH of the solution with dilute NH_4OH to within the range 4–6 before determining the aluminium content spectrophotometrically, using the 'Aluminon' method previously described.

9.3.4 Hydroxyl release

This method is a laboratory adaptation of the field test for identification of allophanic material in soils, developed in New Zealand (Fieldes and Perrott, 1966). The test is carried out by placing a small sample of soil on a filter paper impregnated with phenolphthalein indicator, placing two drops of saturated sodium fluoride solution on to the soil, and noting the intensity of the pink colour then developed. Poorly ordered inorganic material in soils is highly hydrated. The test utilizes the fact that aqueous solutions of fluorides at pH values > 7 react with hydroxyl sites, causing a corresponding increase in pH, which can be related to the relative amounts of allophanic material.

The field test has been adapted for laboratory measurement of the relative amounts of poorly ordered material in soils (Perrott, Smith and Inkson, 1976). The method involves the reaction of 25 mg of soil (< 200 mesh) or clay, at 25°C (thermostatted), with 0.85 M sodium fluoride at pH 6.8. As the exchange between fluoride and hydroxyl ions is pH-dependent, the reaction is carried out using an automatic titrator in the pH stat mode at 6.8. The pH is kept constant by titration with a standard acid. The amount consumed, and thus the amount of hydroxyl released after 25 minutes, is taken as a measure of the hydroxyl release.

This technique has been shown to be very sensitive in detecting small amounts of inorganic gel material in soils. The method is unsuitable for soils with moderate to high organic matter contents (> 8%) and those which contain carbonate must be pretreated with acid or an acidified sodium acetate solution.

Perrott, Smith and Inkson (1976) showed that selected crystalline minerals were non-reactive under the above conditions, whereas gel material was very reactive (Table 9.2). They also used the method to examine 57 freely drained Scottish podzols and brown forest soils, obtaining a mean value of 80 mmol/100 g.

Perrott, Smith and Mitchell (1976) examined the effect of pH on the reaction of a series of mixed inorganic gels with sodium fluoride, and found that the reactions of the gels were pH-dependent. When the pH was raised to 7.6, the silica gel no longer reacted, and when raised to pH 9 the iron gel showed little reaction. Therefore, by varying the pH of the reaction, which should be carried out under a nitrogen atmosphere, it is possible to separate the hydroxyl release values due to silica and iron gels from those due to alumina gel. The values obtained for the gels, however, are less than those at pH 6.8, because that reaction is pH-dependent.

9.3.5 Trimethylsilylation

Soil is formed through the weathering of minerals, and the type of weathering is related to the arrangement of silica units within the crystal

Table 9.2 Hydroxyl release values for minerals and poorly ordered materials (Perrott, Smith and Inkson, 1978)

Material	*Origin*	*Hydroxyl release (mmol/100 g, 25 minutes)*
Quartz	Brazil	1
Goethite	Pribram, Bohemia	9
Kaolinite	Cornwall, UK	12
Hematite	Locality unknown	14
Chlorite	Wanibuchi, Japan	16
Illite	Fithian, Illinois, USA	17
Boehmite	Hungary	20
Montmorillonite	Wyoming, USA	22
Halloysite	Wagon Wheel, Colorado, USA	25
Gibbsite	Guyana	51
Allophane	Taupo, New Zealand	1075
Allophane	Kodonbaru, Japan	1900
Imogolite	Uemura, Japan	1208
Ferric oxide gel	Synthetic	547
Silica gel	Synthetic	1756
Alumina gel	Synthetic	2112

structure. The degree of polymerization of the silica units can be quantitatively assessed by preparing organic trimethylsilyl derivatives, which can be separated and identified. Trimethylsilylation is well known in the field of organic chemistry, and was first used in silicate studies by Lentz (1964). It was further refined by Gotz and Masson (1970), and given a quantitative basis by Gotz and Masson (1971) using an internal standard, and then by Lachowski and Glasser (1973). These authors calibrated the method by a comparison with solutions of pure derivatives isolated by chromatographic separation. The procedure is based on the reaction of silicic acids with a monofunctional organosilicon compound to give stable and volatile organosilyl derivatives of the acids.

Trimethylsilylation has been used in weathering studies of rocks and their derived soils. Smith, Paterson and Mitchell (1982) and Smith, Mitchell and Mackenzie (1983) showed that:

- of the primary minerals, the ferromagnesian ones were the most reactive;
- inorganic gels, particularly those high in alumina were reactive;
- crystalline clay minerals, especially kaolinite and illite were relatively stable (Table 9.3).

Gel material is certainly much more reactive than crystalline clay minerals although not as reactive as some ferromagnesian minerals. The use of the trimethylsilylation method in soil studies can, therefore, give some indication of a non-crystalline complement.

Table 9.3 Monomeric silica released on trimethylsilylation from minerals, clay minerals, naturally occurring and synthetic gels (Smith *el al.*, 1982; Smith, Mitchell and Mackenzie 1983)

Mineral	*Monomeric silica (mg/g)*	*Mineral*	*Monomeric silica (mg/g)*	*Gels*	*Monomeric silica (mg/g)*
Fayalite	247	Chlorite	2.4	Allophane, Misotsuchi, Japan	117
Olivine	207	Halloysite	2.0	Allophane, Mitaka, Japan	72
Lepidomelane	16	Montmorillonite	1.9	Allophane (A), Tirau, New Zealand	59
Hornblende	2.6	Vermiculite	1.5	Allophane (B), Taupo, New Zealand	58
Hydrobiotite	2.0	Nontronite	1.3	SiO_2/Al_2O_3 1:7	38
				SiO_2/Al_2O_3 1.3:1	25
Augite	0.7	Serpentine	1.0	SiO_2/Al_2O_3 1.8:1	23
Albite	0.7	Kaolinite	0.6	SiO_2/Al_2O_3 2.8:1	8.9
Orthoclase	0.4	Illite	0.3	SiO_2/Al_2O_3	2.4
Muscovite	< 0.1				
Quartz	< 0.1				

The method involves treatment of the samples with trimethylchlorosilane (TMCS) in the presence of a small amount of water. The hydrochloric acid produced by hydrolysis of the TMCS leaches the silicate minerals, replacing cations with protons. The silicic acids that are released immediately form derivatives with trimethylsilyl (TMS) groups. The TMS derivatives of the various silica polymers can then be separated, identified and quantified – using standard solutions of pure derivatives – by gas-liquid chromatography. The reaction with a monomeric orthosilicate group is as follows:

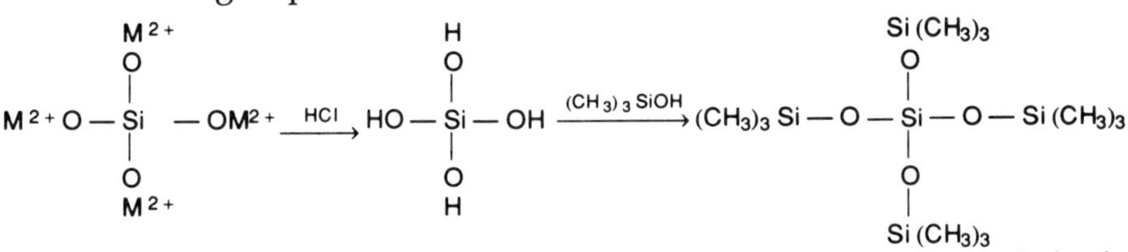

Monomer — Silicic acid — TMS monomer derivative

9.3.5.1 *Trimethylsilylation method*

Reagents Trimethylchlorosilane (Koch-Light).
Hexamethyldisiloxane (Koch-Light).

Isopropyl alcohol – (BDH-AR).
n-Eicosane – $CH_3(CH_2)_{18}CH_3$ (British Drug House, BDH).
Amberlyst 15 (BDH) (cation-exchange resin).

Procedure Weigh a lightly ground sample (25–200 mg) into a glass vial, add 9 ml hexamethyldisiloxane (HMD), 0.2 ml distilled water and 1 ml isopropyl alcohol (containing 2 mg *n*-eicosane ml as an internal standard), stopper the vial and stir the mixture for 5 minutes using a magnetic stirrer. Then add 2 ml trimethylchlorosilane (TMCS) and stir continuously for 16 h (overnight). All reagents, prior to use, are cooled in a refrigerator to minimize any side reactions. In order to hydrolyse the excess TMCS to HMD, add 2 ml distilled water and stir the mixture for a further 1.5 h before centrifuging in 15-ml tubes at 2200 *g* for 5 minutes. Transfer the organic-solution layer to a glass vial, weigh, add *c.* 1 g cation-exchange resin (Amberlyst 15) and stir the mixture for at least another 5 h to ensure complete trimethylsilylation. The organic solution is separated from the exchange resin by filtration, and the TMS derivatives of the silicate anions present are determined by gas-liquid chromatography.

Gas–liquid chromatography A temperature-programmed *Pye 104* chromatograph, fitted with dual flame-ionization detectors and coupled to an integrator, is used to separate and quantify the TMS derivatives. Dual stainless steel columns (3.7 m long × 2 mm diameter) packed with silicon gum rubber (SE 30), 3% on acid-washed diatomite (80–90 mesh), are used for the separation. The carrier gas is nitrogen, the flow rate is 20 ml/minute, and the following conditions are used as standard.

Injection block temperature – 200°C.
Detector block temperature – 350°C.
110°C – isothermal 5 minutes.
110–140°C – 150/minute.
140–220°C – 80/minute.
220–250°C – 30/minute.
250°C isothermal – 10 minutes.

The inert *n*-eicosane is added to the isopropyl alcohol as an internal standard. The greater part of the solvent is dumped by inserting a stream splitter between the column and the detector. Figure 9.1 is a typical chromatograph of the TMS derivatives. The amounts of these present in the solutions are determined from the peak areas, using pure TMS derivatives as internal standards.

The amount of monomeric silica obtained from soil-related materials is always very much larger than that of the dimer; even smaller amounts (if any) are obtained from the higher polymer forms, i.e. ring trimer, ring tetramer and chain trimer.

Since the technique involves the determination of the amount and the progress of silicate dissolution during prolonged and controlled acid attack, it can be used in soil-profile studies, particularly for temperate soils developed in a naturally acidic environment.

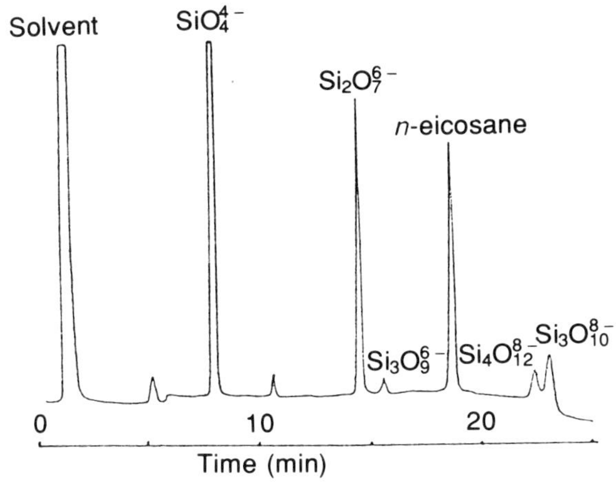

Figure 9.1 Typical gas chromatographic separation of trimethylsilyl derivatives of lower molecular weight silicate anions.

9.3.6 Pyrophosphate extraction

Alexsandrova (1960) showed that pyrophosphate could extract organic matter and associated aluminium and iron from soils. The method was further developed by McKeague (1967) and Bascomb (1968), the main difference between the procedures being that McKeague utilized sodium pyrophosphate and Bascomb utilized potassium pyrophosphate. It was found that with poorly ordered iron oxides and silicates and well-crystallized goethite and hematite there was virtually no reaction. On the other hand, examination of a wide range of organomineral soils showed a good correlation between the extracted iron and aluminium and the amounts of the extracted organic carbon. Pyrophosphate extraction of soils at pH 10 can, therefore, give a measure of organically complexed iron and aluminium. The method is now used as an aid to the classification of soils, in particular spodosols (McKeague, Brydon and Miles, 1971; Loveland and Bullock, 1975), i.e. those soils that are characterized by an illuvial mineral horizon resulting from the accumulation of poorly ordered material. This material was thought to be composed of aluminium and organic matter only, with or without iron. However, as allophane can also be of eluvial origin (Farmer, Russell and Berrow, 1980), acid oxalate is replacing pyrophosphate as the preferred extractant for estimating illuvial materials.

9.3.6.1 Pyrophosphate extraction (Bascomb, 1968)

Reagents Potassium pyrophosphate (0.1 M) – dissolve 38.44 g potassium pyrophosphate $K_4P_2O_7 \cdot 3H_2O$ (AR) in 1 l distilled water.

Sodium acetate–acetic acid buffer, pH 3.8 – dissolve 680 g sodium acetate $NaOAc \cdot 3H_2O$ (AR) (hydrated) in 3 l of distilled water, add 1 l 100% glacial acetic acid (AR), adjust the pH to 3.8 ± 0.05 and dilute to 5 l.
Sodium dithionite $Na_2S_2O_4$ – general purpose reagent (GPR).
Superfloc flocculating agent, 0.2% aqueous.

Extraction Weigh 0.8 g soil (< 2 mm) into a 100-ml polypropylene centrifuge tube, add 80 ml of 0.1 M potassium pyrophosphate solution and shake the mixture for 16 h (overnight). Add five drops flocculating agent (Schuppli, Ross and McKeague, 1983), centrifuge for 15 minutes at 2800 *g*, and siphon off the supernatant liquid into a 200-ml volumetric flask. Wash the residual soil with 50 ml distilled water, recentrifuge and add to the volumetric flask. Dilute to volume, reserving the solution for the determination of extractable aluminium, iron and carbon. The residue is used for the determination of the residual iron, after dithionite treatment. A reagent blank should be carried through the complete extraction procedure.

(i) Residual iron

Add 80 ml sodium acetate buffer at pH 3.8 and 3.2 g sodium dithionite to the soil residue in the centrifuge tube, and shake the mixture for 16 h (overnight). After centrifuging as in the last section, siphon off the supernatant into a 200-ml volumetric flask, wash the residue with 50 ml distilled water, centrifuge and add the supernatant again to the solution in the flask. Dilute to the mark with distilled water. The residual iron content is then measured using the 1,10-phenanthroline method.

(ii) Pretreatment for Al and Fe determinations

Treat a suitable aliquot (5–10 ml) of the pyrophosphate extract, in a conical flask, with 1 ml 69% nitric acid and two drops sulphuric acid (1 : 1), and evaporate to dryness on a hot-plate until white fumes of sulphuric acid appear. If the residue is still brown in colour, cool and treat again with nitric acid. Cool and add 30 ml distilled water and 2 ml of 1 M hydrochloric acid to dissolve the residue, and heat on a water-bath for 2 h. Then cool the solution and reserve for the determination of iron as detailed previously, and the determination of aluminium by xylenol orange (Pritchard, 1967).

(iii) Extractable carbon

Reagents Potassium dichromate (0.166 M) – dissolve 49.0 g $K_2Cr_2O_7$ (AR) in 1 l distilled water and standardize against standard ferrous sulphate solution.
Ferrous sulphate (1 M) – dissolve 278.0 g $FeSO_4 \cdot 7H_2O$ (AR) in 1 l distilled water containing 10 ml 98% H_2SO_4.
Diphenylamine indicator – dissolve 1.60 g barium diphenylamine sulphonate by boiling in 500 ml distilled water containing 15 ml 98% sulphuric acid. Allow mixture to cool and settle, and decant the clear supernatant liquid into a reagent bottle.

Evaporate 25 ml of the pyrophosphate extract to dryness at 80°C in a 500-ml conical flask. Then add 10 ml of 0.166 M potassium dichromate to the residue, carefully add 20 ml 98% sulphuric acid, mix thoroughly and allow to stand for

30 minutes. Dilute the mixture to 200 ml with distilled water, add 10 ml 85% orthophosphoric acid and 20 drops of diphenylamine indicator solution, and titrate rapidly with 1 M ferrous sulphate until the colour changes to green. Then add a further 0.50 ml potassium dichromate, which will cause the blue colour to reappear, continue the ferrous sulphate titration slowly until the end-point, noting the amount of titrant consumed. A blank titration is carried out on 10 ml of standard potassium dichromate in a similar manner to that detailed previously, to measure the amount of standard dichromate consumed in the oxidation of the extractable carbon.

$$1 \text{ ml of } 0.166 \text{ M } K_2Cr_2O_7 \equiv 3 \text{ mg C}.$$

Only a 77% recovery can be obtained, so that 1 ml of 0.166 M $K_2Cr_2O_7$ is assumed to be equivalent to 3.9 mg C reduced.
Then

$$\%C = \frac{390 \times \text{volume (ml) } 0.166 \text{ M dichromate}}{\text{wt of sample (mg)}}.$$

9.3.6.2 *Aluminium – Xylenol orange (Pritchard, 1967)*

Reagents Xylenol orange (0.15%) aqueous.
Sodium acetate buffer, pH 3.8 – dissolve 136 g $NaOAc \cdot 3H_2O$ in water, adjust the pH to 3.8 and dilute to 1 l.
EDTA (ethylenediamenetetra-acetic acid) (0.05 M) – dissolve 18.6 g EDTA (disodium salt) in 1 l distilled water.
Cresol red indicator – dissolve 0.1 g cresol red in 100 ml ethanol.
Al standard (stock) – dissolve 1.7584 g potassium aluminium sulphate (AR) in 1 l distilled water (1 ml = 0.1 mg Al).
Al standard (working) – dilute 100 ml stock Al standard to 1 l (1 ml = 10 μg Al).

Procedure Pipette a suitable aliquot (0–60 μg Al) of the treated pyrophosphate extract into a 100-ml conical flask. Add 20 ml distilled water and four drops 1 : 1 sulphuric acid, and hydrolyse the residue by digesting in a boiling water-bath for 4 h. After cooling, add two drops of cresol red indicator and adjust the pH to *c*. 2 with dropwise addition of 1 : 1 M NH_4OH (to a yellow end-point). Then transfer quantitatively to a 100-ml standard flask, add 25 ml buffer solution and 10 ml xylenol orange, and place in a water-bath at 40°C for 1.5 h. Cool the solutions, add 5 ml EDTA, dilute to 100 ml and measure the absorptions at 550 nm after 1 h in a 4-cm cell. The results are compared with those of a range of standard aluminium solutions prepared in a similar manner (0–60 μg Al).

9.3.7 Evaluation of selective dissolution techniques

The use of differential X-ray diffraction (DXRD) (Campbell and Schwertmann, 1985; Childs, Matsue and Yoshinaga, 1991) and difference infrared (DIR) (Farmer and Russell, 1990) should be adopted to evaluate the effects of dissolution techniques on soils and minerals. These methods give a means of following the processes involved – phases removed and products altered or produced. This is achieved by an XRD or IR comparison of the sample before and after dissolution treatments. An inert standard, either internal (quartz) or artificially added (α-alumina) can be used to normalize the peaks. The patterns obtained are then computer-subtracted to show what has been removed or altered.

9.3.8 Other methods

Other less commonly used techniques for the characterization of poorly ordered material in soils may be mentioned. Most can be classed as selective dissolution methods, e.g. treatment with sodium EDTA at pH 10 (Borggard, 1976), alkaline tiron at pH 10.5 (Biermans and Baert, 1977; Kodama and Ross, 1991), and sodium hydroxide-tetraborate (Bruckert, 1979), monitoring Al, Fe and Si in the respective extracts.

Aomine and Jackson (1959) proposed a method for the estimation of allophanic material in soils by measuring the 'cation-exchange capacity delta value'. This involved treating two separate samples of peroxidized soil, one with 2% Na_2CO_3 for 60 minutes and the other with boiling sodium acetate for 15 minutes, and measuring the difference in exchange capacity between the two samples: the CEC delta value. This was shown to be very high for allophanic clays, large for halloysitic and smectitic clays but very low for the other crystalline clay minerals.

9.4 COMPARISON OF METHODS

Alkali extraction by 0.5 M Na_2CO_3 (Follett *et al.*, 1965) has been shown to remove poorly ordered aluminosilicate material from soils and to have little or no effect on well-crystallized minerals. These authors considered that the extraction method with 0.5 M NaOH (Hashimoto and Jackson, 1960) resulted in some instances in considerable attack of crystalline clay minerals. However, if the ratio of extractant to soil is high enough to prevent saturation of the extract with respect to Si and Al, and if the time of reaction is kept to a minimum, then the attack on the crystalline components can be greatly reduced. As the pH of these extracts is high (*c.* 11), iron is not removed from the clay or soil. The amounts of alkali-soluble material, particularly in B horizons of podzols and brown podzolic soils, are quite large, indicating a high content of poorly ordered aluminosilicates in these horizons. Improved dispersion has been noted after alkali treatment. This is probably due to the removal of poorly ordered aluminosilicates that cement the primary clay particles into aggregates or due to aluminosilica gel coatings on individual particle surfaces or perhaps to both. The amounts of aluminium and silicon extracted by alkali show a highly significant correlation with values obtained for hydroxyl release from both freely drained and poorly drained soils (Perrott, Smith and Inkson, 1976).

Dithionite extraction removes 'free' poorly ordered and crystalline iron oxides from soils, with a minimal effect on crystalline clay minerals. Silicon and aluminium are found in the extracts, normally in smaller amounts, and are thought to derive from poorly crystalline aluminosilicate material in the soil. Dithionite extraction provides a method for determining free (non-silicate) iron in soils, dissolving finely divided hematite and goethite. However, coarse-grained hematite and magnetite are not attacked, or at least only very slowly.

Ammonium oxalate dissolves poorly ordered inorganic- and organic-complexed iron, silicon and aluminium from soils. It has a minimal effect on crystalline silicate minerals, goethite and hematite, but does dissolve poorly ordered lepidocrocite. Oxalate extraction has been used in Canada (McKeague and Day, 1966) in the classification of soils, and in New Zealand (Saunders, 1968) to relate the changes in amorphous iron and aluminium to the degree of weathering and leaching in soils. More recently, Farmer, Russell and Berrow (1980) and Childs, Parfitt and Lee (1983) have used oxalate-soluble silicon and aluminium to support the proposal that movement of aluminium in the podzolization process can be attributed to inorganic complexes with silicon.

Hydroxyl release values can give a measure of the relative amounts of poorly ordered material in soils (Perrott, Smith and Inkson, 1976), although the absolute amounts must remain uncertain. A major part of the reactivity can be attributed to the clay fraction, in particular to gel material. The highest reactivities for soils occur in the Bs horizons of podzols, whereas for brown earths and gley soils there is little variation between the reactivities of the genetic horizons within a profile.

An estimation of the degree of polymerization of silicon in silicates in soils that react to acid attack can be obtained by trimethylsilylation. Crystalline clay minerals are relatively resistant, whereas poorly ordered aluminosilicate material is highly reactive, particularly gels with a high ratio of Al : Si. A good overall correlation exists (Smith, Mitchell and Mackenzie, 1983) between the amount of silica released during trimethylsilylation and hydroxyl release by fluoride exchange.

Organic-complexed iron and aluminium can be extracted from soils by pyrophosphate. The treatment has very little effect on silicate minerals, or on well-crystallized aluminium and iron oxides, but some dissolution of poorly crystalline forms has been observed. This technique has been widely used as a chemical basis for identifying podzolic B horizons. Schuppli, Ross and McKeague (1983) highlighted the fact that pyrophosphate can cause peptization of ferruginous material, which, in some cases, will not centrifuge out easily unless a flocculant is used. This may have led to erroneously high values for organically bound iron and aluminium being reported in the past.

More recently, Parfitt and Childs (1988) showed that extraction with pyrophosphate is a reliable method for estimating organically bound Al.

Table 9.4 Characterization of poorly ordered material in soils

	Poorly ordered material					
Methods	Alkali dissolution	Ammonium oxalte	Pyrophosphate	Dithionite	Trimethylsilylation	Fluoride exchange
Elements	Al, Si	Al, Fe, Mn, Si	Al, Fe_{ext}, Fe_{res}, C	Al, Fe, Si	Si polymers	Relative amounts or poorly ordered material
Components detected	Poorly ordered inorganic aluminosilicates	Poorly ordered inorganic and organic	Poorly ordered organic	Poorly ordered inorganic and organic, and crystalline Fe	Acid-reactive crystalline and poorly ordered silicates	Poorly ordered inorganic material

In the case of Fe the Fe_{py} values do not relate to any particular form of Fe in the soil and should not be used to measure Fe in humus complexes.

Pyrophosphate extraction can give reproducible Fe and Al values if careful use is made of the method. However if the soils contain ferrihydrite or goethite these can be dispersed by pyrophosphate causing high Fe_{py} values which are not attributable to organically bound iron.

A combination of various methods can give an estimation of the different forms of iron and aluminium in soils:

- crytalline Fe oxides = dithionite iron – oxalate iron ($Fe_d - Fe_{ox}$);
- poorly ordered iron and aluminium = oxalate soluble – pyrophosphate soluble iron and aluminium ($Fe_{ox} - Fe_{py}$ and $Al_{ox} - Al_{py}$);
- organically bound iron and aluminium = pyrophosphate soluble iron and aluminium (Fe_{py} and Al_{py}).

The role of the different selective methods in soils is summarized in Table 9.4.

- ferrihydrite = oxalate Fe × 1.7 ($Fe_{ox} \times 1.7$). In some cases more than one oxalate extraction may be necessary. In fact oxalate may extract Fe species other than ferrihydrite (e.g. hisingerite).

Selective chemical dissolution techniques play a very important role in the characterization of clay materials, but must not be considered in isolation. The various instrumental techniques and selective chemical methods are complementary. As succinctly expressed by Mackenzie (1957) 'every method, whatever its nature, will add its own quota of information and fill in some of the shadows or emphasize some of the highlights of the picture'.

REFERENCES

Alexsandrova, L. N. (1960) The use of pyrophosphate for isolating free humic substances and their organic-mineral compounds from the soil. *Soviet Soil Science*, **1960**, 190–197.

Aomine, S. and Jackson, M. L. (1959) Allophane determinations in Ando soils by cation exchange delta value. *Soil Science Society of America, Proceedings*, **23**, 210–214.

Bascomb, C. L. (1968) Distribution of pyrophosphate-extractable iron and organic carbon in soils of various groups. *Journal of Soil Science*, **19**, 251–268.

Biermans, V. and Baert, L. (1977) Selective extraction of the amorphous Al, Fe and Si oxides using an alkaline tiron solution. *Clay Minerals*, **12**, 127–135.

Borggard, O. K. (1976) Selective extraction of amorphous iron oxide by EDTA from a mixture of amorphous iron oxide, goethite and hematite. *Journal of Soil Science*, **27**, 478–486.

Bruckert, S. (1979) Classification des sols bruns ocreux et cryptopodzoliques par analyse des extraits tetraborates tamponnés à pH 9.5. *Geoderma*, **22**, 205–212.

Bruckert, S. and Souchier, B. (1975) Mise au point d'un test de differentiation chimique des horizons cambiques et spodiques. *Compte Rendus Hebdomadaires des Seances de l'Academie des Sciences*, **280**, 1361–1364.

Campbell, A. S. and Schwertmann, U. (1985) Evaluation of selective dissolution extractants in soil chemistry and mineralogy by differential X-ray diffraction. *Clay Minerals*, **20**, 515–519.

Childs, C. W., Matsue, N. and Yoshinaga, N. (1991) Ferrihydrite in volcanic ash soils of Japan. *Soil Science and Plant Nutrition*, **37**, 299–311.

Childs, C. W., Parfitt, R. L. and Lee, R. (1983) Movement of aluminium as an inorganic complex in some podzolised soils, New Zealand. *Geoderma*, **29**, 139–155.

Cradwick, P. D. G., Farmer, V. C., Russell, J. D. *et al.* (1972) Imogolite, a hydrated aluminium silicate of tubular structure. *Nature*, **240**, 187–189.

Deb, B. C. (1950) The estimation of free iron oxides in soils and clays and their removal. *Journal of Soil Science*, **1**, 212–220.

De Endredy, A. S. (1963) Estimation of free iron oxides in soils and clays by a photolytic method. *Clay Minerals Bulletin*, **5**, 209–217.

Farmer, V. C. and Russell, J. D. (1990) The structure and genesis of allophanes and imogolite; their distribution in non-volcanic soils, in *Soil Colloids and their Associations in Soil Aggregates* (eds M. F. de Boodt, M. H. B. Hayes and A. Herbillon). Proceedings, NATO Advanced Studies Workshop, Ghent, 1985, Springer-Verlag, New York, pp. 165–178.

Farmer, V. C., Russell, J. D. and Berrow, M. L. (1980) Imogolite and protoimogolite allophanes in spodic horizons: evidence of a mobile aluminium silicate complex. *Journal of Soil Science*, **31**, 673–684.

Farmer, V. C., Russell, J. D. and Smith, B. F. L. (1983) Extraction of inorganic forms of translocated Al_2O_3, Fe_2O_3 and SiO_2 from a podzol Bs horizon. *Journal of Soil Science*, **34**, 571–576.

Farmer, V. C., Smith, B. F. L. and Tait, J. M. (1977) Alteration of allophane and imogolite by alkaline digestion. *Clay Minerals*, **12**, 195–198.

Farmer, V. C., McHardy, W. J., Palmieri, F., Violante, A. and Violante, P. (1991) Synthetic allophanes formed in calcareous environments: nature conditions of formation and transformations. *Soil Science Society of America, Journal*, **55**, 1162–1166.

Fieldes, M. (1955) Clay mineralogy of New Zealand soils. Part II. Allophane and related mineral colloids. *New Zealand Journal of Science and Technology*, **B37**, 336–350.

Fieldes, M. and Perrott, K. W. (1966) Nature of allophane in soils. III. Rapid field and laboratory test for allophane. *New Zealand Journal of Science*, **9**, 623–629.

Follett, E. A. C., McHardy, W. J., Mitchell, B. D. and Smith, B. F. L. (1965) Chemical dissolution techniques in the study of soil clays. *Clay Minerals Bulletin*, **6**, 23–34.

Gotz, J. and Masson, C. R. (1970) Trimethylsilyl derivatives for the study of silicate structures. Part I: a direct method of trimethylsilylation. *Journal of the Chemical Society (A)*, 2683–2686.

Gotz, J. and Masson, C. R. (1971) Trimethylsilylation of silicate anions: a method of studying the structure of crystalline silicates and glasses, in *Science and Technical Committee, 9th International Conference on Glass, Versailles*, 27 September–2 October, pp. 261–275.

Guillet, B. and Souchier, B. (1982) Amorphous and crystalline oxyhydroxides and oxides in soils (iron, aluminium manganese, silicon), in *Constituents and Properties of Soils* (eds M. Bonneau and B. Souchier). Academic Press, London.

Hashimoto, I. and Jackson, M. L. (1960) Rapid dissolution of allophane and kaolinite–halloysite after dehydration in *Proceedings, 7th Conference on Clays and Clay Minerals* (ed. A. Swineford). Pergamon Press, Oxford, pp. 102–113.

Hendricks, S. B. and Fry, W. H. (1930) The results of X-ray and microscopical examinations of soil colloids. *Soil Science*, **29**, 457–479.

Hoyos, A. and Pino, C. (1958) A comparative study of soils from the Canary Islands and from Spanish Guinea. *Agrochimica*, **2**, 218–235.

Jackson, M. L., Pennington, R. P. and Mackie, W. Z. (1950) Crystal chemistry of soils. I. The fundamental structure groups and families of silicate minerals. *Soil Science Society of America, Proceedings*, **13**, 139–145.

Jones, R. C. and Uehura, G. (1973) Amorphous coatings on mineral surfaces. *Soil Science Society of America, Proceedings*, **37**, 792–798.

Jorgensen, S. S., Birnie, A. C., Smith, B. F. L. and Mitchell, B. D. (1970) Assessment of gibbsitic material in soil clays by differential thermal analysis and alkali dissolution methods. *Journal of Thermal Analysis*, **2**, 277–286.

Kirkman, J. H., Mitchell, B. D. and Mackenzie, R. C. (1966) Distribution in some Scottish soils of an inorganic gel system related to 'allophane'. *Transactions, Royal Society of Edinburgh*, **66**, 393–418.

Kodama, H. and Ross, G. J. (1991) Tiron dissolution method used to remove and characterize inorganic components in soils. *Soil Science Society of America, Journal*, **55**, 1180–1187.

Lachowski, E. E. and Glasser, F. P. (1973) Application of gas chromatography to mineral chemistry: aluminium-silicon ordering in melilites. *Mineralogical Magazine*, **39**, 412–419.

Lentz, C. W. (1964) Silicate minerals as sources of trimethylsilyl silicates and silicate structure analysis of sodium silicate solutions. *Inorganic Chemistry*, **3**, 574–579.

Loveland, P. J. and Bullock, P. (1975) Crystalline and amorphous components of the clay fractions in brown podzolic soils. *Clay Minerals*, **10**, 451–469.

Mackenzie, R. C. (1957) Modern methods for studying clays. *Agrochimica*, **1**, 305–327.

Mackenzie, R. C. (1963) *De Natura Lutorum*, in *Proceedings, 11th National Conference on Clays and Clay Minerals, 1962* (ed. W. F. Bradley). Pergamon Press, Oxford, pp. 11–28.

McKeague, J. A. (1967) An evaluation of 0.1 M pyrophosphate and pyrophosphate-dithionite in comparison with oxalate as extractants of the accumulation products in podzols and some other soils. *Canadian Journal of Soil Science*, **47**, 95–99.

McKeague, J. A. and Day, J. H. (1966) Dithionite and oxalate extractable Fe and Al as aids in differentiating various classes of soils. *Canadian Journal of Soil Science*, **46**, 13–22.

McKeague, J. A., Brydon, J. E. and Miles, N. M. (1971) Differentiation of forms of extractable iron and aluminium in soils. *Soil Science Society of America, Proceedings*, **34**, 33–38.

Mehra, O. P. and Jackson, M. L. (1960) Iron oxide removal from soils and clays by a dithionite–citrate system buffered by sodium bicarbonate, in *Proceedings, 7th National Conference on Clays and Clay Minerals* (ed. A. Swineford), National Academy of Science, Natural Research Council Publication, pp. 317–327.

Mitchell, B. D. and Farmer, V. C. (1962) Amorphous clay minerals in some Scottish soil profiles. *Clay Minerals Bulletin*, **5**, 128–144.

Mitchell, B. D. and Mackenzie, R. C. (1954) Removal of free iron oxide from clays. *Soil Science*, **77**, 173–184.

Mitchell, B. D. and Smith, B. F. L. (1974) The removal of organic matter from soil extracts by bromine oxidation. *Journal of Soil Science*, **25**, 239–241.

Mitchell, B. D., Farmer, V. C. and McHardy, W. J. (1964) Amorphous inorganic material in soils. *Advances in Agronomy*, **16**, 327–383.

Mitchell, B. D., Smith, B. F. L. and de Endredy, A. S. (1971) The effect of buffered sodium dithionite solution and ultrasonic agitation on soil clays. *Israel Journal of Chemistry*, **9**, 45–52.

Parfitt, R. L. (1989) Optimum conditions for extraction of Al, Fe and Si from soils with acid oxalate. *Communications in Soil Science and Plant Analysis*, **20**, 801–816.

Parfitt, R. L. (1990) Allophane in New Zealand – a review. *Australian Journal of Soil Research*, **28**, 343–360.

Parfitt, R. L. and Childs C. W. (1988) Estimation of forms of Fe and Al: a review and analysis of contrasting soils by dissolution and Mössbauer methods. *Australian Journal of Soil Research*, **26**, 121–144.

Parfitt, R. L. and Henmi, T. (1982) Comparison of an oxalate-extraction method and an infrared spectroscopic method for determining allophane in soil clays. *Soil Science and Plant Nutrition*, **28**, 183–190.

Perrott, K. W., Smith, B. F. L. and Inkson, R. E. (1976) The reaction of fluoride with soils and soil minerals. *Journal of Soil Science*, **27**, 58–67.

Perrott, K. W., Smith, B. F. L. and Mitchell, B. D. (1976) Effect of pH on the reaction of sodium fluoride with hydrous oxides of silicon, aluminium and iron and with poorly ordered aluminosilicates. *Journal of Soil Science*, **27**, 348–356.

Pritchard, D. T. (1967) Spectrophotometric determination of aluminium in soil extracts with xylenol orange. *Analyst*, **92**, 103–106.

Rich, C. I. and Thomas, G.W. (1960) The clay fraction of soils. *Advances in Agronomy*, **12**, 1–39.

Robertson, G. (1950) The colorimetric determination of aluminium in silicate materials. *Journal of the Science of Food and Agriculture*, **1**, 59–63.

Ross, C. S. and Kerr, P. F., (1934) Halloysite and allophane. *US Geological Survey Professional Paper*, **185-G**, 135–148.

Saunders, W. M. H. (1968) Soils of New Zealand. Part 2, *New Zealand Soil Bureau Bulletin*, **26**, 109–114.

Schuppli, P. A., Ross, G. J. and McKeague, J. A. (1983) The effective removal of suspended material from pyrophosphate extractions of soils from tropical and temperate regions. *Soil Science Society of America, Journal*, **47**, 1026–1032.

Schwertmann, U. (1964) The differentiation of iron oxide in soils by a photochemical extraction with acid ammonium oxalate. *Zeitschrift fuer Pflanzenemachrung, Duengung, Bodenkunde*, **105**, 194–201.

Schwertmann, U. (1973) Use of oxalate for Fe extraction from soils. *Canadian Journal of Soil Science*, **53**, 244–246.

Schwertmann, U. (1978) Non-crystalline and accessory minerals, in *Proceedings, International Clay Conference, 1978, Oxford* (eds M. M. Mortland and V. C. Farmer), *Developments in Sedimentology*, **27**. Elsevier, Amsterdam.

Smith, B. F. L. (1984) The determination of silicon in ammonium oxalate extracts of soils. *Communications in Soil Science and Plant Analysis*, **15**, 199–204.

Smith, B. F. L., Mitchell, B. D. and Mackenzie, R. C. (1983) Susceptibility to weathering of some Scottish rocks and their derived soils. *Transactions of the Royal Society of Edinburgh: Earth Sciences*, **73**, 191–203.

Smith, B. F. L., Paterson, E. and Mitchell, B. D. (1982) Trimethylsilylation of commonly occurring primary and secondary minerals in soil. *Journal of Soil Science*, **33**, 115–124.

Tait, J. M., Yoshinaga, N. and Mitchell, B. D. (1978) The occurrence of imogolite in some Scottish soils. *Soil Science and Plant Nutrition*, **24**, 145–151.

Tamm, O. (1922) Eine methode zur Bestimmung der anorganischen Komponenten des Gelkomplexes. *Meddelelser Skogsforsoksanst Stockholm*, **19**, 385–404.

Thomsen, J., Johnson, K. and Petty, R. (1983) Determination of reactive silicate in seawater by flow injection analysis. *Analytical Chemistry*, **55**, 2378–2382.

Van Bemmelen, J. M. (1910) The different modes of weathering of silicates in the Earth's crust. *Zeitschrift fuer Anorganische und Allgemeine Chemie*, **66**, 322–357.

Wang, C. and Schuppli, P. A. (1986) Determining ammonium oxalate-extractable Si in soils. *Canadian Journal of Soil Science*, **66**, 751–755.

Weaver, R. M., Syers, J. K. and Jackson, M. L. (1968) Determination of silica in citrate-bicarbonate extracts of soils. *Soil Science Society of America, Proceedings*, **32**, 497–501.

Webber, A. and Wilson, A. L. (1969) The absorptiometric determination of silicon in water. Part VII. Improved method for determining the total silicon content of high purity water. *Analyst*, **94**, 110–120.

Yoshinaga, N. (1966) Chemical composition and some thermal data of eighteen allophanes from Ando soils and weathered pumices. *Soil Science and Plant Nutrition*, **12**, 47–54.

Yoshinaga, N. and Aomine, S. (1962) Imogolite in some Ando soils. *Soil Science and Plant Nutrition*, **8**, 22–29.

Yoshinaga, N., Tait, J. M. and Soong, R. (1973) Occurrence of imogolite in some volcanic ash soils of New Zealand. *Clay Minerals*, **10**, 127–130.

Index